Lecture Notes in Artificial Intelligence 8417

Subseries of Lecture Notes in Computer Science

More information about this series at http://www.springer.com/series/1244

Yukiko Nakano · Ken Satoh
Daisuke Bekki (Eds.)

New Frontiers
in Artificial Intelligence

JSAI-isAI 2013 Workshops,
LENLS, JURISIN, MiMI, AAA, and DDS
Kanagawa, Japan, October 27–28, 2013
Revised Selected Papers

 Springer

Editors
Yukiko Nakano
Seikei University
Musashino-shi, Tokyo
Japan

Daisuke Bekki
Ochanomizu University
Tokyo
Japan

Ken Satoh
National Institute of Informatics
    Research Division
Tokyo
Japan

ISSN 0302-9743   ISSN 1611-3349 (electronic)
ISBN 978-3-319-10060-9   ISBN 978-3-319-10061-6 (eBook)
DOI 10.1007/978-3-319-10061-6

Library of Congress Control Number: 2014954575

LNCS Sublibrary: SL7 – Artificial Intelligence

Springer Cham Heidelberg New York Dordrecht London

Printed on acid-free paper

Springer is part of Springer Science+Business Media (www.springer.com)

Preface

The JSAI-isAI (JSAI International Symposium on Artificial Intelligence) 2013 was the 5th international symposium on AI supported by the Japanese Society of Artificial Intelligence (JSAI). JSAI-isAI 2013 was successfully held during October 27th to 28th at Keio University in Kanagawa, Japan; 161 people from 16 countries participated. The symposium took place after the JSAI SIG joint meeting. As the total number of participants for these two co-located events was over 400, it was the second-largest JSAI event in 2013 after the JSAI annual meeting.

The JSAI-isAI 2013 included 6 workshops, where 9 invited talks and 48 papers were presented. This volume, *New Frontiers in Artificial Intelligence*: JSAI-isAI 2013Workshops, is the post-proceedings of JSAI-isAI 2013. From 5 of the 6 workshops (LENLS10, JURISIN2013, MiMI2013, AAA2013, and DDS2013), 26 papers were carefully selected and revised according to the comments of the workshop Program Committees. About 40% of the total submissions were selected for inclusion in the conference proceedings.

- LENLS (Logic and Engineering of Natural Language Semantics) is an annual international workshop on formal semantics and pragmatics. LENLS10 was the 10th event in the series, and it focused on the formal and theoretical aspects of natural language. The workshop was chaired by Shunsuke Yatabe (West Japan Railway Company).
- JURISIN (Juris-Informatics) 2013 was the 7th event in the series, organized by Katsumi Nitta (Tokyo Institute of Technology). The purpose of this workshop was to discuss fundamental and practical issues for juris-informatics, bringing together experts from a variety of relevant backgrounds, including law, social science, information and intelligent technology, logic, and philosophy (including the area of AI and law).
- MiMI (Multimodality in Multiparty Interaction) 2013 was organized by Mayumi Bono (National Institute of Informatics) and Yasuyuki Sumi (Future University Hakodate). The topics covered in this workshop spanned interaction studies, communication studies, conversation analysis, and workplace studies, as well as their applications in other research fields.
- AAA (Argument for Agreement and Assurance) 2013 focused on the theoretical foundations of argumentation in AI, and the application of argumentation to various fields such as agreement formation and assurance. The organizers were Yoshiki Kinoshita (National Institute of Advanced Industrial Science and Technology), Kazuko Takahashi (Kwansei Gakuin University), Hiroyuki Kido (The University of Tokyo), and Kenji Taguchi (National Institute of Advanced Industrial Science and Technology).
- DDS (Data Discretization and Segmentation for Knowledge Discovery) 2013 was organized by Akihiro Yamamoto (Kyoto University), Hiroshi Sakamoto (Kyushu Institute of Technology), and Tetsuji Kuboyama (Gakushuin University). This workshop

discussed segmentation methods for various types of data, such as graphs, trees, strings, and continuous data, and their applications in the areas of machine learning and knowledge discovery.

It is our great pleasure to be able to share some highlights of these fascinating workshops in this volume. We hope this book will introduce readers to the state-of-the-art research outcomes of JSAI-isAI 2013, and motivate them to participate in future JSAI-isAI events.

April 2014   Yukiko Nakano
                                                         Ken Satoh
                                                         Daisuke Bekki

Logic and Engineering of Natural Language Semantics (LENLS) 10

Shunsuke Yatabe

Department of Letters, Kyoto University
shunsuke.yatabe@gmail.com

The Workshop

Between October 27 and 28, 2013 the Tenth InternationalWorkshop of Logic and Engineering of Natural Language Semantics (LENLS 10) took place at Raiousha Building, Keio University, Kanagawa, Japan. This was held as a workshop of the Fifth JSAI International Symposia on AI (JSAI-isAI 2013), sponsored by The Japan Society for Artificial Intelligence (JSAI).

LENLS is an annual international workshop focusing on topics in formal semantics, formal pragmatics, and related fields. This year the workshop featured invited talks by Nicholas J.J. Smith (the University of Sydney) on "Vagueness, Counting and Cardinality" and Richard Dietz (the University of Tokyo) on "The Possibility of Vagueness". In addition there were 21 presentations of talks selected by the program committee from the abstracts submitted for presentation.

LENLS workshops do not only focus on formal accounts of specific empirical linguistic phenomena, but also attempts to tackle broader theoretical, logical, philosophical and coverage issues. Topics discussed at the workshop included issues as conditionals, plurals, speech acts exhaustivity, lexicon-semantics interface, semantic similarity, an opinion classification task on a French corpus, rhetorical questions, and politeness, as well as more specific issues involving multiple constraints in ACG, type-theoretic approaches to linguistics, ontology on natural language processing, as well as general issues of Dummett's philosophy of language, modal logics, quantum linguistics and many-valued semantics,

In addition to the workshop, on October 26th, a Tutorial Lecture by Nicholas J. J. Smith was held at the Ochanomizu University in Tokyo. The title of the lecture is "Vagueness and Fuzzy Logic", and he gave a positive appraisal of the prospects for a fuzzy logic based solution to the problems of vagueness. He also explained his challenge of integrating the fuzzy theory of vagueness into the wider theoretical landscape in the field of degrees of belief.

Papers

The submitted papers in the LENLS part of the present volume are as follows:

"A Type-Theoretic Account of Neg-Raising Predicates in Tree Adjoining Grammars" by Laurence Danlos, Philippe De Groote and Sylvain Pogodalla, which provides a type theoretic semantics for TAG which derives the NR and non-NR readings of NR predicates in a compositional way.

"Semantic similarity: foundations" by Nicholas Asher, Cedric Degremont and Antoine Venant, which takes up the interesting and useful problem of "semantic similarity" between two discourses, proposes several metrics and investigates their formal properties.

"World history ontology for reasoning truth/falsehood of sentences: Event classification to fill in the gaps between knowledge resources and natural language texts" by Ai Kawazoe, Yusuke Miyao, Takuya Matsuzaki, Hikaru Yokono and Noriko Arai, which provides an ontology for the task of truth/falsehood judgement of simple historical descriptions in university-level history entrance examinations.

"Hypersequent calculi for modal logics extending S4" by Hidenori Kurokawa, which provides hypersequent calculi for S4.2 and S4.3 in addition to S4 and S5 from a uniform perspective.

"Discourse-level Politeness and Implicature" by Elin McCready and Nicholas Asher, which proposes a game-theoretic analysis of politeness, especially from the perspective of descriptive set theory.

"Bare Plurals in the Left Periphery in German and Italian" by Yoshiki Mori and Hitomi Hirayama, which discusses the interpretation possibilities of bare plurals in the left periphery in German and Italian that cannot be predicted by the so-called neo-Carlsonian approaches to the semantics of nominals.

"Analyzing Speech Acts based on Dynamic Normative Logic" by Yasuo Nakayama, which proposes an alternative approach for dynamic epistemic logic for a dynamic aspect of normative change.

"Constructive Generalized Quantifiers Revisited" by Ribeka Tanaka, Yuki Nakano and Daisuke Bekki, which introduces generalized quantifiers in the type theoretic approach to natural language along the line of Sundholm.

"Argumentative insights from an opinion classification task on a French corpus" by Marc Vincent and Gre'goire Winterstein, which reports on the production of a corpus and the results of using certain automatic techniques for classification of opinion and the interpretation of the task results in a particular semantic/pragmatic framework.

"Exhaustivity through the Maxim of Relation" by Matthijs Westera, which clearly shows that taking a view of proposition as a set of sets of worlds and employing attentive semantics with its entailment relation enables us to derive the exhaustiveness of answers as a conversational implicature with the help of the maxim of relation.

"First-Order Conditional Logic and Neighborhood-Sheaf Semantics for Analysis of Conditional Sentences" by Hanako Yamamoto and Daisuke Bekki, which shows the equivalence between Neighborhood-Sheaf Semantics and Kripke-sheaf semantics with respect to a first-order conditional logic.

Acknowledgements

Let me acknowledge some of those who helped with the workshop. The program committee and organisers, in addition to myself, were Daisuke Bekki (Ochanomizu University/National Institute of Informatics), Alastair Butler (PRESTO JST/Tohoku University), Elin McCready

(Aoyama Gakuin University), Koji Mineshima (Keio University), Yoshiki Mori (University of Tokyo), Yasuo Nakayama (Osaka University), Katsuhiko Sano (Japan Advanced Institute of Science and Technology), Katsuhiko Yabushita (Naruto University of Education), Tomoyuki Yamada (Hokkaido University), and Kei Yoshimoto (Tohoku University). Daisuke Bekki was liaison with JSAI and together with Kei Yoshimoto organised and mentored many aspects of the workshop. Finally, the organisers would like to thank JSAI for giving us the opportunity to hold the workshop.

International Workshop on Juris-Informatics (JURISIN 2013)

Katsumi Nitta

Tokyo Institute of Technology
4259 Nagatsuta, Midori-ku, Yokohama 226-8502, Japan
nitta@dis.titech.ac.jp

TheWorkshop

JURISIN is the International Workshop on Juris-Informatics. The purpose of JURISIN is to discuss both the fundamental and practical issues for jurisinformatics from various backgrounds such as law, social science, information and intelligent technology, logic and philosophy, including the conventional "AI and law" area. JURISIN 2013 was held on 27 and 28, 2013, in association with the Fifth JSAI International Symposia on AI (JSAI-isAI 2013) supported by the Japanese Society for Artificial Intelligence (JSAI).

From submitted papers, we accepted ten papers. They cover various topics such as legal reasoning systems, formal argumentation theory, legal text processing, and so on.

As guest speakers, we invited Professor Davide Grossi from University of Liverpool, UK, Professor Martin Caminada from University of Aberdeen, UK, and Professor Kotaro Takagi from Aoyamagakuin University, Japan. Professor Davide Grossi and Professor Martin Caminada are leading scientists in the field of argumentation theory. And Professor Kotaro Takagi is one of leading scientists in the field of legal communication from the view of social informatics. And, as the guest speaker of AAA 2013, Professor Tim Kelly from University of York, UK, and Professor Thomas Agotnes from University of Bergen, Norway gave talks.

Papers

According to discussions of JURISIN 2013, authors revised their papers. The program committee reviewed these revised papers, and selected five papers.

"Requirements of legal knowledge management systems to aid normative reasoning in specialist domains" (Alessio Antonini, et. al.) focuses on the interplay between industry/professional standards and legal norms, the information gap between legal and specialist domains and the need for interpretation at all stages of compliance. They propose extensions to the Eunomos legal knowledge management tool to help address the information gap, with particular attention to aligning norms with operational procedures, and the use of domain-specific specialist ontologies from multiple domains to help users understand and reason with norms on specialist topics.

"ArgPROLEG: A Normative Framework for The JUF Theory" (Zohreh Shams, et. al.) proposes ArgPROLEG, a normative framework for legal reasoning based on

PROLEG, an implementation of the the Japanese "theory of presupposed ultimate facts" (JUF). This theory was mainly developed with the purpose of modelling the process of decision making by judges in the court.

"Answering Yes/No Questions in Legal Bar Exams" (Mi-Young Kim, et. al.) develops a QA approach to answer yes/no questions relevant to civil laws in legal bar exams. The first step is to identify legal documents relevant to the exam questions; the second step is to answer the questions by analyzing the relevant documents. Their experimental results show reasonable performance, which improves the baseline system, and outperforms an SVM-based supervised machine learning model.

"Answering Legal Questions by Mining Reference Information" (Oanh Thi Tran, et. al.) presents a study on exploiting reference information to build a question answering system restricted to the legal domain. To cope with referring to multiple documents, they propose a novel approach which exploits the reference information among legal documents to find answers. he experimental results showed that the proposed method is quite effective and outperform a traditional QA method, which does not use reference information.

"Belief Re-revision in Chivalry Case" (Pimolluck Jirakunkanok, et. al.) proposes a formalization of legal judgment revision in terms of dynamic epistemic logic, with two dynamic operators; commitment and permission. In order to demonstrate their formalization, they analyze judge's belief change in Chivalry Case in which a self-defense causes a misconception.

Acknowledgement

JURISIN 2013 was held in conjunction with JSAI-isAI 2013 supported by JSAI. We thank all staffs of JSAI-isAI 2013 and JSAI for their suppots.

And also JURISIN 2013 was supported by members of the steering committee, the program committee and advisory committee. We really appreciate their support.

Multimodality in Multiparty Interaction (MiMI2013)

Mayumi Bono[1,2] and Yasuyuki Sumi[3]

[1] Digital Content and Media Sciences Research Division,
National Institute of Informatics
[2]Department of Informatics, School of Multidisciplinary Sciences,
Graduate University of Advanced Studies (SOKENDAI)
[3] Department of Complex and Intelligent Systems, Future University Hakodate

The Workshop

The International Workshop on Multimodality in Multiparty Interaction (MiMI2013) took place at Keio University, Kanagawa, on October 28, 2013. This was held as part of the Japan Society for Artificial Intelligence (JSAI) International Symposia on Artificial Intelligence (JSAI-isAI 2013), sponsored by JSAI and Innovation for Interdisciplinary Approaches across the Humanities and Social Sciences, of the Japan Society for the Promotion of Science (JSPS).

In this workshop, we tried to cover a broad range of perspectives related to interaction studies, communication studies, conversation analysis, and workplace studies and their application to other research fields including, but not limited to, human–computer interaction (HCI). Moreover, we tried to provide a space where HCI researchers who have created original work can start collaborative projects with interaction and communication analysts to evaluate their products and upgrade their perspectives on human interaction in our daily lives. Recently, the interest of linguists, interaction analysts, and conversation analysts has turned increasingly toward observing interactional practices in the material world (Streeck et al., 2011). To turn to new domains of communication, we need to focus not only on the systematic structure of dyadic dialogue, i.e., two-party interaction, but also on the complexity of conversations involving more than three. Our daily communication is not limited to dyadic dialogue, but open to multi-party interactions.

Multimodality is a research concept that emerges from the history of traditional language research that has treated only verbal and text information of human language. Human social interaction involves the intertwined interaction of different modalities, such as talk, gesture, gaze, and posture. Human–computer interactions involve studying, planning, and designing interactions between humans and computers. Traditionally, HCI researchers have adopted the methodologies of experimental psychology to evaluate their products by measuring human behaviors and human knowledge under experimental conditions. However, we believe that experimental settings are limited in their ability to study human daily interactions.

Since 2012, we have been conducting the Ido-Robo project as one of the grand challenge projects at the National Institute of Informatics (NII-Today, 2014). Can a robot engage in communicative activities such as gossiping beside the well? Ido-bata kaigi (congregate at the side of a well) is a concept in Japanese that reflects how women living in villages used to chat, circulate gossip, and exchange community information as they gathered beside a well and washed clothes and pumped water from the well. Now, that phrase refers to spontaneous congregations that serve as hubs for the communicative, intellectual, and political life of Japanese people. Such a phenomenon is not yet possible even for a robot manufactured with the latest technology.

Figuratively speaking, we hope to build an infrastructure that will enable robots to congregate and engage in small talk, which is based on an interdisciplinary research framework involving scholars in linguistics, cognitive science, information science, sociology, and robotics. In this workshop, we discussed how a marriage between interaction studies and informatics could affect developments in both research fields.

Papers and Future plans

This one-day workshop included two invited talks and five general papers. The first invited talk was a curious talk entitled 'Social Robotics in Classrooms' by Prof. Fumihide Tanaka from the perspective of engineering research. Then Prof. Morana Alač gave an insightful talk entitled 'Just a Robot: Haptic Interaction and the Thingness of the Social Robot' from the perspective of the social sciences. These two speakers once collaborated at the University of California, San Diego (UCSD). From their different perspectives, they outlined how they conducted their research projects in classrooms using social robotics. These invited talks were a good example of the marriage between interaction studies and informatics.

The general paper sessions discussed several aspects of multi-party interactions: listener's nonverbal behaviors in Bibliobattle; the sequential structure of improvisation in Robot-Human Theater; listener's behaviors during table talk; conversation during table cooking; and the home position of gestures in multi-party conversation. We accepted one paper from the invited session and one paper from the general session for the post-proceedings.

Currently, we are preparing for a special session of Multimodality in Multiparty Interactions (MiMI) with Social Robots: Exploring Human–Robot Interaction (HRI) in the Real World (MiMI2014), for the 23rd IEEE International Symposium on Robot and Human Interactive Communication (IEEE RO-MAN 2014), which will be held in Edinburgh, Scotland, on August 25-29, 2014. As we discussed numerous aspects of multi-party interactions at MiMI2013, we will try to focus on a number of aspects of interactions with social robots. Through such activities, we hope to continue to be able to discuss the main issues of MiMI.

References

1. Streeck, J., Goodwin, C., and LeBaron, C.: Embodied Interaction Language and Body in the Material World. Cambridge University Press (2011)
2. NII Today: Can a Robot Join an Idobata Kaigi?. NII Today No.48 (2014)

Argument for Agreement and Assurance (AAA)

Hiroyuki Kido[1], Yoshiki Kinoshita[2], Kenji Taguchi[3], and Kazuko Takahashi[4]

[1] The University of Tokyo
[2] Kanagawa University
[3] National Institute of Advanced Industrial Science and Technology
[4] Kwansei Gakuin University

Preface

Lattice theoretical and combinatorial analysis of argumentation has now become an established field in Artificial Intelligence and there is much hope for its application to agreement formation and consensus building. On the other hand, there is a growing interest in assurance cases in Assurance Engineering, where the logical analysis of arguments by Toulmin is much appreciated. Both of these two activities aim at analysis of arguments, but it seems they have had rather few interaction with each other. The aim of this workshop is to encourage exchange of idea between these two fields. To that end, we called for submissions of the work in the following topics.

- Abstract and structured argumentation systems including studies of frameworks, proof-theories, semantics and complexity.
- Dialogue systems for persuasion, negotiation, deliberation, eristic and information-seeking dialogues.
- Applications of argumentation and dialogue systems to various fields such as agreement technologies, systems assurance, safety engineering, multi-agent systems, practical reasoning, belief revision, learning and semantic web.
- Agreement and assurance technologies through arguments including safety cases, assurance cases and dependability cases.
- Tools for argumentation systems, dialogue systems, argument-based stakeholders' agreement, argument-based accountability achievement, argument-based open systems dependability and argument-based verification and validation.

We have seven contributed talks and one invited lecture by Professor Tim Kelly in the workshop. Moreover, AAA 2013 have made an agreement with JURISIN 2013 that participants of AAA are encouraged to attend the JURISIN invited lecture by Dr. Caminada.

Workshop on Data Discretization and Segmentation for Knowledge Discovery (DDSS13)

Akihiro Yamamoto[1], Tetsuji Kuboyama[2], and Hiroshi Sakamoto[3]

[1] Graduate School of Informatics, Kyoto University, Japan
akihiro@i.kyoto-u.ac.jp
[2] Computer Centre, Gakushuin University, Japan
ori-ds2013@tk.cc.gakushuin.ac.jp
[3] Graduate School of Computer Science and Systems Engineering,
Kyushu Institute of Technology, Japan
hiroshi@ai.kyutech.ac.jp

The Workshop

The Workshop on Data Discretization and Segmentation for Knowledge Discovery (DDS13) was held on October 27th at Keio University in the fifth JSAI International Symposia on AI (JSAI-isAI 2013), sponsored by the Japan Society for Artificial Intelligence (JSAI).

DDS13 is the first workshop on subjects related to discretization and segmentation of data in developing methods for discovering knowledge from large-scale data. Originally, decomposing one datum into a set of several small data is found to play a crucial role in computer science. Decomposition usually indicates an operation to consecutive data structure, but in the context of Machine Learning and Knowledge Discovery, it can include, for example, segmentation of sentences in natural languages and segmentation of time series. Moreover decomposition can be extended so that it may include discretization of continuous data, data compression, and algebraic methods for Knowledge Discovery. We focus this workshop on decomposition methods of various types of data, such as graphs, trees, strings, and continuous data, and their applications to Machine Learning and Knowledge Discovery. We welcomed scientific results based on, but not restricted to, decomposition of data and its applications to bioinformatics, natural language processing, social network analysis, and other related areas.

We first organized the program committee consisting of 14 researchers concerning with subjects in the workshop scope, and announced a call for papers. As the result of review by the PC members, 10 submitted papers were accepted. More information on DDS13 is available at the workshop homepage[1]. The proceedings were published from JSAI[2].

---

[1] https://sites.google.com/site/dds13workshop
[2] ISBN 978-4-915905-59-9 C3004(JSAI).

Post-Workshop Proceedings

Five papers in those presented in the workshop were submitted after revision to this post-workshop proceedings. Each of them was peer reviewed by three PC members, which consists of two PC members previously assigned plus another, and eventually the PC selected three papers.

Two of the papers are focused on trees. Kernel functions for structured data such as trees are in the scope of the workshop because most of them are defined by decomposing input data along the structure. The third paper treats mixed-type data, which contain both discrete and continuous features. In order to treat such types of data in formal concept analysis (FCA), an algebraic method for analyzing data, the author discretizes continuous features. The abstracts of the three papers are following. Hamada et al. introduced a kernel for tree data based on counting all of the agreement subtree mappings, and designed an efficient algorithm to compute the kernel value for unordered leaf-labeled full binary trees. Then they applied it to analysis of nucleotide sequences for A (H1N1) influenza viruses. They also showed that the problem of counting all of the agreement subtree mappings is #P-complete unless the trees are full binary.

Shin et al. provided comprehensive research on various tree kernels proposed previously, in order to choose good ones for analyzing data. They picked up 32 algorithms for tree kernels under two different parameter settings, and showed that three of the 64 tree kernels are superior to the others with proving statistically significant through t-tests.

Sugiyama introduced a method for detecting outliers in mixed-type data, based on FCA. In this method, a lattice of concepts which represents a hierarchy of clusters is constructed, and outliers are clusters highly isolated in the hierarchy. Though continuous features are discretized for the method, he showed by experiments that the method detects outliers more effectively than other popular distance-based methods.

Acknowledgments

DDS13 was closed successfully. We are grateful for the great support received from the program committee members: Hiroki Arimura, Basabi Chakraborty, Kouichi Hirata, Yoshinobu Kawahara, Nobuhiro Kaji, Noriaki Kawamae, Tesuhiro Miyahara, Yoshiaki Okubo, Takeshi Shinohara, and Tomoyuki Uchida. Most of the PC members are steering members of the Special Interest Group on Fundamental Problems in AI (SIG-FPAI), chaired by Kuboyama. We are thankful to Prof. Yukiko Nakano for her organization of JSAI-isAI 2013. We also thank Prof. Ken Satoh and Prof. Daisuke Bekki for their arrangement to publish the LNAI volume of these post-workshop proceedings. Finally, we thank all speakers and all audiences who attended the workshop.

Contents

LENLS

A Type-Theoretic Account of Neg-Raising Predicates in Tree Adjoining Grammars

Laurence Danlos[1,2,3], Philippe de Groote[4,5,6], and Sylvain Pogodalla[4,5,6(✉)]

[1] Université Paris Diderot (Paris 7), 75013 Paris, France
[2] ALPAGE, INRIA Paris-Rocquencourt, 75013 Paris, France
[3] Institut Universitaire de France, 75005 Paris, France
laurence.danlos@inria.fr
[4] INRIA, 54600 Villers-lès-Nancy, France
[5] Université de Lorraine, LORIA, UMR 7503, 54500 Vandœuvre-lès-Nancy, France
[6] CNRS, LORIA, UMR 7503, 54500 Vandœuvre-lès-Nancy, France
{philippe.degroote,sylvain.pogodalla}@inria.fr

Abstract. Neg-Raising (NR) verbs form a class of verbs with a clausal complement that show the following behavior: when a negation syntactically attaches to the matrix predicate, it can semantically attach to the embedded predicate. This paper presents an account of NR predicates within Tree Adjoining Grammar (TAG). We propose a lexical semantic interpretation that heavily relies on a Montague-like semantics for TAG and on higher-order types.

Keywords: Tree Adjoining Grammar · Semantics · λ-calculus · Type theory · Neg-raising

1 Introduction

Neg-Raising (NR) verbs form a class of verbs with a clausal complement that show the following behavior: when a negation syntactically attaches to the matrix predicate, it can semantically attach to the embedded predicate, as the implication of (1c) by (1b) shows. This corresponds to the NR reading of this predicate.

(1)   a.   Marie pense que Pierre partira
           Mary thinks that Peter will leave

       b.   Marie ne pense pas   que Pierre partira
           Mary does not think that Peter will leave

       c.   Marie pense que Pierre ne partira pas
           Mary thinks that Peter will not leave

This work has been supported by the French agency Agence Nationale de la Recherche (ANR-12-CORD-0004).

Y. Nakano et al. (Eds.): JSAI-isAI 2013, LNAI 8417, pp. 3–16, 2014.
DOI: 10.1007/978-3-319-10061-6_1

Such an implication does not always hold. Some contexts make it impossible to consider the negation as having scope over the embedded predicate only [1, 2]. This corresponds to the non-NR reading of the predicate.

This paper aims at providing an account of NR predicates within Tree Adjoining Grammar (TAG) [3]. We propose a lexical semantic interpretation that heavily relies on a Montague-like semantics for TAG and on higher-order types. As a base case, our approach lexically provides both NR and non-NR readings to NR predicates. We implement our proposal in the Abstract Categorial Grammar (ACG) framework [4] as it offers a fairly standard interface to logical formal semantics for TAG. However, our approach could be implemented in other synchronous frameworks such as Synchronous TAG [5–7].

Not all sentence-embedding predicates show a NR behavior [8,9]. In order to test whether a predicate is NR, we can look at its interaction with Negative Polarity Items (NPI). Such items need to be in the scope of a negative operator in order for the utterance to be felicitous as the contrast between (2a) and (2b) shows. This contrast also shows up when the NPI occurs within the *positive* clause embedded under a *negative* matrix clause as in (3).[1] On the other hand, non-NR predicates do not allow NPI[2] in a positive embedded clause even if it is itself in a negative context (4b).

(2)   a.   Pierre n'est pas dans son assiette
           Peter   doesn't feel good

       b.   *Pierre est dans son assiette
           Peter   feels good

(3)   a.   Marie pense que Pierre n'est pas dans son assiette
           Mary   thinks that Peter   doesn't feel good

       b.   Marie ne pense pas que Pierre soit dans son assiette
           Marie   doesn't think that Peter   feels good

       c.   *Marie pense que Pierre est/soit dans son assiette
           Mary   thinks that Peter   feels good

(4)   a.   Marie affirme que Pierre n'est pas dans son assiette
           Mary   claims   that Peter   doesn't feel good

       b.   *Marie n'affirme pas que Pierre est/soit dans son assiette
           Mary   doesn't claim that Peter   feels good

The modeling we propose takes into account several constraints or properties of NR-predicates. First, the availability of the different readings should not introduce spurious ambiguities, in particular when no negation occur. To achieve this, we make use of fine-grained typing.

---

[1] We do not discuss here the use of subjunctive in the embedded clause when the matrix predicate is in negative form as it does not add to the constraints we describe.

[2] At least some NPI. Some other NPI seems to perfectly occur in a similar position. See [2].

Second, we want to give an account of *NR cyclicity*. This phenomenon occurs when a NR predicate embeds another NR predicate: a negation at the matrix level will semantically cycle down to the most embedded NR predicate, giving rise to several possible interpretations as (5) shows with the interpretations (5a–b). We achieve this effect by making use of higher-order types. In particular, the clausal argument is type-raised so that it can further be modified. Note however that according to [2,10] we may want to block NR cyclicity when a NR desire predicate embeds a NR belief predicate. This amounts to force non-NR readings of the predicate.

(5) Marie ne pense pas que Jeanne croie que Pierre partira
    Mary doesn't think that Jane believes Peter will leave

   a. $\neg$(**think** (**believe** (**leave p**) **j**) **m**)

   b. **think** ($\neg$(**believe** (**leave p**) **j**)) **m**

   c. **think** (**believe** ($\neg$(**leave p**)) **j**) **m**

A similar effect on forcing specific readings occurs with NPI. When a NPI occurs in a positive embedded clause, it forces a NR reading of the (negated) matrix predicate [11]. Then, the negation has scope over the embedded sentence, but not over the whole sentence.

On the other hand, some adverbial discourse connectives (ADCs) force a NR reading. [12] discusses the syntax-semantics mismatches of arguments of an ADC in French and introduces three principles. It illustrates in particular the phenomena with complex sentences that include an ADC (e.g. *par contre (however)*) in the matrix clause and shows that this ADC can have scope: over the whole sentence (principle 1) as in (6), over the embedded clause only (principle 2) as in (7), or over the negation of the embedded clause with a NR reading of the matrix predicate (principle 3) as in (8).

(6) Fred ira à Dax pour Noel. Jeanne pense, par contre, qu'il
    Fred will go to Dax for Christmas. Jane thinks, however, that he
    n'ira pas.
    will not go.

(7) Fred ira à Dax pour Noel. Jeanne pense, par contre, que
    Fred will go to Dax for Christmas. Jane thinks, however, that
    Pierre n'ira pas.
    Peter will not go.

(8) Fred ira à Dax pour Noel. Jeanne ne pense pas, par contre,
    Fred will go to Dax for Christmas. Jane doesn't think, however,
    que Pierre ira.
    that Peter will go.

Our long term goal is to provide a semantic and discourse analysis of these phenomena within TAG and D-STAG [13] that would account for these interactions. However, in this article, we only deal with the NR and non-NR readings of NR predicates. We propose a simply typed λ-calculus approach at the semantic level, while we take the standard analysis of NR predicates as auxiliary trees allowing for adjunction at the syntactic level.

2   Verbs with Clausal Arguments in TAG

In TAG, verbs that have sentential arguments usually are represented as auxiliary trees (see [14, Sect. 6.7] for English and [15, Chap. 3, Sect. 1.2] for French) in order to allow for describing long distance dependencies with multiple embeddings as in (9). Negation is analyzed with adjunction as well[3]. Figure 1(a) and (b) show the TAG analysis of (1b) and Fig. 1(c) shows the corresponding derivation tree γ_0: the auxiliary tree of *ne pas* is adjoined to the V node of the auxiliary tree of *pense que*, and the latter is adjoined to the S node of the initial tree of *partira*. The initial trees of *Marie* and *Pierre* are substituted to the NP nodes of the trees of *pense que* and *partira* resp.

(9)   a.   Quelle fille Paul pense-t-il que Bob sait que Jean aime ?   [15, p. 234]

      b.   What did Bill tell Mary that John said?   [14, p. 43]

3   Type-Theoretic Perspective on TAG

3.1   Abstract Categorial Grammars

ACGs provide a framework in which several grammatical formalisms may be encoded. They generate languages of linear λ-terms, which generalize both string and tree languages. A key feature is to provide the user direct control over the parse structures of the grammar, the *abstract language*, which allows several grammatical formalisms to be defined in terms of ACG, in particular TAG [16]. In this perspective, derivation trees of TAG are straightforwardly represented as terms of the abstract language, while derived trees (and yields) are represented by terms of the object language. We refer the reader to [4, 17] for the details and introduce here only few relevant definitions and notations.

A *higher-order linear signature* defines a finite set of atomic types and a finite set of typed (possibly with complex types $\alpha \to \beta$) constants. It is also called a *vocabulary*. $\Lambda(\Sigma)$ is the set of λ-terms built on Σ, and for $t \in \Lambda(\Sigma)$ such that t has type α, we note $t : \alpha$.

An *abstract categorial grammar* is a quadruple $\mathscr{G} = \langle \Sigma, \Xi, \mathscr{L}, s \rangle$ where Σ and Ξ are two higher-order linear signatures, which are called the *abstract vocabulary* and the *object vocabulary* respectively. $\mathscr{L} : \Sigma \longrightarrow \Xi$ is a lexicon from the abstract

---

[3] For sake of clarity, we use a simplified version here.

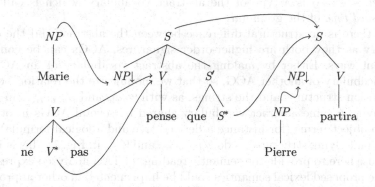

(a) Tree operations

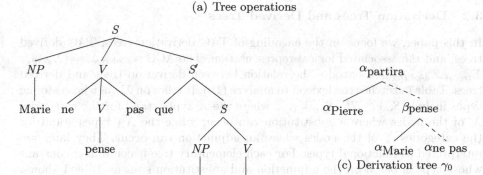

(b) Derived tree c_0

(c) Derivation tree γ_0

Fig. 1. TAG analysis of *Marie ne pense pas que Pierre partira*

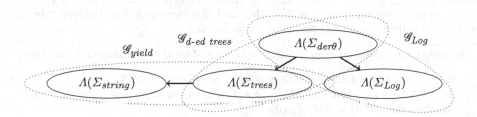

Fig. 2. ACG architecture for TAG

vocabulary to the object vocabulary. It is a homomorphism.[4] We note $t := u$ if $\mathscr{L}(t) = u$. $s \in \mathscr{T}_\Sigma$ is a type of the abstract vocabulary, which is called the *distinguished type* of the grammar.

Since there is no structural difference between the abstract and the object vocabulary as they both are higher-order signatures, ACGs can be combined in different ways. Either by making the abstract vocabulary of an ACG the object vocabulary of another ACG, so that we modularize the relation between the derivation structures and the strings, as with $\mathscr{G}_{yield}$ and $\mathscr{G}_{d\text{-}ed\ trees}$ in Fig. 2. Or by having a same abstract vocabulary shared by several ACGs in order to make two object terms (for instance a derived tree and a logical formula) share the same underlying structure, as do $\mathscr{G}_{d\text{-}ed\ trees}$ and $\mathscr{G}_{Log}$ in Fig. 2. This is indeed what we use here to provide the semantic readings of TAG analyzed expressions, even if the proposed lexical semantics could be implemented in other approaches, in particular synchronous approaches [5,6,18].

3.2   Derivation Trees and Derived Trees

In this paper, we focus on the encoding of TAG derivation trees, TAG derived trees, and the associated logical representation. The ACG $\mathscr{G}_{d\text{-}ed\ trees} = \langle \Sigma_{der\theta},$ $\Sigma_{trees}, \mathscr{L}_{d\text{-}ed\ trees}, S \rangle$ encodes the relation between derivation trees and derived trees. Table 1 sketches the lexicon to analyze (1b). It relies on $\Sigma_{der\theta}$ whose atomic types include S, V,[5] NP, S_A, V_A... where the X types stand for the categories X of the nodes where a substitution can occur while the X_A types stand for the categories X of the nodes where an adjunction can occur. They later are interpreted as functional types. For each elementary tree it contains a constant whose type is based on the adjunction and substitution sites as Table 1 shows. It additionally contains constants $\beta_X^{Id} : X_A$ that are meant to provide a fake auxiliary tree on adjunction sites where no adjunction actually takes place in a TAG derivation.

The other signature, Σ_{trees}, has τ the type of trees as unique atomic type. Then, for any X of arity n belonging to the ranked alphabet describing the elementary trees of the TAG, we have a constant $X_n : \underbrace{\tau \to \cdots \to \tau}_{n \text{ times}} \to \tau$.

The relation between these vocabularies is given by the lexicon $\mathscr{L}_{d\text{-}ed\ trees}$ where $\mathscr{L}_{d\text{-}ed\ trees}(X_A) = \tau \to \tau$ and for any other type X, $\mathscr{L}_{d\text{-}ed\ trees}(X_A) = \tau$. Then, the derivation tree γ_0 of Fig. 1(c) and the corresponding derived tree c_0 of Fig. 1(b) are represented and related as follows:

---

[4] In addition to defining $\mathscr{L}$ on the atomic types and on the constants of Σ, we have:

- if $\alpha \to \beta$ is a type build on Σ then $\mathscr{L}(\alpha \to \beta) = \mathscr{L}(\alpha) \to \mathscr{L}(\beta)$;
- if $x \in \Lambda(\Sigma)$ (resp. $\lambda x.t \in \Lambda(\Sigma)$ and $t\,u \in \Lambda(\Sigma)$) then $\mathscr{L}(x) = x$ (resp. $\mathscr{L}(\lambda x.t) = \lambda x.\mathscr{L}(t)$ and $\mathscr{L}(t\,u) = \mathscr{L}(t)\,\mathscr{L}(u)$);

with the proviso that for any constant $c : \alpha$ of Σ we have $\mathscr{L}(c) : \mathscr{L}(\alpha)$.

[5] We follow [15] and we do not use VP categories. Using it would not change our analysis.

$\gamma_0 = \alpha_{\text{partira}} \, (\beta_{\text{pense}} \, \beta_S^{Id} \, \beta_{\text{ne pas}} \, \alpha_{\text{Marie}}) \, \beta_V^{Id} \, \alpha_{\text{Pierre}}$

$c_0 = \mathscr{L}_{d\text{-}ed\ trees}(\gamma_0)$

$\quad = S_3(NP_1 \text{ Marie})(V_3 \text{ ne}(V_1 \text{ pense})\text{pas}) \, (S'_2 \text{ que}(S_2(NP_1 \text{ Pierre})(V_1 \text{ partira})))$

Parallel (or synchronous) to this interpretation as derived trees, we can also interpret terms representing derivation trees as logical formulas representing the associated meanings.

Table 1. TAG as ACG: the $\mathscr{L}_{d\text{-}ed\ trees}$ lexicon

Abstract constants of $\Sigma_{der\theta}$	Their images by $\Sigma_{der\theta}$ and the corresponding TAG trees
$\alpha_{\text{Marie}} : NP$	$c_{Marie} = NP_1 \text{ Marie} : \tau$ $\gamma_{Marie} = \begin{array}{c} NP \\ \mid \\ \text{Marie} \end{array}$
$\alpha_{\text{partira}} : S_A \to V_A \to NP \to S$	$c_{partira} = \lambda^\circ a \, v \, s.s \ (S_2 \, s \, (v \, (V_1 \text{ partira})))$ $: (\tau \multimap \tau) \multimap (\tau \multimap \tau) \multimap \tau \multimap \tau$ $\gamma_{partira} = \begin{array}{c} S \\ \overbrace{\quad\quad} \\ NP{\downarrow} \quad V \\ \mid \\ \text{partira} \end{array}$
$\beta_{\text{ne pas}} : V_A$	$c_{ne\ pas} = \lambda^\circ x. V_3 \text{ ne } x \text{ pas} : \tau \to \tau$ $\gamma_{ne\ pas} = \begin{array}{c} V \\ \overbrace{\quad\quad\quad} \\ \text{ne} \quad V^* \quad \text{pas} \end{array}$
$\beta_{\text{pense}} : S_A \to V_A \to NP \to S_A$	$c_{pense\ que} = \lambda^\circ s \, v \, x \, y.s \ (S_3 \, x(v \, (V_1 \text{ pense})) \, (S'_2 \text{ que } y))$ $: (\tau \to \tau) \to (\tau \to \tau) \to \tau \to (\tau \to \tau)$ $\gamma_{pense\ que} = \begin{array}{c} S \\ \overbrace{\quad\quad\quad} \\ NP{\downarrow} \quad V \quad S' \\ \mid \quad\quad \overbrace{\quad} \\ \text{pense} \quad \text{que} \quad S^* \end{array}$

3.3 Building Semantic Representations

In order to define the translation of terms denoting derivation trees into a logical formula with the ACG $\mathscr{G}_{Log} = \langle \Sigma_{der\theta}, \Sigma_{Log}, \mathscr{L}_{Log}, S \rangle$, we need to define the interpretation of each atomic type and of each constant, and then to consider the homomorphic extension of this interpretation. In other worlds, we have to define the semantic recipe of each lexical item.

The higher-order signature Σ_{Log} for the logical representation defines the following typed constants:

$$\mathbf{m} : e \qquad \neg : t \to t \qquad \mathbf{leave} : e \to t \qquad \mathbf{think} : t \to e \to t$$

And we consider the following interpretation $\mathscr{L}_{Log}$:

$$
\begin{aligned}
S &:= t & \alpha_{\text{Marie}} &:= \mathbf{m} \\
NP &:= e & \alpha_{\text{partira}} &:= \lambda p\, v\, s.p\, (\lambda f.f\, (v\, (\mathbf{leave}\, s))) \\
V &:= e \to t & \beta_X^{Id} &:= \lambda x.x \\
V_A &:= t \to t & \beta_{\text{ne pas}} &:= \lambda p.\neg\, p \\
S_A &:= ((t \to t) \to t) & \beta_{\text{pense}} &:= \lambda p\, v\, s\, c.p\, (\lambda f.f\, (v\, (\mathbf{think}\, (c\, (\lambda x.x))\, s))) \\
&\quad \to (t \to t) \to t & \beta'_{\text{pense}} &:= \lambda p\, v\, s\, c.p\, (\lambda f.\mathbf{think}\, (f\, (c\, v))\, s)
\end{aligned}
$$

We model the NR and non-NR readings with two possible auxiliary trees for NR predicates[6]: one with the non-NR reading (β_{pense}) and one with the NR reading (β'_{pense}), both with the same type $S_A \to V_A \to NP \to S_A$.

Remark 1. An important point is the type interpreting S_A. In previous works [17], S_A was interpreted with a $t \to t$ type, a function from truth values to truth values where the parameter corresponded to the truth value associated with the embedded clause. This gave the following interpretation:

$$
\beta'_{\text{pense}} := \lambda p\, v\, s\, c.p\, (\mathbf{think}\, (v\, c)\, s)
$$

where the c parameter conveys the meaning of the embedded clause. However, in this setting, this meaning cannot further be subject to changes, in particular by a negation provided by the v parameter (interpreting the V_A adjunction site of the verb *pense*). While this would provide us with a reading where the negation scopes over the embedded clause only, it cannot cycle down to a clause more deeply embedded under another NR predicate.

In order to model the NR cyclicity we then need that the semantic argument of the negation can occur arbitrarily far away from the matrix verb as in (5). With $S_A := ((t \to t) \to t) \to ((t \to t) \to t)$, the modification stipulated by the second argument v of $[\![\beta'_{\text{pense}}]\!]$ (which will be replaced by the semantics of the auxiliary tree adjoined to the V_A node of β'_{pense}, typically the negation) will be given as argument to the "raised" clausal argument c so that it can possibly cycle down if c itself represents a NR predicate.

In $[\![\beta_{\text{pense}}]\!]$, v directly has scope over the **think** predicate and c, applied to the identity $\lambda x.x$, is not modified.

Remark 2. Another difference in the interpretation of type modeling the adjunction occurs in the interpretation of $[\![V_A]\!]$. We usually have $V_A := (e \to t) \to (e \to t)$ for V modifiers interpretations. It should not be surprising that we need to actually consider two V adjunction sites since their semantic contribution definitely differ. Technically, it amounts to duplicate the V node to create two adjunction sites in the elementary trees for NR predicates:

---

[6] We use the $\beta_{\text{NR predicate}}$ notation for constants to be interpreted with the non-NR reading, and $\beta'_{\text{NR predicate}}$ for those to be interpreted with the NR reading.

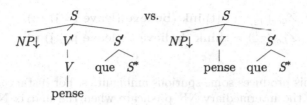

where one of the sites is dedicated to the negation and the other one to usual V adjunctions such as auxiliaries. We would have

$$\beta'_{\text{pense}} : \quad S_A \to V'_A \to V_A \to NP \to S_A$$

$$\beta'_{\text{pense}} := \lambda p \; v'_1 \; v_2 \; s \; c.p \; (\lambda f.(v'_1 \; (\lambda y.\textbf{think} \; (f \; (c \; v_2)) \; y) \; s))$$

This would prevent the meaning of auxiliaries (represented by the parameter v'_1) to modify the meaning of the embedded clauses, contrary to the meaning of the negation (represented by the parameter v_2). However, for sake of simplicity, we drop this additional adjunction site as we do not provide any example using it in this paper.

We now are in position to give examples of interpretations. (1b) can be given the derivation trees and the associated meanings:

$$\gamma_0 = \alpha_{\text{partira}}(\beta_{\text{pense}} \; \beta_S^{Id} \; \beta_{\text{ne pas}} \; \alpha_{\text{Marie}})\beta_V^{Id} \; \alpha_{\text{Pierre}}$$

$$\gamma_1 = \alpha_{\text{partira}}(\beta'_{\text{pense}} \; \beta_S^{Id} \; \beta_{\text{ne pas}} \; \alpha_{\text{Marie}})\beta_V^{Id} \; \alpha_{\text{Pierre}}$$

$$\mathscr{L}_{Log}(\gamma_0) = \neg(\textbf{think} \; (\textbf{leave} \; \textbf{p}) \; \textbf{m})$$

$$\mathscr{L}_{Log}(\gamma_1) = \textbf{think} \; (\neg(\textbf{leave} \; \textbf{p})) \; \textbf{m}$$

Extending in a straightforward way the given lexicon with other NR predicates, we can give (5) the following derivation trees:[7]

$$\gamma'_0 = \alpha_{\text{partira}}(\beta_{\text{croie}}(\beta_{\text{pense}}\beta_S^{Id}\beta_{\text{ne pas}}\alpha_{\text{Marie}})\beta_V^{Id}\alpha_{\text{Jeanne}})\beta_V^{Id} \; \alpha_{\text{Pierre}}$$

$$\gamma'_1 = \alpha_{\text{partira}}(\beta_{\text{croie}}(\beta'_{\text{pense}}\beta_S^{Id}\beta_{\text{ne pas}}\alpha_{\text{Marie}})\beta_V^{Id}\alpha_{\text{Jeanne}})\beta_V^{Id} \; \alpha_{\text{Pierre}}$$

$$\gamma'_2 = \alpha_{\text{partira}}(\beta'_{\text{croie}}(\beta_{\text{pense}}\beta_S^{Id}\beta_{\text{ne pas}}\alpha_{\text{Marie}})\beta_V^{Id}\alpha_{\text{Jeanne}})\beta_V^{Id} \; \alpha_{\text{Pierre}}$$

$$\gamma'_3 = \alpha_{\text{partira}}(\beta'_{\text{croie}}(\beta'_{\text{pense}}\beta_S^{Id}\beta_{\text{ne pas}}\alpha_{\text{Marie}})\beta_V^{Id}\alpha_{\text{Jeanne}})\beta_V^{Id} \; \alpha_{\text{Pierre}}$$

and the associated meanings:

$$\mathscr{L}_{Log}(\gamma'_0) = \neg(\textbf{think} \; (\textbf{believe} \; (\textbf{leave} \; \textbf{p}) \; \textbf{j}) \; \textbf{m})$$

$$\mathscr{L}_{Log}(\gamma'_1) = \textbf{think} \; (\neg \; (\textbf{believe} \; (\textbf{leave} \; \textbf{p}) \; \textbf{j})) \; \textbf{m}$$

---

[7] In the described architecture, the semantic ambiguities are derived from "derivation ambiguities". We can avoid this in considering an intermediate level between Σ_{derv} and Σ_{trees}. β_{pense} and β'_{pense} would map on the same term of this intermediate level, and the latter would be considered as the actual derivation tree representation level. The upper part would then be considered as lying within the semantic device. However this intermediate level would not provide any additional modeling capability so we do not consider it here.

$$\mathscr{L}_{Log}(\gamma_2') = \neg(\textbf{think}\ (\textbf{believe}\ (\textbf{leave p})\ \textbf{j})\ \textbf{m})$$
$$\mathscr{L}_{Log}(\gamma_3') = \textbf{think}\ (\textbf{believe}\ (\neg(\textbf{leave p}))\ \textbf{j})\ \textbf{m}$$

However, this produces some spurious ambiguities. For instance, assigning a NR reading to the intermediary NR predicate when the matrix NR predicate has a non-NR reading (as for γ_2') is useless. Similarly, when no negation occur in the matrix predicate, using the NR readings yields the same result as using the non-NR one as the following possible derivations and their interpretations for (1a) shows:

$$\gamma_0'' = \alpha_{\text{partira}}(\beta_{\text{pense}}\ \beta_S^{Id}\ \beta_V^{Id}\ \alpha_{\text{Marie}})\beta_V^{Id}\ \alpha_{\text{Pierre}}$$
$$\gamma_1'' = \alpha_{\text{partira}}(\beta_{\text{pense}}'\ \beta_S^{Id}\ \beta_V^{Id}\ \alpha_{\text{Marie}})\beta_V^{Id}\ \alpha_{\text{Pierre}}$$
$$\mathscr{L}_{Log}(\gamma_0'') = \textbf{think}\ (\textbf{leave p})\ \textbf{m}$$
$$\mathscr{L}_{Log}(\gamma_1'') = \textbf{think}\ (\textbf{leave p})\ \textbf{m}$$

4 Improvements

4.1 Avoiding Spurious Ambiguities

In order to avoid spurious ambiguities, we refine the types for S and V adjunction sites. There are several ways to present this refinement using records and dependent types as proposed in [19,20] that are rather similar to feature structures. As it would involve additional notations, we prefer to keep the atomic types we have used so far, but instead of the atomic type S_A and V_A we now have $S_A[\text{Neg} = no]$, $S_A[\text{Neg} = yes]$, $V_A[\text{Neg} = no]$, and $V_A[\text{Neg} = yes]$ as atomic types.

With these types, an auxiliary tree for a NR predicate will have type:

$$\beta : S_A[\text{Neg} = m] \rightarrow V_A[\text{Neg} = n] \rightarrow NP \rightarrow S_A[\text{Neg} = r] \text{ with } (m,n,p) \in \{\text{yes}, \text{no}\}^3$$

While in principle we could instantiate these auxiliary trees with all the possible combinations, removing some of them will prevent us from getting unwanted readings.

The accepted combinations for non-NR readings are given in Table 2(a) and the ones for NR readings in Table 2(b). The tables show the resulting type $S_A[\text{Neg} = r]$ for each combination of $S_A[\text{Neg} = m]$ and $V_A[\text{Neg} = n]$ values. When a cell is empty, it means there is no term with this type combination.

Table 2. Allowed feature combinations

(a) Combinations for non-NR readings

	$S_A[\text{Neg} = no]$	$S_A[\text{Neg} = yes]$
$V_A[\text{Neg} = no]$	$S_A[\text{Neg} = no]$	$S_A[\text{Neg} = no]$
$V_A[\text{Neg} = yes]$	$S_A[\text{Neg} = no]$	$S_A[\text{Neg} = no]$

(b) Combinations for NR readings

	$S_A[\text{Neg} = no]$	$S_A[\text{Neg} = yes]$
$V_A[\text{Neg} = no]$		$S_A[\text{Neg} = yes]$
$V_A[\text{Neg} = yes]$	$S_A[\text{Neg} = yes]$	$S_A[\text{Neg} = yes]$

We now have four constants for the non-NR reading of *pense que*, but only three for its NR readings:

$$\beta^0_{\text{pense}} : S_A[\text{Neg} = no] \rightarrow V_A[\text{Neg} = no] \rightarrow NP \rightarrow S_A[\text{Neg} = no]$$

$$\beta^1_{\text{pense}} : S_A[\text{Neg} = yes] \rightarrow V_A[\text{Neg} = no] \rightarrow NP \rightarrow S_A[\text{Neg} = no]$$

$$\beta^2_{\text{pense}} : S_A[\text{Neg} = no] \rightarrow V_A[\text{Neg} = yes] \rightarrow NP \rightarrow S_A[\text{Neg} = no]$$

$$\beta^3_{\text{pense}} : S_A[\text{Neg} = yes] \rightarrow V_A[\text{Neg} = yes] \rightarrow NP \rightarrow S_A[\text{Neg} = no]$$

$$\beta'^0_{\text{pense}} : S_A[\text{Neg} = yes] \rightarrow V_A[\text{Neg} = no] \rightarrow NP \rightarrow S_A[\text{Neg} = yes]$$

$$\beta'^1_{\text{pense}} : S_A[\text{Neg} = no] \rightarrow V_A[\text{Neg} = yes] \rightarrow NP \rightarrow S_A[\text{Neg} = yes]$$

$$\beta'^2_{\text{pense}} : S_A[\text{Neg} = yes] \rightarrow V_A[\text{Neg} = yes] \rightarrow NP \rightarrow S_A[\text{Neg} = yes]$$

We also need to adapt the types of the other terms and set the type of β^{Id}_S, β^{Id}_V, and $\beta_{\text{ne pas}}$ to $S_A[\text{Neg} = no]$, $V_A[\text{Neg} = no]$, and $V_A[\text{Neg} = yes]$ respectively. For α_{partira} to accept any kind of parameter, it now comes in four forms $\alpha^{m,n}_{\text{partira}}$ whose types are $S_a[\text{Neg} = m] \rightarrow V_A[\text{Neg} = n] \rightarrow NP \rightarrow S$ with $(m, n) \in \{yes, no\}^2$.

With this setting, the derivation tree analyzing (1a) that uses a NR predicate with two fake adjunctions $\beta^{Id}_{S[\text{Neg}=no]}$ and $\beta^{Id}_{V[\text{Neg}=no]}$ cannot make use of any of the constants associated with the NR reading, but only of β^0_{pense}. So that it now has the single analysis:

$$\gamma'''_0 = \alpha^{no,no}_{\text{partira}}(\beta^0_{\text{pense}}\ \beta^{Id}_{S[\text{Neg}=no]}\ \beta^{Id}_{V[\text{Neg}=no]}\ \alpha_{\text{Marie}})\beta^{Id}_{V[\text{Neg}=no]}\ \alpha_{\text{Pierre}}$$

Similarly, if a negated NR predicate embeds another NR predicate as in (5), and if we have a non-NR reading for the matrix predicate (as for *pense que* in γ'_0 and γ'_2), the type of $\beta^2_{\text{pense}}\ \beta^{Id}_{S[\text{Neg}=no]}\ \beta_{\text{ne pas}}\ \alpha_{\text{Marie}}$ necessarily is $S_A[\text{Neg} = no]$, hence cannot serve as first argument of any NR reading β'^i_{croie} of *croie que* if the latter is not negated. This avoids the $\mathscr{L}_{Log}(\gamma'_2)$ reading of the sentence.

4.2 Enforcing the NR Reading

A similar technique can be used to enforce the NR reading when a NPI occurs in the embedded clause. S_A and NP types are declined as $S_A[\text{NPI} = b]$ and $NP[\text{NPI} = b]$ with $b \in \{yes, no\}$. [NPI = yes] means it licenses a NPI in the clause. Only NR readings should allow for a $S_A[\text{NPI} = yes]$ resulting type:

$$\beta'_{\text{pense}} : S_A[\text{NPI} = m] \rightarrow V_A[\text{Neg} = n] \rightarrow NP[\text{NPI} = p] \rightarrow S_A[\text{NPI} = yes]$$

if at least one of the m, n, and p is set to *yes*[8] while non-NR readings only allow for a $S_A[\text{NPI} = no]$ type.

---

[8] We should be more careful with the licensed combinations, but this is enough to show how to force the reading.

Then, together with the constants:

$$\alpha^0_{\text{est... assiette}} : S_A[\text{NPI} = no] \to V_A[\text{Neg} = yes] \to NP[\text{NPI} = no] \to S$$
$$\alpha^1_{\text{est... assiette}} : S_A[\text{NPI} = yes] \to V_A[\text{Neg} = no] \to NP[\text{NPI} = no] \to S$$
$$\beta^{Id}_{S[\text{NPI}=no]} : S_A[\text{NPI} = no]$$

the (positive) embedded clause *Pierre est/soit dans son assiette* requires the adjunction of an auxiliary tree $\gamma^{(4)}$ of type $S_A[\text{NPI} = yes]$ to be analyzed as:

$$\gamma^{(5)} = \alpha^1_{\text{est... assiette}} \, \gamma^{(4)} \, \beta^{Id}_{V[\text{Neg}=no]} \, \alpha_{\text{Pierre}}$$

Because it has type $S_A[\text{NPI} = yes]$, the $\gamma^{(4)}$ auxiliary tree can only be the result of a β'_{pense} application (NR reading). For instance, the actual analysis of (3b) would be:

$$\gamma^{(6)} = \alpha^1_{\text{est... assiette}} \, (\beta'_{\text{pense}} \, \beta^{Id}_{S[\text{NPI}=no]} \, \beta_{\text{ne pas}} \, \alpha_{\text{Marie}}) \, \beta^{Id}_{V[\text{Neg}=no]} \, \alpha_{\text{Pierre}}$$

This approach could also be used to model the blocking of NR cyclicity when a NR desire predicate embeds a NR belief predicate [2,10], or to model the readings forced by embedded discourse connectives.

5   Related Works

Our approach models NR predicates at the syntax-semantics interface. This contrasts with the semantic/pragmatic approaches that explains NR behaviors from presupposition. [2] unifies different trends from this vein using *soft presuppositions* [21]. Because we aim at articulating the semantics of NR predicates with discourse connectives that are lexically introduced as in [12], the syntax-semantics approach allows us to have them modeled at a same level, at least as a first approximation.

Another syntax-semantics interface perspective in TAG on NR predicates is proposed in [11] for German. Its modeling relies on multicomponent TAG [22] with an underspecified semantic representation [23]. The motivation for using multicomponents does not only depend on the analysis of NR predicates. It is motivated in the first place by the modeling of scrambling and other word-order phenomena in German. As it also shows useful to analyze long-dependencies, it is used for NR predicates as well. Then, both ambiguities and relative scoping are managed at the semantic representation level with the underspecified framework whereas we model them at the syntax-semantics interface level and at the logical representation level respectively.

It is also worth noting that the constraints we model using the fine tuning of types with $[\text{Neg} = m]$ and $[\text{NPI} = n]$ extensions is also present in [11]. The latter implements specific computations on the NEG features of the verbal spine that extend the usual notion of unification of TAG. The different combinations we propose in Table 2(a) and (b) for instance closely relate to this computation.

6   Conclusion

We have presented an account of NR predicates within TAG that relies on a Montague-like semantics for TAG. The different properties of NR predicates are rendered at different levels: the ambiguity of the readings is modeled by lexical ambiguity; the scoping and cyclicity properties are modeled through the lexical semantics and the higher-order interpretation of adjunction nodes; spurious ambiguities are avoided using fine-grained types for terms representing derivation trees. This provides us with a base layer where to account for interactions with discourse connectives and discourse representation.

References

1. Bartsch, R.: "Negative Transportation" gibt es nicht. Linguistische Berichte **27** (1973)
2. Gajewski, J.R.: Neg-raising and polarity. Linguist. Philos. **30**(3), 289–328 (2007)
3. Joshi, A.K., Schabes, Y.: Tree-adjoining grammars. In: Rozenberg, G., Salomaa, A. (eds.) Handbook of Formal Languages, vol. 3. Springer, Berlin (1997)
4. de Groote, P.: Towards abstract categorial grammars. In: Association for Computational Linguistics, 39th Annual Meeting and 10th Conference of the European Chapter, Proceedings of the Conference, pp. 148–155 (2001)
5. Shieber, S.M., Schabes, Y.: Synchronous tree-adjoining grammars. In: Proceedings of the 13th International Conference on Computational Linguistics, Helsinki, Finland, vol. 3, pp. 253–258 (1990)
6. Nesson, R., Shieber, S.M.: Simpler TAG semantics through synchronization. In: Proceedings of the 11th Conference on Formal Grammar, Malaga, Spain, CSLI Publications, 29–30 July 2006
7. Storoshenk, D.R., Frank, R.: Deriving syntax-semantics mappings: node linking, type shifting and scope ambiguity. In: hye Han, C., Satta, G. (eds.) Proceedings of the 11th International Workshop on Tree Adjoining Grammars and Related Framework (TAG+11), pp. 10–18 (2012)
8. Horn, L.R.: Neg-raising predicates: towards an explanation. In: Proceedings of CLS 11 (1975)
9. Horn, L.R.: A Natural History of Negation. University of Chicago Press, Chicago (1989)
10. Horn, L.R.: Negative transportation: unsafe at any speed? In: Proceedings of CLS 7 (1971)
11. Lichte, T., Kallmeyer, L.: Licensing German negative polarity items in LTAG. In: Proceedings of the Eighth International Workshop on Tree Adjoining Grammar and Related Formalisms, Sydney, Australia, pp. 81–90. Association for Computational Linguistics, July 2006
12. Danlos, L.: Connecteurs de discours adverbiaux: Problèmes à l'interface syntaxe-sémantique. Lingvisticæ Investigationes **36**(2) (2013)
13. Danlos, L.: D-STAG: a formalism for discourse analysis based on SDRT and using synchronous TAG. In: de Groote, P., Egg, M., Kallmeyer, L. (eds.) FG 2009. LNCS (LNAI), vol. 5591, pp. 64–84. Springer, Heidelberg (2011)
14. XTAG Research Group: A lexicalized tree adjoining grammar for English. Technical report IRCS-01-03, IRCS, University of Pennsylvania (2001)

15. Abeillé, A.: Une grammaire électronique du français. CNRS Éditions (2002)
16. de Groote, P.: Tree-adjoining grammars as abstract categorial grammars. In: TAG+6, Proceedings of the sixth International Workshop on Tree Adjoining Grammars and Related Frameworks, Università di Venezia, pp. 145–150 (2002)
17. Pogodalla, S.: Advances in abstract categorial grammars: language theory and linguistic modeling. ESSLLI 2009 Lecture Notes, Part II (2009)
18. Shieber, S.M.: Unifying synchronous tree adjoining grammars and tree transducers via bimorphisms. In: Proceedings of the 11th Conference of the European Chapter of the Association for Computational Linguistics (EACL-2006), pp. 377–384 (2006)
19. de Groote, P., Maarek, S.: Type-theoretic extensions of abstract categorial grammars. In: New Directions in Type-Theoretic Grammars, Proceedings of the Workshop, 18–30 (2007). http://let.uvt.nl/general/people/rmuskens/ndttg/ndttg2007.pdf
20. de Groote, P., Maarek, S., Yoshinaka, R.: On two extensions of abstract categorial grammars. In: Dershowitz, N., Voronkov, A. (eds.) LPAR 2007. LNCS (LNAI), vol. 4790, pp. 273–287. Springer, Heidelberg (2007)
21. Abusch, D.: Presupposition triggering from alternatives. J. Semant. 27(1), 37–80 (2010)
22. Weir, D.J.: Characterizing Mildly Context-Sensitive Grammar Formalisms. Ph.D. thesis, University of Pennsylvania (1988)
23. Kallmeyer, L., Romero, M.: Scope and situation binding for LTAG. Res. Lang. Comput. 6(1), 3–52 (2008)

Semantic Similarity: Foundations

Cédric Dégremont$^{(\boxtimes)}$, Antoine Venant, and Nicholas Asher

IRIT, Toulouse, France
{cedric.degremont,antoine.venant,asher}@irit.fr
http://www.irit.fr

Abstract. This paper investigates measures of semantic similarity between conversations from an axiomatic perspective. We abstract away from real conversations, representing them as sequences of formulas, equipped with a notion of semantic interpretation that maps them into a different space. An example we use to illustrate our approach is the language of propositional logic with its classical semantics. We introduce and study a range of different candidate properties for metrics on such conversations, for the structure of the semantic space, and for the behavior of the interpretation function, and their interactions. We define four different metrics and explore their properties in this setting.

Keywords: Semantics · Distance · Metric · Similarity · Lattice · Conversations

1 Introduction

If linguistic behavior is to be analyzed as a form of rational behavior (Grice 1967), it is important to be able to assess the conversational goals of linguistic agents and the extent to which they are fulfilled by any given conversation in a manageable way. Specifying preferences over the set of all possible choices of what to say is clearly intractable for us as theorists and for speakers as practioners. Instead, speakers must be able to group conversations into semantically similar classes and to assess the relative semantic proximity of any two pairs of conversations. The preferences of the agents over different ways of expressing themselves have to do with how close these ways are from satisfying certain positive or negative semantic goals. An elegant way to be able to do this, is to have a metric over conversations that is semantic in nature. The goal of this paper is to identify properties that characterize 'semantic metrics' and to identify reasonable axioms that can help us isolate well-behaved semantic metrics.

A workable definition of semantic distance between texts or conversations is also important for the evaluation of annotations of discourse structure in text and dialogue. It is also crucial to the success of the machine learning of semantic structures from annotated data, as all known algorithms rely on some

---

This research was supported by ERC Grant 269427.

© Springer International Publishing Switzerland 2014
Y. Nakano et al. (Eds.): JSAI-isAI 2013, LNAI 8417, pp. 17–41, 2014.
DOI: 10.1007/978-3-319-10061-6_2

notion of similarity or loss with respect to the target structure. While measures of syntactic similarity like ParseEval (Black et al. 1991) and Leaf Ancestor (Sampson 2000) are well-understood and used in computational linguistics, they yield intuitively wrong results. ParseEval, for instance, places too much importance on the boundaries of discourse constituents, which are often notoriously hard even for expert annotators to agree on. Investigations of distances between semantic interpretations of a text are rarely examined. While a natural equivalence and ordering relation over contents comes from the underlying logic of formal semantic analysis, this only gives a very crude measure. Some have appealed to a language of semantic primitives to exploit the more developed measures of syntactic distance in a more semantic setting. But such an approach depends on the choice of semantic primitives, with no clear consensus on how to go about determining these primitives.

Semantic distances are also relevant in the context of formal theories of belief revision. Lehmann et al. (2001) explores Alchourrón et al. (1985) style postulates that characterize a wide family of belief revision operator based on pseudo-distances on models satisfying only very mild assumptions. Our problem is also closely related to the problem of determining the distance of a scientific theory from the truth. This problem, referred to as the problem of verisimilitude or truthlikeness in philosophy of science (since Popper 1968), is arguably reducible to the problem of having a satisfactory concept of similarity between theories in a formal language.

The aim of this paper is to study semantic metrics for an abstract and simple concept of conversations. Syntactically, we assume that conversations are monoids with respect to concatenation. These conversations are equipped with an interpretation function mapping them into some distinct semantic space. In general, our assumptions about the semantic space and the interpretation function will be as minimal as possible. As far as identifying the axioms that characterize our concept of 'semanticity' for a metric goes, we will not be making any assumption. To analyze candidate axioms that characterize *well-behaved* semantic metrics, it will be interesting to consider the effect of assuming a bit more structure. Specifically we will pay some attention to the case in which the semantic co-domain of the interpretation function is a lattice. As an example, sequences of propositional formulas with their classical interpretation certainly fall under this category. We will moreover consider interpretation functions that satisfy some structural properties, for example assuming that the semantic meaning of a sequence is invariant under stuttering, that is immediate repetition of the same element in a sequence, or even assuming complete invariance under permutation.

To develop semantic metrics for conversations in natural language or for their representations in some formalism suitable for discourse interpretation (like for instance SDRT, Asher and Lascarides 2003) we first need to clarify the space of reasonable axioms and metrics for the simplest and most general representations. We take this first step here.

The paper is organized as follows. We start with a first section, Sect. 2, that contains technical preliminaries and settles the notation. We then describe in Sect. 3 different properties that can be met by an interpretation function, regarding how it interacts with sequences-concatenation or the structure of the semantic space. Section 4 introduces some elementary background about generalized metrics and metrics over subsets of a metric space. Section 5 draws a map of different level of *semanticity* for metrics, corresponding to different requirements on the interaction between the interpretation function and the metric. Section 6 introduces some concrete candidate semantic metrics. Sections 6–8 then describe how these potential measures of semantic similarity fare with respect to different lists of axioms. We show that certain combinations of axioms lead to trivialization results. We then conclude.

2 Preliminaries and Notation

This section contains some technical preliminaries and settles notation. The reader can skip this section on a first reading, and come back to it when needed.

2.1 Sets, Functions, Sequences, Orders and Lattices

Let X, Y be two sets. We let $X \ominus Y$ denote the symmetric difference of X and Y. Let $\mathsf{card}(X)$ denote the cardinality of X. let X^* be the set of finite strings over X. If $f : X \to Y$, we let $f(X)$ and $f[X]$ be alternative notation for the image of X under f, that is $f(X) = f[X] = \{f(x) | x \in X\}$. We let $\mathsf{dom}(f) = X$, $\mathsf{target}(f) = Y$ and $\mathsf{ran}(f) = f(X)$. The kernel of f is the equivalence relation $\sim$, such that $x \sim y$ iff $f(x) = f(y)$. We say that f is isotone whenever for all x, y with $x \leq y$ we have $f(x) \leq f(y)$. We write $f : X \nrightarrow Y$ whenever f is partial function from X to Y, that is there exists a non-empty subset of $A \subseteq Y$ such that $f : A \to Y$.

Given a sequence $\sigma \in X^*$ we let $\mathsf{len}(\sigma)$ be the length of σ. If $k \leq \mathsf{len}(\sigma)$ then we let $\sigma|_k$, be the prefix of σ of length k, and we let $\sigma(k)$ or $\sigma[k]$ be the kth element of σ. We let $\mathsf{ran}(\sigma)$ be the range of σ, that is $\mathsf{ran}(\sigma) = \{\sigma[k] | 1 \leq k \leq \mathsf{len}(\sigma)\}$. We let $\overrightarrow{\epsilon}$ be the empty sequence.

A relation $\leq$ on X is a pre-order on X iff it is a reflexive and transitive relation on X. $(X, \leq)$ is a poset iff $\leq$ is a pre-order on X such that $\leq$ is antisymmetric on X, that is $x \leq y$ and $y \leq x$, implies that $x = y$. A lattice $(X, \leq)$ is a poset such that every two elements $x, y \in X$ have a least upper bound (or join, denoted $x \wedge y$) and a greatest lower bound (or meet, denoted $x \vee y$). A lattice $(X, \leq)$ is bounded whenever X has a least and a greatest element (denoted $\perp$ and $\top$).

2.2 Propositional Languages and Interpretation Functions

Given a language L we let $\varphi, \psi, \chi, \varphi_1, \varphi_2 \ldots$ range over L, and we let $\overrightarrow{\varphi}, \overrightarrow{\psi}, \overrightarrow{\sigma}$, $\overrightarrow{\tau}, \overrightarrow{\upsilon}, \overrightarrow{\sigma_1}, \overrightarrow{\sigma_2} \ldots$ range over L*. Given a finite set PROP $= \{p_1, \ldots, p_n\}$ we let $\mathsf{L_{PROP}}(1)$ be defined as follows:

$$\varphi ::= p | \neg p | \top | \perp$$

where p ranges over PROP. And we define L_{PROP} as follows:

$$\varphi ::= p \,|\, \top \,|\, \bot \,|\, \neg\varphi \,|\, \varphi \wedge \varphi \,|\, \varphi \vee \varphi$$

where p ranges over PROP. We define $\mathsf{sig}(\varphi){:}\mathsf{L} \to \wp(\text{PROP})$ where $p \in \mathsf{sig}(\varphi)$ iff p occurs in φ. Given a sequence $\overrightarrow{\sigma} \in \mathsf{L}^*$ or a subset $A \subseteq \mathsf{L}$, we write $\mathsf{sig}(\overrightarrow{\sigma}) := \bigcup_{\psi \in \mathsf{ran}(\overrightarrow{\sigma})} \mathsf{sig}(\psi)$ or $\mathsf{sig}(A) := \bigcup_{\psi \in A} \mathsf{sig}(\psi)$, respectively.

Classical truth-functional interpretation of L_{prop}. Let $W_{\text{PROP}} = 2^{\text{PROP}}$. Depending on context, will treat a member $V \in W_{\text{PROP}}$ either as a function $V{:}\,\text{PROP} \to \{0,1\}$ or as a subset $V \subseteq \text{PROP}$. These two representations are of course equivalent. We let $[\![\cdot]\!]^t : \mathsf{L} \to 2^{\text{PROP}}$, be the classical truth-functional interpretation function of $\mathsf{L}_{\text{PROP}}(1)$ and L_{PROP}.

3   Properties of Interpretation Functions

In general, we will work with abstract concepts of a language and of an interpretation function. Let L, X be non-empty sets.

Definition 3.1 (Interpretation function). An interpretation function of L into X is a function $\|\cdot\| : \mathsf{L}^* \to \mathsf{X}$.

An interpretation function for L is an interpretation function L into Y for some non-empty set Y.

3.1   Co-Domain

In this paper, we will sometimes assume that the semantic space has some structure. We are always explicit about these assumptions whenever we make them.

Definition 3.2. We say that $\|\cdot\|$ is $(W, \leq)$–pre-order-valued $((W, \leq)$–poset-valued) whenever $\mathsf{target}(\|\cdot\|){=}W$ and $(W, \leq)$ is a pre-order (respectively, a poset).

We say that $\|\cdot\|$ is pre-order–valued, iff it is $(W, \leq)$–pre-order-valued for some pre-ordered set $(W, \leq)$, and similarly for poset–valued.

Definition 3.3. $\|\cdot\|$ is $(W, \leq, \wedge, \vee)$–lattice-valued iff $\mathsf{target}(\|\cdot\|) = W$ and $(W, \leq)$ is a lattice, with $\wedge$ and $\vee$ as its meet and join operator, respectively.

Definition 3.4. $\|\cdot\|$ is $(W, \leq, \wedge, \vee, \bot, \top)$–lattice-valued iff $\|\cdot\|$ is $(W, \leq, \wedge, \vee)$–lattice-valued, and $(W, \leq)$ is a bounded lattice, with $\bot$ and $\top$ as its least and greatest element, respectively.

We say that $\|\cdot\|$ is lattice-valued, iff it is $(W, \leq, \wedge, \vee)$–lattice-valued for some $(W, \leq)$, $\wedge$ and $\vee$. We say that $\|\cdot\|$ is bounded lattice-valued, iff it is $(W, \leq, \wedge, \vee, \bot, \top)$–lattice-valued, for some $(W, \leq)$, $\wedge$, $\vee$, $\bot$ and $\top$.

Definition 3.5. $\| \cdot \|$ is set-valued iff we have $\mathsf{target}(\| \cdot \|) = \wp(W)$ for some non-empty set W.

3.2 Structural Properties for Interpretation Functions

It will sometimes be interesting to restrict ourselves to interpretation functions satisfying certain structure properties. Assume that $\| \cdot \|$ is $\leq$–poset-valued. Below the comma ',' is the concatenation operator. The axiom in the table below, are to be understood as quantifying universally. $\vec{\alpha}, \vec{\beta}$ ranging over $\mathsf{dom}(\| \cdot \|)^*$ and α, β ranging over $\mathsf{dom}(\| \cdot \|)$.

Axiom name	Meaning
contraction	$\| \vec{\alpha}, \varphi, \varphi, \vec{\beta} \| \leq \| \vec{\alpha}, \varphi, \vec{\beta} \|$
expansion	$\| \vec{\alpha}, \varphi, \vec{\beta} \| \leq \| \vec{\alpha}, \varphi, \varphi, \vec{\beta} \|$
exchange	$\| \vec{\alpha}, \varphi, \psi, \vec{\beta} \| = \| \vec{\alpha}, \psi, \varphi, \vec{\beta} \|$
right monotonicity	$\| \vec{\alpha}, \varphi \| \leq \| \vec{\alpha} \|$
left monotonicity	$\| \varphi, \vec{\alpha} \| \leq \| \vec{\alpha} \|$
$\vec{\epsilon}$–⊤	$\| \vec{\alpha} \| \leq \| \vec{\epsilon} \|$
adjunction	If $\| \vec{\alpha} \| \leq \| \vec{\beta} \|$ and $\| \vec{\alpha} \| \leq \| \vec{\gamma} \|$
	then $\| \vec{\alpha} \| \leq \| \vec{\beta} \, \vec{\gamma} \|$
mix	If $\| \vec{\alpha}_1 \| \leq \| \vec{\beta}_1 \|$ and $\| \vec{\alpha}_2 \| \leq \| \vec{\beta}_2 \|$
	then $\| \vec{\alpha}_1 \vec{\alpha}_2 \| \leq \| \vec{\beta}_1 \vec{\beta}_2 \|$

For example, $\| \cdot \|$ satisfies contraction iff for every $\vec{\alpha}, \vec{\beta} \in \mathsf{L}^*$ and $\varphi \in \mathsf{L}$ we have $\| \vec{\alpha}, \varphi, \varphi, \vec{\beta} \| \leq \| \vec{\alpha}, \varphi, \vec{\beta} \|$.

Remark 3.6. If $\| \cdot \|$ satisfies exchange, then $\| \cdot \|$ for every $\vec{\varphi}$ and $\vec{\psi}$ that are equivalent up to permutation we have $\| \vec{\varphi} \| = \| \vec{\psi} \|$. If $\| \cdot \|$ satisfies either right or left monotonicity, then $\| \cdot \|$ satisfies $\vec{\epsilon}$–⊤.

3.3 Stronger Properties

Definition 3.7 (Conjunctive, intersective interpretation)

- $\| \cdot \|$ is conjunctive iff it is lattice-valued and $\forall \vec{\varphi}, \vec{\psi}, \| \vec{\varphi} \, \vec{\psi} \| = \| \vec{\varphi} \| \wedge \| \vec{\psi} \|$.
- $\| \cdot \|$ is intersective iff it is set-valued and $\forall \vec{\varphi}, \vec{\psi}, \| \vec{\varphi} \, \vec{\psi} \| = \| \vec{\varphi} \| \cap \| \vec{\psi} \|$.

Definition 3.8. We let $\| \cdot \|^t$ be the interpretation function for L^* defined by $\| \vec{\varphi} \|^t := [\![\bigwedge_{\varphi \in \mathsf{ran}(\vec{\varphi})}]\!]^t$.

Example 3.1. $\| \cdot \|^t$ is intersective. If $\| \cdot \|$ is intersective, then it is bounded $\subseteq$–lattice-valued.

4   Generalized Metrics

We start by recalling some basic definitions.

Definition 4.1 (Semi-Pseudometric). A semi-pseudometric on a set X is a function $d : (X \times X) \to \mathbb{R}$, such that for all $x, y, z \in X$ we have:

1. $d(x, x) = 0$;
2. $d(x, y) = d(y, x)$.

Definition 4.2 (Pseudometric). A pseudometric on a set X is a semi-pseudometric on X, such that for all $x, y, z \in X$ we have:

3. $d(x, z) \le d(x, y) + d(y, z)$ (triangle inequality).

If d is a (semi-)pseudometric on X, then (X, d) is a (semi-)pseudometric space.

Definition 4.3 (Trivial pseudo-metric). The trivial pseudo-metric over a set A is the function $d : \begin{cases} A \times A \to \mathbb{R} \\ \forall x, y \in A \quad d(x, y) = 0 \end{cases}$

4.1   Metrics on Valuations, Relations and Graphs

Given a finite set PROP Hamming distance on 2^{PROP} is the metric $\delta_{ham} : 2^{\mathrm{PROP}} \times 2^{\mathrm{PROP}} \to \omega$ defined as $\delta_{ham}(V, V') = \mathsf{card}(V \ominus V')$.

4.2   Aggregators

Let δ_i be a pseudo-metric on a set X. We want to study closeness between subsets of X, and so we provide some natural aggregators α associating with δ_i a function $d_\alpha^i : 2^X \times 2^X \to \mathbb{R}$, that may or may not be a pseudo-metric, depending on the particular aggregator.

Definition 4.4 (min aggregator). Let $d_{\min}^i(A, B) = \min_{x \in A, y \in B} d_i(x, y)$.

Definition 4.5 (max aggregator). Let $d_{\max}^i(A, B) = \max_{x \in A, y \in B} d_i(x, y)$.

Definition 4.6 (Hausdorff aggregator). Formally $d_H^i(A, B) = \max\{\max_{x \in A} \min_{y \in B} d_i(x, y), \max_{y \in B} \min_{x \in A} d_i(x, y)\}$.

Definition 4.7 (mean aggregator). Formally $d_{am}^i(A, B) = \sum_{x \in A, y \in B} \frac{1}{\mathsf{card}(A \times B)} d_i(x, y)$.

Remark 4.8. In general max and mean will return a non-zero value for (A, A). Note also that the min aggregator will return 0 for (A, B) whenever $A \cap B \ne \emptyset$.

Let W be a set and let d be a pseudo-metric on W. Let L be a language and let $\| \cdot \|$ be an interpretation function for L such that $\mathsf{target}(\| \cdot \|) = \wp(W)$ for some non-empty set W. Let $d_\alpha^i : \wp(W) \times \wp(W) \to \mathbb{R}$ be an aggregator based on the distance d_i between points of W. We let $d_{\alpha, \mathsf{L}, \| \cdot \|}^i : \mathsf{L}^* \times \mathsf{L}^* \to \mathbb{R}$ be defined by $d_{\alpha, \mathsf{L}, \| \cdot \|}^i(\overrightarrow{\varphi}, \overrightarrow{\psi}) = d_\alpha^i(\| \overrightarrow{\varphi} \|, \| \overrightarrow{\psi} \|)$. When L and $\| \cdot \|$ are clear from context, we will simply write d_α^i for $d_{\alpha, \mathsf{L}, \| \cdot \|}^i$. For instance, $d_{H, \mathsf{L}^{prop}, \| \cdot \|^t}^{ham}$ is sometimes shortened as d_H^{ham} when L^{prop} and $\| \cdot \|^t$ are clear from context.

5   What Is a Semantic Metric?

Now that we have set the stage for our investigations, our first task is to define our object of interest: semantic pseudometrics. Semantic pseudometrics are a subclass of linguistic pseudometrics.

Definition 5.1 (Linguistic (semi-)pseudometric). A linguistic (semi-) pseudometric on a language L is a partial function $d : (L^* \times L^*) \nrightarrow \mathbb{R}$ such that $\mathrm{dom}(d)$ is symmetric and $(\mathrm{dom}(d), d)$ is a (semi-)pseudometric space.

How semantic pseudometrics should be defined is not a fully straightforward matter. A minimal requirement would be the following:

$$\text{If for every} \overrightarrow{\chi_1}, \overrightarrow{\chi_2} \text{ with } \|\overrightarrow{\chi_1}\| = \|\overrightarrow{\chi_2}\|$$
$$\text{we have } d(\overrightarrow{\varphi}, \overrightarrow{\chi_1}) = d(\overrightarrow{\psi}, \overrightarrow{\chi_2}) \quad \text{then} \quad \|\overrightarrow{\varphi}\| = \|\overrightarrow{\psi}\| \qquad \text{(min sem separation)}$$

That is, if two sequences $\overrightarrow{\chi_1}, \overrightarrow{\chi_2}$ are semantically non-equivalent, then there should be two (other) semantically equivalent sequences, that are not pairwise equidistant from $\overrightarrow{\chi_1}$ and $\overrightarrow{\chi_2}$. A stronger, yet reasonable, assumption is:

$$\text{If for every } \overrightarrow{\chi} \text{ we have } d(\overrightarrow{\varphi}, \overrightarrow{\chi}) = d(\overrightarrow{\psi}, \overrightarrow{\chi}) \quad \text{then} \quad \|\overrightarrow{\varphi}\| = \|\overrightarrow{\psi}\|$$
$$\text{(sem separation)}$$

The axiom states that if two sequences of formulas $\overrightarrow{\varphi}$ and $\overrightarrow{\psi}$ are not semantically equivalent, then there is some sequence of formulas $\overrightarrow{\chi}$ that is not at the same distance from both $\overrightarrow{\varphi}$ and $\overrightarrow{\psi}$.

Fact 5.2. *Let d be a semi-pseudometric. If d satisfies* (sem separation), *then it satisfies* (min sem separation).

Finally we consider a stronger axiom:

$$\text{If } d(\overrightarrow{\varphi}, \overrightarrow{\psi}) = 0 \quad \text{then} \quad \|\overrightarrow{\varphi}\| = \|\overrightarrow{\psi}\| \qquad \text{(zero} \Rightarrow \text{sem}\equiv)$$

The axiom is a regularity condition stating, that semantically non-equivalent sequences of formulas, should be at positive distance of each other.

Fact 5.3. *Let d be a semi-pseudometric. If d satisfies* (zero $\Rightarrow$ sem $\equiv$), *then it satisfies* (sem separation).

The two become equivalent if we assume triangle inequality.

Fact 5.4. *Let d be a pseudo-metric. d satisfies* (sem separation) *iff d satisfies* (zero $\Rightarrow$ sem $\equiv$).

The converse of (zero $\Rightarrow$ sem $\equiv$), below, states that semantically equivalent sequences formulas, should be a distance 0 of each other.

$$\text{If } \|\overrightarrow{\varphi}\| = \|\overrightarrow{\psi}\| \quad \text{then} \quad d(\overrightarrow{\varphi}, \overrightarrow{\psi}) = 0 \qquad \text{(sem}\equiv \Rightarrow\text{zero)}$$

Unsurprisingly (sem $\equiv$ $\Rightarrow$ zero) will filter out syntactically driven notions such as δ_{count} or $\delta_{synt,count}$.

Definition 5.5. Given a language L, let $\delta_{count}(\vec{\varphi}, \vec{\psi}) := card(ran(\vec{\varphi}) \ominus ran(\vec{\psi}))$

Definition 5.6. Given a language L, let $\delta_{synt,count}(\vec{\varphi}, \vec{\psi}) := \delta_{count}(\vec{\varphi}, \vec{\psi}) + card(sig(\vec{\varphi}) \ominus sig(\vec{\psi}))$

Fact 5.7. $(L_{PROP}(1), \delta_{count})$ *does not satisfy* (sem $\equiv$ $\Rightarrow$ zero).

Fact 5.8. $(L_{PROP}(1), \delta_{synt,count})$ *does not satisfy* (sem $\equiv$ $\Rightarrow$ zero).

As observed previously, (sem $\equiv$ $\Rightarrow$ zero) rules out a number of aggregators, e.g. :

Fact 5.9. $(L_{PROP}, d_{max}^{ham})$ *does not satisfy* (sem $\equiv$ $\Rightarrow$ zero).

We have seen that (sem$\equiv$ $\Rightarrow$ zero) and the triangle inequality together imply that two semantically equivalent points are equidistant to any other third point. This latter notion of semantic invariance implies in return (sem $\equiv$ $\Rightarrow$ zero) and might be a desirable property as well:

$$\text{If } \|\vec{\varphi}\| = \|\vec{\psi}\| \quad \text{then} \quad \text{for every } \vec{\chi} \text{ we have } d(\vec{\varphi}, \vec{\chi}) = d(\vec{\psi}, \vec{\chi})$$
$$\text{(sem preservation)}$$

Fact 5.10. *Let d be a semi-pseudometric. If d satisfies* (sem preservation) *then d satisfies* (sem $\equiv$ $\Rightarrow$ zero).

Finally, we can require our (semi-)pseudometric to be fully induced by a distance on the co-domain of the interpretation function $\| \cdot \|$, which we define as follows:

$$\text{If } \|\vec{\varphi_1}\| = \|\vec{\psi_1}\| \text{ and } \|\vec{\varphi_2}\| = \|\vec{\psi_2}\| \quad \text{then} \quad d(\vec{\varphi_1}, \vec{\varphi_2}) = d(\vec{\psi_1}, \vec{\psi_2})$$
$$\text{(sem induced)}$$

Fact 5.11. *Let d be a semi-pseudometric. If d satisfies* (sem induced) *then d satisfies* (sem preservation).

Fact 5.12. *Let d be a pseudo-metric. If d satisfies* (sem $\equiv$ $\Rightarrow$ zero), *then it satisfies* (sem induced).

Corollary 5.13. *Let d be a pseudo-metric that satisfies* (sem $\equiv$ $\Rightarrow$ zero). *There exists a pseudo-metric $\dot{d}$ on* ran($\| \cdot \|$) *such that $\dot{d}(\|\vec{\varphi}\|, \|\vec{\psi}\|) = d(\vec{\varphi}, \vec{\psi})$.*

Fact 5.14. *Let d be a pseudo-metric that verifies* (sem $\equiv$ $\Rightarrow$ zero). *Let $\equiv$ be the kernel of $\| \cdot \|$. The following holds:*

1. *If $\vec{\varphi} \equiv \vec{\psi}$ then $\forall \vec{\chi}\, d(\vec{\chi}, \vec{\varphi}) = d(\vec{\chi}, \vec{\psi})$.*

Fact 5.15. *Let d be a pseudo-metric. If d satisfies* (min sem separation) *and* (sem $\equiv$ $\Rightarrow$ zero), *then it satisfies* (sem separation).

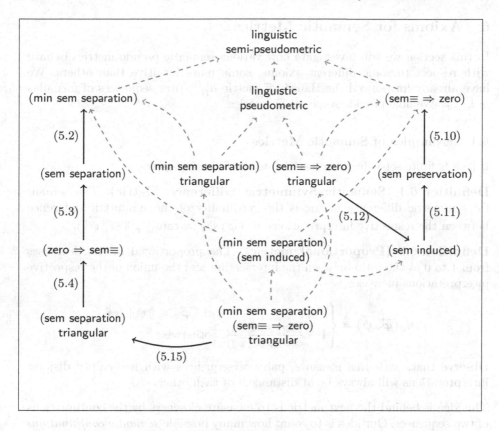

Fig. 1. Summary of the results in Sect. 5. Dashed arrows follow from definitions.

Figure 1, in p. 9, summarizes the relation between the axioms discussed in this section. We are now ready to define our notion of 'semanticity'.

Definition 5.16 (Semantic Pseudometric). A linguistic (semi-)pseudometric is semantic whenever for all $\overrightarrow{\varphi}, \overrightarrow{\psi}$ we have $\|\overrightarrow{\varphi}\| = \|\overrightarrow{\psi}\|$ iff for all $\overrightarrow{\chi}_1, \overrightarrow{\chi}_2$ such that $\|\overrightarrow{\chi}_1\| = \|\overrightarrow{\chi}_2\|$ we have $d(\overrightarrow{\varphi}, \overrightarrow{\chi}_1) = d(\overrightarrow{\psi}, \overrightarrow{\chi}_2)$.

Fact 5.17. *A linguistic semi-pseudometric is semantic iff it satisfies* (min sem separation) *and* (sem induced).

Fact 5.18. *A linguistic pseudometric is semantic iff it satisfies* (min sem separation) *and* (sem $\equiv$ $\Rightarrow$ zero).

Now that we have settled our definition of semantic pseudo metric, which we will use in the sequel, we can tackle our main problem—in brief:

What are reasonable properties of a semantic pseudometric on (a subset of) L^*?

6 Axioms for Semantic Metrics

In this section we will investigate how various semantic pseudo metrics behave with respect to some different axioms, some more intuitive than others. We have already introduced the Hausdorff metric d_H^{ham} over sequences of formulas in $\mathsf{L_{PROP}}$. We now provide some other metrics.

6.1 Examples of Semantic Metrics

If $\| \cdot \|$ is finite set-valued, we can define the following metrics:

Definition 6.1 (Semantic Symmetric Difference Metric). The semantic symmetric difference metric is the cardinality of the symmetric difference between the respective interpretations. $d_{\ominus}(\overrightarrow{\varphi}, \overrightarrow{\psi}) := \mathsf{card}(\| \overrightarrow{\varphi} \| \ominus \| \overrightarrow{\psi} \|)$.

Definition 6.2 (Proportional metric). The proportional metric decreases from 1 to 0 as the ratio between the intersection and the union of the respective interpretations increases.

$$d_{\alpha}(\overrightarrow{\varphi}, \overrightarrow{\psi}) = \begin{cases} 0 & \text{if } \| \overrightarrow{\varphi} \| = \| \overrightarrow{\psi} \| = \emptyset \\ 1 - \dfrac{\mathsf{card}(\| \overrightarrow{\varphi} \| \cap \| \overrightarrow{\psi} \|)}{\mathsf{card}(\| \overrightarrow{\varphi} \| \cup \| \overrightarrow{\psi} \|)} & \text{otherwise} \end{cases}$$

Observe that, with this measure, pairs of sequences with non-empty disjoint interpretations will always be at distance 0 of each other.

The idea is behind the next metric is to measure closeness by the continuations of two sequences. Our idea is to count how many possible *semantic continuations* one sequence has that can not apply to the other. For a set-valued interpretation function, this is given by the cardinal of the symmetric difference between the power sets of the respective interpretations. This should then be normalized by the possible ways to continue these sequences. This is given by the cardinal of the cartesian product of the power sets of the respective interpretations. Moreover we normalize this measure by the number of possible semantic values, $\mathsf{card}(\wp(\{0,1\}^{\text{PROP}})) = 2^{2^{\mathsf{card}(\text{PROP})}}$. Here is the definition:

Definition 6.3 (Continuation-based pseudometric)

$$d_{\mathcal{C}}(\overrightarrow{\varphi}, \overrightarrow{\psi}) = \frac{\mathsf{card}(2^{\| \overrightarrow{\varphi} \|} \ominus 2^{\| \overrightarrow{\psi} \|})}{\mathsf{card}(2^{\| \overrightarrow{\varphi} \|} \times 2^{\| \overrightarrow{\psi} \|})} \cdot \frac{1}{2^{2^{\mathsf{card}(\text{PROP})}}}$$

In the simplistic setting of building a semantic metric for $\mathsf{L_{PROP}^*}$ with the classical truth functional $\| \cdot \|^t$ interpretation, a natural notion of understanding *possible continuation*, is one of *consistent continuation*. A set $\| \overrightarrow{\varphi} \|^t$ is a possible continuation of $\overrightarrow{\psi}$ just in case $\| \overrightarrow{\varphi} \overrightarrow{\psi} \|^t = \| \overrightarrow{\varphi} \|^t \cap \| \overrightarrow{\psi} \|^t \neq \emptyset$. Following this definition, a possible continuation of $\overrightarrow{\varphi}$ which is not a possible continuation of $\overrightarrow{\psi}$ is uniquely decomposed into a part of $\| \overrightarrow{\varphi} \|^t \setminus \| \overrightarrow{\psi} \|^t$ and a set of valuations that are neither in $\| \overrightarrow{\varphi} \|^t$ nor $\| \overrightarrow{\psi} \|^t$.

Fact 6.4. *For any set-valued interpretation, the preceding definition is equivalent to the following:*

$$d_C(\overrightarrow{\varphi}, \overrightarrow{\psi}) = \frac{2^{\mathsf{card}(\|\overrightarrow{\varphi})\|} + 2^{\mathsf{card}(\|\overrightarrow{\psi}\|)} - 2 \cdot 2^{\mathsf{card}(\|\overrightarrow{\varphi}\| \cap \|\overrightarrow{\psi}\|)}}{2^{\mathsf{card}(\|\overrightarrow{\varphi}\|) + \mathsf{card}(\|\overrightarrow{\psi}\|)}}$$

6.2   Shortest Paths in Covering Graphs

Monjardet (1981) summarizes interesting results concerning metrics on posets and lattices. We will make use of two of these results to shed a different light on the semantic metrics defined in the previous section. In what follows, let $\langle W, \vee, \top, \leq \rangle$ be a bounded semi-lattice with $\top$ as greatest element. For all $x, y \in W$, we say that y covers x iff $x < y$ and $\forall z, x < z \leq y \to y = z$. Define also inductively the *rank* of an element, by setting all $<$-minimal elements of rank 0, and for every y covering a x of rank n, setting y of rank $n + 1$.

Definition 6.5 (Covering graph of a semi-lattice). Let $\langle W, \top, \leq \rangle$ be a semi-lattice with $\top$ as greatest element. The covering graph $G(W) = \langle V, E \rangle$ of W is such that $V = W$ and $(x, y) \in E$ iff x covers y or y covers x.

Definition 6.6. An upper valuation is an isotone map $v : W \to \mathbb{R}$ such that $\forall z\, z \leq x, y \to v(x) + v(y) \geq v(x \vee y) + v(z)$.

Let $G(W) = \langle W, E \rangle$ be the covering graph of W and let v be an upper valuation on W. For each edge $(x, y) \in E$, let the weight function $\omega_v : E \to \mathbb{R}$ induced by v be defined by $\omega(x, y) = |v(x) - v(y)|$. Moreover, let $\pi(x, y)$ be the set of paths from x to y. We make use of two results exposed in Monjardet (1981):

Fact 6.7 (Monjardet 1981**).** *Let v be an isotone upper valuation. We have:*

1. *the function $d_v(x, y) = 2v(x \vee y) - v(x) - v(y)$ is positive and verifies the triangle inequality.*
2. *$d_v(x, y) = \delta_v(x, y) := \min_{p_{xy} \in \pi(x,y)} \sum_{(z_1, z_2) \in p_{xy}} \omega(x, y)$.*

Definition 6.8. If v is an isotone, positive, upper valuation that assigns 0 to minimal elements in the semi-lattice, then the normalized distance is defined by

$$d_v^n(x, y) = \begin{cases} \frac{d_v(x,y)}{v(x \vee y)} & \text{if } v(x \vee y) \neq 0, \\ 0 & \text{otherwise.} \end{cases}$$

Fact 6.9. *If v is an isotone, positive, upper valuation that assigns 0 to minimal elements in the semi-lattice, then the normalized distance $d_v^n \geq 0$ and d_v^n verifies the triangle inequality.*

This offers a new perspective on the Symmetric difference and Proportional metrics as metrics defined by minimal-weighted paths in the lattice:

Fact 6.10. *Consider the (semi-)lattice* $\langle 2^{prop}, \subseteq, \mathrm{PROP} \rangle$*, and the mapping* v_0 *that assigns to each* $V \in 2^{prop}$ *its rank. We have*

$$d_{\ominus}(\overrightarrow{\varphi}, \overrightarrow{\psi}) = \delta_{v_0}(\|\overrightarrow{\varphi}\|, \|\overrightarrow{\psi}\|) = \mathsf{card}(\|\overrightarrow{\varphi}\| \ominus \|\overrightarrow{\psi}\|)$$

Fact 6.11. *Consider the (semi-)lattice* $\langle 2^{prop}, \subseteq, \mathrm{PROP} \rangle$*, and the mapping* v_0 *that assigns to each* $V \in 2^{prop}$ *its rank. We have*

$$d_{\alpha}(\overrightarrow{\varphi}, \overrightarrow{\psi}) = \delta_{v_0}^n(\|\overrightarrow{\varphi}\|, \|\overrightarrow{\psi}\|)$$

When the lattice is the lattice of subsets of some set A, the rank of $X \subseteq A$ coincide with its cardinal, hence the two facts above. This suggest d_{v_0} and $d_{v_0}^n$ as natural generalisations of $d_{\ominus}$ and d_{α} for more general semi-lattices.

The continuations-based metric is also expressible in these minimal-weighted paths terms, but this requires a little more work:

Fact 6.12. *Let* $v : W \to \mathbb{R}$ *be an isotone, mapping. The mapping* $w : W \to \mathbb{R}$ *such that* $w : x \mapsto -2^{-v(x)}$ *is isotone as well. Moreover, if* v *is an upper valuation, then so is* w.

Corollary 6.13. *Consider the (semi-)lattice* $\langle 2^{prop}, \subseteq, \mathrm{PROP} \rangle$*, and the mapping* v_0 *that assigns to each* $V \in 2^{prop}$ *its rank, and* w_0 *defined by* $w_0(x) = -2^{-v_0(x)}$. *We have*

$$d_{\mathcal{C}}(\overrightarrow{\varphi}, \overrightarrow{\psi}) = \delta_{w_0}(\|\overrightarrow{\varphi}\|, \|\overrightarrow{\psi}\|)$$

Corollary 6.14. $d_{\ominus}, d_{\alpha}$ *and* $d_{\mathcal{C}}$ *are all pseudo-metrics.*

6.3   Stronger Semantic Axioms

We next move to axioms that differentiate between our metrics. We start by considering the following axiom:

$$d(\overrightarrow{\varphi}, \overrightarrow{\varphi}\,\overrightarrow{\psi}) \le d(\overrightarrow{\varphi}, \overrightarrow{\psi}) \qquad\qquad \text{(rebar property)}$$

From the four metrics we have introduced, only d_H^{ham} does not satisfy it.

Fact 6.15. *Let* $\| \cdot \|$ *be intersective.* $d_{\ominus}, d_{\alpha}$ *and* $d_{\mathcal{C}}$ *satisfy* (rebar property).

Fact 6.16. *Let* $\| \cdot \| = \| \cdot \|^t$. $(L_{\mathrm{PROP}}, d_H^{ham})$ *does not satisfy* (rebar property).

Assume that $\| \cdot \|$ is $\preceq$-poset-valued. The following axiom is very mild:

$$\text{If } \|\overrightarrow{\varphi}\| \preceq \|\overrightarrow{\psi}\| \quad \text{then} \quad d(\overrightarrow{\varphi}, \overrightarrow{\varphi}\,\overrightarrow{\psi}) \le d(\overrightarrow{\varphi}, \overrightarrow{\psi}) \qquad \text{(antitonicity)}$$

Fact 6.17. *Let* $\| \cdot \|$ *satisfy* adjunction *and* right weakening *and let* d *be a semi-pseudometric. If* d *satisfies* (sem $\equiv$ $\Rightarrow$ zero)*, then* d *satisfies* (antitonicity).

6.4   Domain Axioms for Semantic Pseudometrics

Domain axioms require the metric to be well-defined on large portions of the language. Given an interpretation function $\| \cdot \| : L^* \to D$ of these sequences of sentences into some co-domain, we could expect to have:

$$\text{If } \vec{\varphi}, \vec{\psi} \in L^*, \text{then } d(\vec{\varphi}, \vec{\psi}) \in \mathbb{R}. \qquad \text{(linguistic domain)}$$

It should be realized that the preceding axiom is relatively strong. Consider for example d_H^{ham}.

Fact 6.18. *But* $(L_{\text{PROP}}, d_H^{ham})$ *does not satisfy* (linguistic domain).

Fact 6.19. $d_\ominus$, d_α, $d_\mathcal{C}$ *verify* (linguistic domain)

Weakening the preceding axiom, without dropping it entirely, can be done if the co-domain of $\| \cdot \|$ is a bounded poset. Recall that $\perp$ denote the least element of a bounded poset.

$$\text{If } \|\vec{\varphi}\| \neq \perp \text{ and } \|\vec{\psi}\| \neq \perp, \text{ then } d(\vec{\varphi}, \vec{\psi}) \in \mathbb{R}. \qquad \text{(consistent domain)}$$

Fact 6.20. $(L_{\text{PROP}}, d_H^{ham})$ *satisfies* (consistent domain).

6.5   Axioms for Set-Valued Semantic Pseudometrics

If $\| \cdot \|$ is set-valued and $\text{target}(\| \cdot \|) = \wp(W)$ for some W and (W, δ) is a metric space then we can investigate axioms like the following one considered in Eiter and Mannila (1997):

$$\text{whenever } \|\vec{\varphi}\| = \{w\} \text{ and } \|\vec{\psi}\| = \{v\} \text{ then } d(\vec{\varphi}, \vec{\psi}) = \delta(w, v) \qquad \text{(EM)}$$

A semantic metric defined as an aggregator of the values of the distance between points in the interpretation of either sequences, will satisfy the preceding axiom.

6.6   Signature Invariance Axioms

The next condition states that the relative proximity of conversations should not depend on irrelevant aspects pertaining to the choice of signature.

$$\text{If } \vec{\varphi}, \vec{\psi}, \vec{\chi} \in L'^* \text{ and } L' \subseteq L, \text{ then we have} \qquad \text{(weak sig inv)}$$
$$d_L(\vec{\varphi}, \vec{\chi}) \leq d_L(\vec{\psi}, \vec{\chi}) \text{ iff } d_{L'}(\vec{\varphi}, \vec{\chi}) \leq d_{L'}(\vec{\psi}, \vec{\chi})$$

Fact 6.21. $(L_{\text{PROP}}(1), \delta_{count})$ *and* $(L_{\text{PROP}}(1), \delta_{synt,count})$ *satisfy* (weak sig inv).

Fact 6.22. d_H^{ham}, $d_\ominus$, d_α, $d_\mathcal{C}$ *satisfy* (weak sig inv)

7 Preservation Axioms

7.1 Uniform Preservation Axioms

The following axiom states that extending conversations with a given piece of information should not change the relative proximity of conversations. Formally:

$$\text{If } d(\overrightarrow{\varphi}, \overrightarrow{\chi}) \leq d(\overrightarrow{\psi}, \overrightarrow{\chi}) \text{ then } d(\overrightarrow{\varphi}\varphi_0, \overrightarrow{\chi}\varphi_0) \leq d(\overrightarrow{\psi}\varphi_0, \overrightarrow{\chi}\varphi_0)$$

(uniform preservation)

But such an axiom can lead to triviality.

Fact 7.1. *Let* $\| \cdot \|$ *satisfy exchange, contraction and expansion and let* d *be a pseudometric. If* d *satisfies* (uniform preservation) *and* (sem $\equiv$ $\Rightarrow$ zero), *then whenever* $d(\varphi, \chi) \leq d(\psi, \chi)$ *then* $d(\psi\chi, \varphi\psi\chi) = 0$.

The next corollary is slightly technical. Let us introduce a bit of notation. Let $o : \mathsf{L}^* \to \mathbb{R}$ be defined by $o(\overrightarrow{\varphi}) = d(\overrightarrow{\epsilon}, \overrightarrow{\varphi})$, and let $\sim$ be the kernel of o. Let $\leq_o$ be the total pre-order induced by o, with $\overrightarrow{\varphi} \leq_o \overrightarrow{\psi}$ iff $o(\overrightarrow{\varphi}) \leq o(\overrightarrow{\psi})$. Let $[\overrightarrow{\varphi}]$ be the equivalence class of $\overrightarrow{\varphi}$ in $\mathsf{L}^* / \sim$. Let $\downarrow[\overrightarrow{\varphi}] = \{\overrightarrow{\psi} \in \mathsf{L}^* | o(\overrightarrow{\psi}) \leq o(\overrightarrow{\varphi})\}$ and let $\downarrow[\overrightarrow{\varphi}]^*$ be the reflexive transitive closure of $\downarrow[\overrightarrow{\varphi}]$.

Corollary 7.2. *Let* $\| \cdot \|$ *satisfy exchange, contraction, expansion and* $\overrightarrow{\epsilon} - \top$ *and let* d *be a pseudometric. If* d *satisfies* (uniform preservation) *and* (sem $\equiv$ $\Rightarrow$ zero), *then for every* $\overrightarrow{\varphi}, \overrightarrow{\chi} \in [\overrightarrow{\varphi}], \overrightarrow{\psi}_1, \overrightarrow{\psi}_2 \in \downarrow[\overrightarrow{\varphi}]^*$ *we have* $d(\overrightarrow{\varphi}, \overrightarrow{\psi}_1\overrightarrow{\chi}\overrightarrow{\psi}_2) = 0$.

Corollary 7.3. *Let* $\| \cdot \|$ *satisfy exchange, contraction, expansion and* $\overrightarrow{\epsilon} - \top$ *and let* d *be a pseudometric. If* d *satisfies* (uniform preservation), (sem $\equiv$ $\Rightarrow$ zero) *and* (min sem separation), *then for every* $\overrightarrow{\varphi}, \overrightarrow{\chi} \in [\overrightarrow{\varphi}], \overrightarrow{\psi}_1, \overrightarrow{\psi}_2 \in \downarrow[\overrightarrow{\varphi}]^*$ *we have* $\|\overrightarrow{\varphi}\| = \|\overrightarrow{\psi}_1\overrightarrow{\chi}\overrightarrow{\psi}_2\|$.

Hence (uniform preservation) comes with very disputable consequences. The converse is even more problematic:

$$\text{If } d(\overrightarrow{\varphi}\varphi_0, \overrightarrow{\chi}\varphi_0) \leq d(\overrightarrow{\psi}\varphi_0, \overrightarrow{\chi}\varphi_0) \text{ then } d(\overrightarrow{\varphi}, \overrightarrow{\chi}) \leq d(\overrightarrow{\psi}, \overrightarrow{\chi})$$

(uniform anti-preservation)

These two axioms are quite demanding.

Fact 7.4. *Let* $\| \cdot \| = \| \cdot \|^t$. d_H^{ham} *satisfy neither* (uniform preservation), *nor* uniform anti-preservation.

Fact 7.5. *Let* $\| \cdot \| = \| \cdot \|^t$. d_α, $d_\ominus$ *and* $d_{\mathcal{C}}$ *satisfy neither* (uniform preservation), *nor* (uniform anti-preservation).

But the situation is much more radical for (uniform anti-preservation): if the interpretation satisfies very mild conditions: such as contraction, expansion and exchange, then the only semi-pseudometric satisfying (uniform anti-preservation) and (sem $\equiv$ $\Rightarrow$ zero), is the trivial metric.

Fact 7.6. *Let* $\| \cdot \|$ *be an interpretation satisfying* contraction, expansion *and* exchange *and let* d *be a semi-pseudometric on* target($\| \cdot \|$). *The following are equivalent:*

1. *d satisfies* (uniform anti-preservation) *and* (sem $\equiv$ $\Rightarrow$ zero)
2. *d is the trivial metric on* target($\| \cdot \|$)

This result is a very strong argument against the reasonableness of (uniform anti-preservation).

7.2   Preservation Axioms: Close Information

The preceding axioms considered extensions of two sequences with the same sequence of formulas. As we have seen, they are too demanding. What if instead, we are interested in the relative effect of extending with a sequences that might be more or less similar to the sequence it is extending. We could expect, that the closer that new sequence is from the original one, the closer the resulting conversation will be from the original one. Or at least that a reverse in respective orderings cannot occur. Formally,

$$\text{If } d(\overrightarrow{\varphi}, \psi_1) < d(\overrightarrow{\varphi}, \psi_2) \text{ then } d(\overrightarrow{\varphi}, \overrightarrow{\varphi}\psi_1) \leq d(\overrightarrow{\varphi}, \overrightarrow{\varphi}\psi_2) \qquad \text{(action pref)}$$

Fact 7.7. d_H^{ham}, $d_\ominus$, d_α, d_C *do not satisfy* (action pref)

Conversely, we can require the deviation of the resulting sequence to be smaller, whenever the original sequence is closer to the new one by which it is extended.

$$\text{If } d(\overrightarrow{\varphi}, \chi) < d(\overrightarrow{\psi}, \chi) \text{ then } d(\overrightarrow{\varphi}, \overrightarrow{\varphi}\chi) < d(\overrightarrow{\psi}, \overrightarrow{\psi}\chi) \qquad \text{(coherent deviation)}$$

Fact 7.8. *Let* $\| \cdot \| = \| \cdot \|^t$. d_α *satisfy neither* (action pref) *nor* (coherent deviation).

Fact 7.9. *Let* d *be a semi-pseudometric on* target($\| \cdot \|$). *The following are equivalent:*

1. *d satisfies* (coherent deviation) *and the* triangle inequality
2. *d is the* trivial metric *on* target($\| \cdot \|$)

As we will show, the respective converses of the two preceding axioms are certainly unreasonable.

$$\text{If } d(\overrightarrow{\varphi}, \overrightarrow{\varphi}\psi_1) \leq d(\overrightarrow{\varphi}, \overrightarrow{\varphi}\psi_2) \text{ then } d(\overrightarrow{\varphi}, \psi_1) \leq d(\overrightarrow{\varphi}, \psi_2)$$
$$\text{(converse strong action pref)}$$

Fact 7.10. *Let* d *be a pseudometric on* target($\| \cdot \|$). *The following are equivalent:*

1. *d satisfies* (converse strong action pref)
2. *d is the trivial metric on* target($\| \cdot \|$)

$$\text{If } d(\overrightarrow{\varphi}, \overrightarrow{\varphi}\chi) \leq d(\overrightarrow{\psi}, \overrightarrow{\psi}\chi) \text{ then } d(\overrightarrow{\varphi}, \chi) \leq d(\overrightarrow{\psi}, \chi)$$
$$\text{(converse coherent deviation)}$$

Fact 7.11. *Let d be a pseudometric on $\mathsf{target}(\|\cdot\|)$. The following are equivalent:*

1. *d satisfies (converse coherent deviation)*
2. *d is the trivial metric on $\mathsf{target}(\|\cdot\|)$.*

8 Conjunction and Disjunction Axioms

8.1 Conjunction Axioms

Assume that $\|\cdot\|$ is lattice-valued. The following axiom regulates the behavior of the distance with respect to the meet. But as we will see, it is much too demanding.

$$\text{If } \|\overrightarrow{\varphi_1}\| \wedge \|\overrightarrow{\varphi_2}\| \leq \|\overrightarrow{\varphi_1}\| \wedge \|\overrightarrow{\varphi_3}\| \quad \text{then} \quad \delta(\overrightarrow{\varphi_1}, \overrightarrow{\varphi_2}) \geq \delta(\overrightarrow{\varphi_1}, \overrightarrow{\varphi_3}) \quad \text{(strong } \wedge \text{ rule)}$$

Fact 8.1. *Let $\|\cdot\|$ be lattice-valued and let d be a semi-pseudometric on $\mathsf{target}(\|\cdot\|)$. If d satisfies (strong $\wedge$ rule), then whenever $\|\overrightarrow{\varphi}\| \leq \|\overrightarrow{\psi}\|$, we have $d(\overrightarrow{\chi}, \overrightarrow{\varphi}) \geq d(\overrightarrow{\chi}, \overrightarrow{\psi})$ for any $\overrightarrow{\chi}$.*

Corollary 8.2. *Let $\|\cdot\|$ be lattice-valued and let d be a semi-pseudometric on $\mathsf{target}(\|\cdot\|)$. If d satisfies (strong $\wedge$ rule), then whenever $\|\overrightarrow{\varphi}\| \leq \|\overrightarrow{\psi}\|$, we have $d(\overrightarrow{\varphi}, \overrightarrow{\psi}) = d(\overrightarrow{\psi}, \overrightarrow{\varphi}) = 0$.*

Corollary 8.3. *Let $\|\cdot\|$ be a lattice-valued interpretation satisfying $(\overrightarrow{\epsilon}-\top)$ and let d be a semi-pseudometric on $\mathsf{target}(\|\cdot\|)$. The following are equivalent:*

1. *d satisfies (strong $\wedge$ rule) and the triangle inequality.*
2. *d is the trivial metric on $\mathsf{target}(\|\cdot\|)$.*

The above facts follow from the equality case in (strong $\wedge$ rule): for any sequences $\overrightarrow{\varphi}, \overrightarrow{\psi_1}, \overrightarrow{\psi_2}$, if $\overrightarrow{\psi_1} \wedge \overrightarrow{\varphi} = \overrightarrow{\psi_2} \wedge \overrightarrow{\varphi}$ then $\overrightarrow{\psi_1}$ and $\overrightarrow{\psi_2}$ have to be equidistant from $\overrightarrow{\varphi}$. Removing this assumption yields a weakening of (strong $\wedge$ rule) which no longer support the trivialisation result above:

$$\text{If } \|\overrightarrow{\varphi_1}\| \wedge \|\overrightarrow{\varphi_2}\| \lneq \|\overrightarrow{\varphi_1}\| \wedge \|\overrightarrow{\varphi_3}\| \quad \text{then} \quad \delta(\overrightarrow{\varphi_1}, \overrightarrow{\varphi_2}) \geq \delta(\overrightarrow{\varphi_1}, \overrightarrow{\varphi_3}) \quad \text{(weak } \wedge \text{ rule)}$$

Fact 8.4. *d_H^{ham}, $d_\ominus$ and d_α do not satisfy (weak $\wedge$ rule).*

Fact 8.5. *For any set-valued $\|\cdot\|$, d_C verifies (weak $\wedge$ rule).*

Corollary 8.6. *$\|\cdot\|^t$ is an intersective interpretation which yield a d_C that verifies (weak $\wedge$ rule), the triangle inequality and is not trivial.*

8.2 Disjunction Axioms

Assume that $\| \cdot \|$ is lattice-valued. The following axiom is very mild.

$$\text{If } \|\overrightarrow{\psi}\| = \|\overrightarrow{\varphi_1}\| \vee \|\overrightarrow{\varphi_2}\| \text{ then } d(\overrightarrow{\varphi}_1, \overrightarrow{\psi}) \leq d(\overrightarrow{\varphi}_1, \overrightarrow{\varphi}_2) \qquad (\vee \text{ rule})$$

Fact 8.7. *Assume that* $\mathsf{target}(\| \cdot \|) = \wp(W)$ *for some non-empty W and that* $\| \cdot \|$ *is lattice-valued. Let δ be a metric on W. d_H^δ satisfies ($\vee$ rule).*

Fact 8.8. $d_\ominus, d_\alpha$ *and* d_C *also satisfy ($\vee$ rule).*

9 Future Directions

So far we have focused on isolating an abstract concept of semanticity for metrics. We have explored general axioms that help us express components of this concept. We have also identified more specific axioms that were candidates at defining the contour of a notion of 'good behavior' for semantic metrics, and thus at being criterion for evaluating such metrics. We have done this at an abstract level, considering conversations as sequences of formulas where one conversational agent plays a sequence of formulas after the other. The conversation thus has the structure of a (syntactic) monoid with a syntactic composition operation of concatenation. Corresponding to sequences of formulas is their abstract interpretation in a different, semantic space; the generic notion of a semantic interpretation furnishes the correspondence, mapping these sequences into the semantic space.

Coming back to the goals outlined in the introduction of this paper, the next step of our work is to extend this perspective to structures that represent real conversations. We mention a few directions here, each of which can be explored independently. In order to do this, we need to fill in this abstract framework with notions that capture aspects of conversational content at various levels of detail. A first step is to refine the notion of sequence of formulas into something that preserves more of the logical form of conversations. Most models of discourse interpretation assume a more structured representation of conversations, e.g., trees or graphs, in which elementary discourse units are linked together via discourse relations to form more complex discourse units. Using such structures to represent conversations would require us to adapt the structural properties of interpretation functions considered in Sect. 3.2 to be able to reflect the semantics of discourse relations and the units they link together. Second we would need to revisit the axioms that make use of concatenation, replacing the latter with a notion of a graph update or graph extension.

Furthermore, to deal adequately with some natural language phenomena such as questions, commands, agreements and disagreements among speakers, explicit or implicit corrections, it is natural to assume additional structure for semantic spaces, on top of that provided by general lattices. This additional semantic structure could also serve to refine some of our axioms, in particular those making hypotheses on lattice-theoretic relations between the semantic interpretations of two conversations.

Different notions of semantic interpretation carry different amounts of the initial syntactic structure into the semantic space. The classical notion of information content for a discourse erases all structural information, mapping discourses such as *Jane fell because John pushed her* and *John pushed Jane so she fell* into the exact same semantic interpretation (either a set of possible worlds or a set of world assignment pairs as in SDRT and other dynamic semantic theories). Differently structured discourses, even when they share the same meaning, however, may exhibit different semantic and pragmatic behavior, concerning the possibility of future coreferences and of ways to extend the conversation. Intuitions dictate that these features are important for a notion of conversational similarity. It will therefore be important to test metrics defined on more structurally-conservative spaces, for instance conserving some aspects of the conversational graph. These metrics should match intuitions as to how far two real conversations are from each other.

10   Conclusions

Our first task was to explore the concept of a semantic metric by identifying a certain number of reasonable axioms that characterize the idea of 'semanticity' for a distance. We clarified the relation between these different axioms and the triangle inequality, and we mapped out a lattice of axioms in terms of their logical strength. Next, we explored a structured list of candidate axioms or desirable properties for any semantic metric. We found several to be too demanding, in the sense that under some structural constraints on the interpretation function and on the distance, they could only be satisfied by the trivial metric. These axioms divide into a certain number of categories. First, we considered a certain number of axioms pertaining to general properties of semantic metrics, including arguably mild assumptions about their structure, their domain and their insensitivity to the choice of signature. Then, we considered preservation axioms that carry a general idea of coherence between the relative proximity of sequences and of their extensions. Finally we considered axioms that are more specific to a lattice- or a set-theoretic approach.

We concentrated on the foundational case of conversations as sequences of propositional formulae with a classical truth functional interpretation by studying four semantically induced metrics that looked intuitively promising (based respectively on the ideas of symmetric semantic difference, semantic proportionality, Hausdorff metric and on possible continuations). We now have a clear picture of their different behavior. Overall however, these metrics satisfy only few of our axioms that do not lead to a triviality result. One reason for this are the very strong structural hypotheses behind the set-theoretic, classical interpretation of the language of the propositional calculus. A further exploration of these axioms in the context of interpretation into structures like lattices with fewer structural hypotheses and of more general families of metrics remains to be done. We hope that the abstract setting that we have set up in this paper can serve a first step towards achieving this goal.

A   Appendix: Selected Proofs

(PROOF OF FACT 5.2). Assume that for every $\overrightarrow{\chi_1}, \overrightarrow{\chi_2}$ with $\|\overrightarrow{\chi_1}\| = \|\overrightarrow{\chi_2}\|$ we have $d(\overrightarrow{\varphi}, \overrightarrow{\chi_1}) = d(\overrightarrow{\psi}, \overrightarrow{\chi_2})$ (i). Now take some $\overrightarrow{\chi}$. We have $\|\overrightarrow{\chi}\| = \|\overrightarrow{\chi}\|$. Hence by (i), we have $d(\overrightarrow{\varphi}, \overrightarrow{\chi}) = d(\overrightarrow{\psi}, \overrightarrow{\chi})$. But $\overrightarrow{\chi}$ was arbitrary, hence $\forall \chi \quad d(\overrightarrow{\varphi}, \overrightarrow{\chi}) = d(\overrightarrow{\psi}, \overrightarrow{\chi})$. It follows, by (sem separation), that $\|\overrightarrow{\varphi}\| = \|\overrightarrow{\psi}\|$.   QED

(PROOF OF FACT 5.3). Assume that $\forall \overrightarrow{\chi}, d(\overrightarrow{\varphi}, \overrightarrow{\chi}) = d(\overrightarrow{\psi}, \overrightarrow{\chi})$. In particular $d(\overrightarrow{\varphi}, \overrightarrow{\psi}) = d(\overrightarrow{\psi}, \overrightarrow{\psi}) = 0$. Hence by (zero $\Rightarrow$ sem $\equiv$), $\|\overrightarrow{\varphi}\| = \|\overrightarrow{\psi}\|$.   QED

(PROOF OF FACT 5.4). The right to left direction follows from Fact 5.3. For the left to right direction, assume that $d(\overrightarrow{\varphi}, \overrightarrow{\psi}) = 0$ (i). Take any $\overrightarrow{\chi}$. By triangle inequality, $d(\overrightarrow{\chi}, \overrightarrow{\varphi}) \leq d(\overrightarrow{\chi}, \overrightarrow{\psi}) + d(\overrightarrow{\psi}, \overrightarrow{\varphi})$. Hence, by (i) we have $d(\overrightarrow{\chi}, \overrightarrow{\varphi}) \leq d(\overrightarrow{\chi}, \overrightarrow{\psi})$. Similarly we have $d(\overrightarrow{\chi}, \overrightarrow{\psi}) \leq d(\overrightarrow{\chi}, \overrightarrow{\varphi})$. Hence $d(\overrightarrow{\chi}, \overrightarrow{\psi}) = d(\overrightarrow{\chi}, \overrightarrow{\varphi})$. But χ was arbitrary, hence for all χ we have $d(\overrightarrow{\chi}, \overrightarrow{\psi}) = d(\overrightarrow{\chi}, \overrightarrow{\varphi})$. By (sem separation), it follows that $\|\overrightarrow{\varphi}\| = \|\overrightarrow{\psi}\|$.   QED

(PROOF OF FACT 5.7). $\|p\neg p\| = \|q\neg q\|$ but $\delta_{count}(p\neg p, q\neg q) = 4$.   QED

(PROOF OF FACT 5.8). $\|p\neg p\| = \|q\neg q\|$ but $\delta_{synt,count}(p\neg p, q\neg q) = 6$.   QED

(PROOF OF FACT 5.9). Take some φ such that $\mathsf{card}(\|\varphi\|) \geq 2$.   QED

(PROOF OF FACT 5.10). Assume that $\|\overrightarrow{\varphi}\| = \|\overrightarrow{\psi}\|$. By (sem preservation) we have $d(\overrightarrow{\varphi}, \overrightarrow{\chi}) = d(\overrightarrow{\psi}, \overrightarrow{\chi})$. In particular we have $d(\overrightarrow{\varphi}, \overrightarrow{\psi}) = d(\overrightarrow{\psi}, \overrightarrow{\psi}) = 0$ QED

(PROOF OF FACT 5.11). Assume that $\|\overrightarrow{\varphi_1}\| = \|\overrightarrow{\psi_1}\|$. Take some $\overrightarrow{\chi}$. We have $\|\overrightarrow{\chi}\| = \|\overrightarrow{\chi}\|$. Hence by (sem induced) we have $d(\overrightarrow{\varphi_1}, \overrightarrow{\chi}) = d(\overrightarrow{\psi_1}, \overrightarrow{\chi})$.   QED

(PROOF OF FACT 5.12). Assume that $\|\overrightarrow{\varphi_1}\| = \|\overrightarrow{\psi_1}\|$ (i) and $\|\overrightarrow{\varphi_2}\| = \|\overrightarrow{\psi_2}\|$ (ii). By triangle inequality we have:

$$d(\overrightarrow{\varphi_1}, \overrightarrow{\varphi_2}) \leq \underbrace{d(\overrightarrow{\varphi_1}, \overrightarrow{\psi_1})}_{0,\ \text{by}(\text{sem}\equiv\ \Rightarrow\ \text{zero})} + d(\overrightarrow{\psi_1}, \overrightarrow{\varphi_2})$$

$$d(\overrightarrow{\psi_1}, \overrightarrow{\varphi_2}) \leq d(\overrightarrow{\psi_1}, \overrightarrow{\psi_2}) + \underbrace{d(\overrightarrow{\psi_2}, \overrightarrow{\varphi_2})}_{0,\ \text{by}(\text{sem}\equiv\ \Rightarrow\ \text{zero})}$$

Hence, $d(\overrightarrow{\varphi_1}, \overrightarrow{\varphi_2}) \leq d(\overrightarrow{\psi_1}, \overrightarrow{\psi_2})$. Similarly, we have $d(\overrightarrow{\varphi_1}, \overrightarrow{\varphi_2}) \geq d(\overrightarrow{\psi_1}, \overrightarrow{\psi_2})$.   QED

(PROOF OF COROLLARY 5.13). By Fact 5.12, d satisfies (sem induced), hence for every $\overrightarrow{\varphi_1}, \overrightarrow{\varphi_2}, \overrightarrow{\psi_1}, \overrightarrow{\psi_2}$ with $\|\overrightarrow{\varphi_1}\| = \|\overrightarrow{\psi_1}\|$ and $\|\overrightarrow{\varphi_2}\| = \|\overrightarrow{\psi_2}\|$ we have

$$d(\overrightarrow{\varphi_1}, \overrightarrow{\varphi_2}) = d(\overrightarrow{\psi_1}, \overrightarrow{\psi_2})$$

It follows that $\dot{d}(\|\overrightarrow{\varphi}\|, \|\overrightarrow{\psi}\|) := d(\overrightarrow{\varphi}, \overrightarrow{\psi})$ is well-defined. Moreover for any $\overrightarrow{\varphi}$, $\dot{d}(\|\overrightarrow{\varphi}\|, \|\overrightarrow{\varphi}\|) = d(\overrightarrow{\varphi}, \overrightarrow{\varphi}) = 0$. Triangle inequality is proven similarly.   QED

(Proof of Fact 5.14). First observe, that by triangle inequality, we have

$$d(\vec{\chi}, \vec{\varphi}) \leq d(\vec{\chi}, \vec{\psi}) + d(\vec{\psi}, \vec{\varphi})$$

Now, assume that $\vec{\varphi} \equiv \vec{\psi}$. By (sem $\equiv$ $\Rightarrow$ zero) we have $d(\vec{\psi}, \vec{\varphi}) = 0$, hence $d(\vec{\chi}, \vec{\varphi}) \leq d(\vec{\chi}, \vec{\psi})$. Similarly, $d(\vec{\chi}, \vec{\psi}) \leq d(\vec{\chi}, \vec{\varphi})$ which proves (1).   QED

(Proof of Fact 5.15). Assume that $\forall \chi$ we have $d(\varphi, \chi) = d(\psi, \chi)$ (i). Take some $\vec{\chi_1}, \vec{\chi_2}$ with $\|\vec{\chi_1}\| = \|\vec{\chi_2}\|$. By (sem $\equiv$ $\Rightarrow$ zero) we have $d(\vec{\chi_1}, \vec{\chi_2}) = 0$ (ii). By triangle inequality we have:

$$d(\vec{\varphi}, \vec{\chi_1}) \leq d(\vec{\varphi}, \vec{\chi_2}) + \underbrace{d(\vec{\chi_2}, \vec{\chi_1})}_{0, \text{ by}(ii)}$$

$$d(\vec{\varphi}, \vec{\chi_2}) \leq d(\vec{\varphi}, \vec{\chi_1}) + \underbrace{d(\vec{\chi_1}, \vec{\chi_2})}_{0, \text{ by}(ii)}$$

Hence $d(\vec{\varphi}, \vec{\chi_1}) = d(\vec{\varphi}, \vec{\chi_2})$. Moreover by (i) we have $d(\vec{\varphi}, \vec{\chi_2}) = d(\vec{\psi}, \vec{\chi_2})$. Hence $d(\vec{\varphi}, \vec{\chi_1}) = d(\vec{\psi}, \vec{\chi_2})$. Since $\vec{\chi_1}, \vec{\chi_2}$ were arbitrary, it follows by (min sem separation), that $\|\vec{\varphi}\| = \|\vec{\psi}\|$.   QED

(Proof of Fact 6.12). Let v be an isotone upper valuation and $z \leq x, y$. Assume without loss of generality that $v(y) \geq v(x)$. We can write, for $0 \leq \alpha \leq 1$:

$$-\alpha \cdot 2^{-v(y)} \geq -2^{-v(y)} \geq -2^{-v(x)},$$

which ensures that

$$-\alpha \cdot 2^{-v(y)} - \frac{1}{\alpha} 2^{-v(x)} \leq -2^{-v(y)} - 2^{-v(x)}.$$

Instantiating this result for $\alpha = 2^{v(z)-v(x)}$ yields after development

$$-2^{v(z)-v(x)-v(y)} - 2^{-v(z)} \leq -2^{-v(y)} - 2^{-v(x)}.$$

Since v is an upper-valuation, we have $-v(x \vee y) \geq v(z) - v(x) - v(y)$ and thus

$$-2^{-v(x \vee y)} - 2^{-v(z)} \leq -2^{v(z)-v(x)-v(y)} - 2^{-v(z)} \leq -2^{-v(y)} - 2^{-v(x)}$$

i.e., $w(x \vee y) + w(z) \leq w(x) + w(y)$ which concludes the proof. The case $v(x) \geq v(y)$ is symmetrically dealt with.   QED

(Proof of Fact 6.16). Let $\vec{\varphi} := (p_1 \wedge p_2) \vee (\neg p_1 \wedge \neg p_2 \wedge \neg p_3)$ and $\vec{\psi} := (p_1 \wedge (p_2 \rightarrow p_3))$, and assume some intersective interpretation of concatenation. We have $\|\vec{\varphi}\vec{\psi}\| = \|p_1 \wedge p_2 \wedge p_3\|$. $d_H^{ham}(\vec{\varphi}, \vec{\varphi}\vec{\psi}) = 3$, but $d_H^{ham}(\vec{\varphi}, \vec{\psi}) = 1$.   QED

(Proof of Fact 6.18). For any φ, d_H^{ham} is neither well-defined for $(\varphi, \perp)$ nor for $(\perp, \varphi)$.   QED

(PROOF OF FACT 6.21). Adding a new propositional letter that does not occur in either sequence will not affect the symmetric difference of the range of formulas, nor the symmetric difference of the respective signature. Allowing for the negation of the propositional letter that was previously forbidden will not change the sets either.   QED

(PROOF OF FACT 7.1). Assume that $d(\varphi, \chi) \leq d(\psi, \chi)$ then $d(\varphi\psi\chi, \chi\psi\chi) \leq d(\psi\psi\chi, \chi\psi\chi)$. By exchange, contraction, expansion, and (sem $\equiv$ $\Rightarrow$ zero), we have $d(\psi\psi\chi, \chi\psi\chi) = 0$. Hence $d(\varphi\psi\chi, \chi\psi\chi) = 0$. By exchange, contraction, expansion, and (sem $\equiv$ $\Rightarrow$ zero), we have $d(\varphi\psi\chi, \chi\psi) = 0$. Concluding our proof.   QED

(PROOF OF COROLLARY 7.2). We only give the idea of the proof. The idea of the proof is to define a linear order on L^* compatible with $\leq_o$. By induction, using Fact 7.1 we first show the claim for formulas in the same o-equivalence class, then we show that the claim propagate downward, that is for every $\vec{\psi} \in\downarrow [\vec{\varphi}]$. Finally we show that the claim propagates with transitive closure.   QED

(PROOF OF COROLLARY 7.3). Direct from Fact 5.15 and Corollary 7.2.   QED

(PROOF OF FACT 7.4). Let $k \geq 2$, $n = 2k$. Now let

$$\varphi := p_1 \rightarrow (\neg p_2 \wedge \ldots \wedge \neg p_n) \wedge \neg p_1 \rightarrow (p_2 \wedge \ldots \wedge p_n),$$
$$\psi := p_1 \wedge \neg p_2 \wedge \ldots \wedge p_{2k-1} \wedge \neg p_{2k},$$

$\chi := p_1 \wedge \ldots \wedge p_n$ and $\varphi_0 := p_1$. Since $k \geq 2$ we have

$$1 = d_H^{ham}(\varphi, \chi) < d_H^{ham}(\varphi, \chi) = k, \text{ and,}$$
$$n - 1 = d_H^{ham}(\varphi\varphi_0, \chi\varphi_0) > d_H^{ham}(\varphi\varphi_0, \chi\varphi_0) = k = n/2$$

Concluding our proof.   QED

(PROOF OF FACT 7.6). (1 $\Rightarrow$ 2). Take some $\vec{\varphi}$, $\vec{\psi}$. By contraction, expansion and exchange we have

$$\|\vec{\varphi}\,\vec{\varphi}\,\vec{\psi}\| = \|\vec{\psi}\,\vec{\varphi}\,\vec{\psi}\|$$

Hence by (sem $\equiv$ $\Rightarrow$ zero), $d(\vec{\varphi}\,\vec{\varphi}\,\vec{\psi}, \vec{\psi}\,\vec{\varphi}\,\vec{\psi}) = 0$ (i). Hence $d(\vec{\varphi}\,\vec{\varphi}\,\vec{\psi}, \vec{\psi}\,\vec{\varphi}\,\vec{\psi})$ $\leq d(\vec{\psi}\,\vec{\varphi}\,\vec{\psi}, \vec{\psi}\,\vec{\varphi}\,\vec{\psi})$ (i). Now, let $\vec{\chi} = \vec{\psi}$ and $\vec{\varphi_0} = \vec{\varphi}\,\vec{\psi}$. By (i) and (uniform anti-preservation) we have

If $d(\vec{\varphi}\,\vec{\varphi}\,\vec{\psi}, \vec{\psi}\,\vec{\varphi}\,\vec{\psi}) \leq d(\vec{\psi}\,\vec{\varphi}\,\vec{\psi}, \vec{\psi}\,\vec{\varphi}\,\vec{\psi})$ then $d(\vec{\varphi}, \vec{\psi}) \leq d(\vec{\psi}, \vec{\psi}) = 0$

Hence by (i), $d(\vec{\varphi}, \vec{\psi}) = 0$. Concluding the proof for this direction. The other direction is trivial.   QED

(PROOF OF FACT 7.8). Let $n \in \omega$ be such that $n > 5$. Moreover let: $\varphi := (p_3 \wedge \ldots \wedge p_n) \wedge \neg(p_1 \wedge p_2)$, $\psi_1 := (p_1 \vee \ldots \vee p_n) \wedge \neg(p_2 \wedge \ldots \wedge p_n)$ and $\psi_2 := (p_2 \wedge \ldots \wedge p_n)$. We have $d_\alpha(\varphi, \psi_1) = 1 - \frac{2}{n+3} = \frac{n+1}{n+3} > d_\alpha(\varphi, \psi_2) = 1 - \frac{1}{4} = \frac{3}{4}$. But we have $d_\alpha(\varphi, \varphi\psi_1) = 1 - \frac{2}{3} = \frac{1}{3} < d_\alpha(\varphi, \varphi\psi_2) = 1 - \frac{1}{3} = \frac{2}{3}$.   QED

(PROOF OF FACT 7.9). Take some $\overrightarrow{\varphi}, \overrightarrow{\psi}$. By (coherent deviation) we have

$$\text{If } d(\overrightarrow{\varphi}, \overrightarrow{\epsilon}) < d(\overrightarrow{\psi}, \overrightarrow{\epsilon}) \text{ then } d(\overrightarrow{\varphi}, \overrightarrow{\varphi}\,\overrightarrow{\epsilon}) < d(\overrightarrow{\psi}, \overrightarrow{\psi}\,\overrightarrow{\epsilon})$$

$$\text{If } d(\overrightarrow{\psi}, \overrightarrow{\epsilon}) < d(\overrightarrow{\varphi}, \overrightarrow{\epsilon}) \text{ then } d(\overrightarrow{\psi}, \overrightarrow{\psi}\,\overrightarrow{\epsilon}) < d(\overrightarrow{\varphi}, \overrightarrow{\varphi}\,\overrightarrow{\epsilon})$$

Since $\overrightarrow{\varphi}, \overrightarrow{\psi}$ were arbitrary, it follows that for any $\overrightarrow{\varphi}, \overrightarrow{\psi}, d(\overrightarrow{\varphi}, \overrightarrow{\epsilon}) = d(\overrightarrow{\psi}, \overrightarrow{\epsilon})$. In particular $d(\overrightarrow{\varphi}, \overrightarrow{\epsilon}) = d(\overrightarrow{\psi}, \overrightarrow{\epsilon}) = d(\overrightarrow{\epsilon}, \overrightarrow{\epsilon}) = 0$. Hence by (triangle inequality) we have $\forall \overrightarrow{\varphi}, \overrightarrow{\psi} d(\overrightarrow{\varphi}, \overrightarrow{\psi}) \leq d(\overrightarrow{\varphi}, \overrightarrow{\epsilon}) + d(\overrightarrow{\epsilon}, \overrightarrow{\psi}) = 0$. $\qquad$ QED

(PROOF OF FACT 7.10). $(1 \Rightarrow 2)$. Take some $\overrightarrow{\varphi}, \overrightarrow{\psi}$. We have

$$0 = d(\overrightarrow{\varphi}, \overrightarrow{\varphi}\,\overrightarrow{\epsilon}) \leq d(\overrightarrow{\varphi}, \overrightarrow{\varphi}\,\overrightarrow{\psi})$$

Hence by (converse strong action pref) $d(\overrightarrow{\varphi}, \overrightarrow{\epsilon}) \leq d(\overrightarrow{\varphi}, \overrightarrow{\psi})$. But ψ was arbitrary, hence, in particular $d(\overrightarrow{\varphi}, \overrightarrow{\epsilon}) \leq d(\overrightarrow{\varphi}, \overrightarrow{\varphi}) = 0$. But φ was arbitrary as well, hence $\forall \chi d(\overrightarrow{\varphi}, \overrightarrow{\epsilon}) = 0$. Hence by triangle inequality for any formula $\overrightarrow{\varphi}, \overrightarrow{\psi}$ we have $d(\overrightarrow{\varphi}, \overrightarrow{\psi}) \leq d(\overrightarrow{\varphi}, \overrightarrow{\epsilon}) + d(\overrightarrow{\epsilon}, \overrightarrow{\psi}) = 0$. Concluding our proof. $\qquad$ QED

(PROOF OF FACT 7.11). $(1 \Rightarrow 2)$. Take some $\overrightarrow{\varphi}$ and $\overrightarrow{\psi}$. We have

$$0 = d(\overrightarrow{\varphi}, \overrightarrow{\varphi}\,\overrightarrow{\epsilon}) \leq d(\overrightarrow{\epsilon}, \overrightarrow{\epsilon}\,\overrightarrow{\epsilon}) = 0$$

By (converse coherent deviation) $d(\overrightarrow{\varphi}, \overrightarrow{\epsilon}) \leq d(\overrightarrow{\epsilon}, \overrightarrow{\epsilon}) = 0$ (i). Similarly, we have $d(\overrightarrow{\psi}, \overrightarrow{\epsilon}) = 0$ (ii). By (i), (ii) and triangle inequality we have $d(\overrightarrow{\varphi}, \overrightarrow{\psi}) \leq d(\overrightarrow{\varphi}, \overrightarrow{\epsilon}) + d(\overrightarrow{\epsilon}, \overrightarrow{\psi}) = 0$. Concluding our proof. $\qquad$ QED

(PROOF OF FACT 8.1). Take some $\overrightarrow{\chi}$ and assume that $\|\overrightarrow{\varphi}\| \leq \|\overrightarrow{\psi}\|$. Since target($\|\cdot\|$) is a lattice. We have $\|\overrightarrow{\chi}\| \wedge \|\overrightarrow{\varphi}\| \leq \|\overrightarrow{\chi}\| \wedge \|\overrightarrow{\psi}\|$. Hence by (strong $\wedge$ rule) $\delta(\overrightarrow{\chi}, \overrightarrow{\varphi}) \geq \delta(\overrightarrow{\chi}, \overrightarrow{\psi})$. $\qquad$ QED

(PROOF OF COROLLARY 8.2). Assume that $\|\overrightarrow{\varphi}\| \leq \|\overrightarrow{\psi}\|$. Since d satisfies (strong $\wedge$ rule), we have by Fact 8.1 we have in particular $0 = \delta(\overrightarrow{\varphi}, \overrightarrow{\varphi}) \geq \delta(\overrightarrow{\varphi}, \overrightarrow{\psi}) = \delta(\overrightarrow{\psi}, \overrightarrow{\varphi})$. $\qquad$ QED

(PROOF OF COROLLARY 8.3). $(1 \Rightarrow 2)$. Take two arbitrary $\overrightarrow{\varphi}, \overrightarrow{\psi}$. By $(\overrightarrow{\epsilon}\text{-}\top)$ we have $\|\overrightarrow{\varphi}\| \leq \|\overrightarrow{\epsilon}\|$ and $\|\overrightarrow{\psi}\| \leq \|\overrightarrow{\epsilon}\|$. Since d satisfies (strong $\wedge$ rule), it follows by Corollary 8.2 that $d(\overrightarrow{\varphi}, \overrightarrow{\epsilon}) = d(\overrightarrow{\epsilon}, \overrightarrow{\psi}) = 0$. By triangle inequality it follows that $d(\overrightarrow{\varphi}, \overrightarrow{\psi}) = 0$. The $(2 \Rightarrow 1)$ direction is trivial. $\qquad$ QED

(PROOF OF FACT 8.5). Take $\overrightarrow{\varphi}, \overrightarrow{\psi_1}, \overrightarrow{\psi_2}$ with $\overrightarrow{\varphi} \cap \overrightarrow{\psi_1} \subsetneq \overrightarrow{\varphi} \cap \overrightarrow{\psi_2}$. And consider the cardinalities assigned in Fig. 2. Let $X_1 := \alpha_1 + \eta_0$ and let $X_2 := \alpha_2 + \eta_0$. Using these cardinalities and inserting them in the expression of the distance, gives us:

$$d_C(\varphi, \psi_1) = \frac{2^{\mathsf{card}(\|\overrightarrow{\varphi})\|} + 2^{\mathsf{card}(\|\overrightarrow{\psi_1}\|)} - 2 \cdot 2^{\mathsf{card}(\|\overrightarrow{\varphi}\| \cap \|\overrightarrow{\psi_1}\|)}}{2^{\mathsf{card}(\|\overrightarrow{\varphi}\|) + \mathsf{card}(\|\overrightarrow{\psi_1}\|)}}$$

$$= \frac{2^{X_\varphi + \eta_1 + \eta_2} + 2^{X_1 + \eta_1} - 2 \cdot 2^{\eta_1}}{2^{X_\varphi + X_1 + 2\eta_1 + \eta_2}}$$

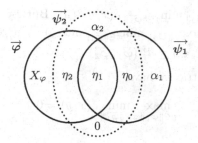

Fig. 2. Assigning cardinalities to the respective intersections.

Similarly:

$$d_C(\varphi, \psi_2) = \frac{2^{X_\varphi + \eta_1 + \eta_2} + 2^{X_2 + \eta_1 + \eta_2} - 2 \cdot 2^{\eta_1 + \eta_2}}{2^{X_\varphi + X_2 + 2\eta_1 + 2\eta_2}}$$

From the two previous expression, after simplifications we find:

$$d_C(\varphi, \psi_1) - d_C(\varphi, \psi_2) = \frac{2^{X_\varphi}(2^{X_2 + \eta_2} - 2^{X_1}) + 2(2^{X_1} - 2^{X_2})}{2^{X_\varphi + X_1 + X_2 + \eta_1 + \eta_2}}$$

From the assumption that $\overrightarrow{\varphi} \cap \overrightarrow{\psi_1} \subsetneq \overrightarrow{\varphi} \cap \overrightarrow{\psi_2}$, it follows that $\eta_2 \geq 1$, hence:

$$2^{X_\varphi}(2^{X_2 + \eta_2} - 2^{X_1}) + 2(2^{X_1} - 2^{X_2}) \geq 2^{X_2 + \eta_2} - 2^{X_2 + 1} + 2^{X_1 + 1} - 2^{X_1} \geq 0$$

which concludes the proof.   QED

(PROOF OF FACT 8.7). Take $\overrightarrow{\varphi}_1, \overrightarrow{\varphi}_2$. Assume that $\|\overrightarrow{\psi}\| = \|\overrightarrow{\varphi_1}\| \vee \|\overrightarrow{\varphi_2}\|$ (i). By definition, we have

$$d_H^\delta(\overrightarrow{\varphi}_1, \overrightarrow{\varphi}_2) = \max\{\max_{x \in \|\overrightarrow{\varphi_1}\|} \min_{y \in \|\overrightarrow{\varphi_2}\|} \delta(x, y),$$

$$\max_{y \in \|\overrightarrow{\varphi_2}\|} \min_{x \subset \|\overrightarrow{\varphi_1}\|} \delta(x, y)\}$$

Hence, we are in one of two cases.

(1) $\max_{x \in \|\overrightarrow{\varphi_1}\|} \min_{y \in \|\overrightarrow{\varphi_2}\|} \delta(x, y) = d_H^\delta(\overrightarrow{\varphi_1}, \overrightarrow{\varphi_2})$, or,

(2) $\max_{y \in \|\overrightarrow{\varphi_2}\|} \min_{x \in \|\overrightarrow{\varphi_1}\|} \delta(x, y) = d_H^\delta(\overrightarrow{\varphi_1}, \overrightarrow{\varphi_2})$.

Case 1. There are two subcases.
Subcase 1a. Assume that $d_H^\delta(\overrightarrow{\varphi}_1, \overrightarrow{\psi}) = \max_{x \in \|\overrightarrow{\varphi_1}\|} \min_{y \in \|\overrightarrow{\psi}\|}$ (a). By (i),

$$\max_{x \in \|\overrightarrow{\varphi_1}\|} \min_{y \in \|\overrightarrow{\psi}\|} \leq \max_{x \in \|\overrightarrow{\varphi_1}\|} \min_{y \in \|\overrightarrow{\varphi}_2\|}$$

By (a) and (1) we have $d_H^\delta(\overrightarrow{\varphi}_1, \overrightarrow{\psi}) \leq d_H^\delta(\overrightarrow{\varphi_1}, \overrightarrow{\varphi_2})$.
Subcase 1b. $d_H^\delta(\overrightarrow{\varphi_1}, \overrightarrow{\psi}) = \max_{y \in \|\overrightarrow{\psi}\|} \min_{x \in \|\overrightarrow{\varphi_1}\|} \delta(x, y)$ (b). Since δ is a metric we have $\max_{y \in \|\overrightarrow{\varphi_1}\|} \min_{x \in \|\overrightarrow{\varphi_1}\|} \delta(x, y) = 0$. Hence by (i) and (b) we have

$d_H^\delta(\overrightarrow{\varphi_1}, \overrightarrow{\psi}) = \max_{y \in \|\overrightarrow{\varphi_2}\|} \min_{x \in \|\overrightarrow{\varphi_1}\|} \delta(x,y)$ (ii). But by (1) and definition of $d_H^\delta(\overrightarrow{\varphi_1}, \overrightarrow{\varphi_2})$ we have $\max_{y \in \|\overrightarrow{\varphi_2}\|} \min_{x \in \|\overrightarrow{\varphi_1}\|} \delta(x,y) \leq d_H^\delta(\overrightarrow{\varphi_1}, \overrightarrow{\varphi_2})$ (iii). By (ii) and (iii) we have $d_H^\delta(\overrightarrow{\varphi_1}, \overrightarrow{\psi}) \leq d_H^\delta(\overrightarrow{\varphi_1}, \overrightarrow{\varphi_2})$.

Case 2. Since δ is a metric we have

$$\max_{y \in \|\overrightarrow{\varphi_1}\|} \min_{x \in \|\overrightarrow{\varphi_1}\|} \delta(x,y) = 0$$

Hence by (2), we have

$$\max_{y \in \|\overrightarrow{\psi}\|} \min_{x \in \|\overrightarrow{\varphi_1}\|} \delta(x,y) = d_H^\delta(\overrightarrow{\varphi_1}, \overrightarrow{\varphi_2})$$

$$= \max_{y \in \|\overrightarrow{\varphi_2}\|} \min_{x \in \|\overrightarrow{\varphi_1}\|} \delta(x,y)$$

$$= d_H^\delta(\overrightarrow{\varphi_1}, \overrightarrow{\varphi_2}) \qquad \text{(iv)}$$

We now consider two subcases.

Subcase 2a. $d_H^\delta(\overrightarrow{\varphi}_1, \overrightarrow{\psi}) = \max_{x \in \|\overrightarrow{\varphi_1}\|} \min_{y \in \|\overrightarrow{\psi}\|}$ (a). But since δ is a metric we have $\max_{x \in \|\overrightarrow{\varphi_1}\|} \min_{y \in \|\overrightarrow{\psi}\|} \leq \max_{x \in \|\overrightarrow{\varphi_1}\|} \min_{y \in \|\overrightarrow{\varphi_2}\|}$ (v). But by definition $d_H^\delta(\overrightarrow{\varphi_1}, \overrightarrow{\varphi_2})$ and (2) we have $\max_{x \in \|\overrightarrow{\varphi_1}\|} \min_{y \in \|\overrightarrow{\varphi_2}\|} \leq d_H^\delta(\overrightarrow{\varphi}_1, \overrightarrow{\varphi}_2)$ (vi). By (v), (vi) and (a), it follows that $d_H^\delta(\overrightarrow{\varphi}_1, \overrightarrow{\psi}) \leq d_H^\delta(\overrightarrow{\varphi_1}, \overrightarrow{\varphi_2})$.

Subcase 2b. Assume that. $d_H^\delta(\overrightarrow{\varphi_1}, \overrightarrow{\psi}) = \max_{y \in \|\overrightarrow{\psi}\|} \min_{x \in \|\overrightarrow{\varphi_1}\|} \delta(x,y)$ (b). By (iv), it follows that $d_H^\delta(\overrightarrow{\varphi_1}, \overrightarrow{\psi}) = d_H^\delta(\overrightarrow{\varphi_1}, \overrightarrow{\varphi_2})$.

Hence in all cases $d_H^\delta(\overrightarrow{\varphi}_1, \overrightarrow{\psi}) \leq d_H^\delta(\overrightarrow{\varphi_1}, \overrightarrow{\varphi_2})$. Concluding our proof.   QED

References

Alchourrón, C.E., Gärdenfors, P., Makinson, D.: On the logic of theory change: partial meet contraction and revision functions. J. Symbolic Logic **50**(2), 510–530 (1985)

Asher, N., Lascarides, A.: Logics of Conversation. Cambridge University Press, Cambridge (2003)

Black, E., Abney, S.P., Flickenger, D., Gdaniec, C., Grishman, R., Harrison, P., Hindle, D., Ingria, R., Jelinek, F., Klavans, J.L., Liberman, M., Marcus, M.P., Roukos, S., Santorini, B., Strzalkowski, T.: A procedure for quantitatively comparing the syntactic coverage of english grammars. In: Proceedings of a Workshop held at Pacific Grove, Speech and Natural Language, California, USA, 19–22 February 1991, Morgan Kaufmann (1991)

Eiter, T., Mannila, H.: Distance measures for point sets and their computation. Acta Informatica **34**(2), 109–133 (1997)

Grice, H.P.: Logic and conversation. Studies in the Way of Words, pp. 22–40. Harvard University Press, Cambridge, MA (1967)

Lehmann, D., Magidor, M., Schlechta, K.: Distance semantics for belief revision. J. Symbolic Logic **66**(1), 295–317 (2001)

Monjardet, B.: Metrics on partially ordered sets - a survey. Discrete Math. **35**(13), 173–184 (1981). Special Volume on Ordered Sets

Popper, K.R.: Conjectures and Refutations: The Growth of Scientific Knowledge. Harper and Row, New York (1968)

Sampson, G.: A proposal for improving the measurement of parse accuracy. Int. J. Corpus Linguist. **5**, 53–68 (2000)

World History Ontology for Reasoning Truth/Falsehood of Sentences: Event Classification to Fill in the Gaps Between Knowledge Resources and Natural Language Texts

Ai Kawazoe[✉], Yusuke Miyao, Takuya Matsuzaki,
Hikaru Yokono, and Noriko Arai

National Institute of Informatics, 2-1-2 Hitotsubashi, Chiyoda-ku,
Tokyo 101-8430, Japan
{zoeai,yusuke,takuya-matsuzaki,yokono,arai}@nii.ac.jp

Abstract. This paper introduces a world history ontology that supports reasoning of truth/falsehood of historical descriptions in natural languages. The core of the ontology includes an event classification according to certain basic properties such as necessary/sufficient conditions for the existence of events in the real world. We will discuss how this ontology functions in solving world history problems in Japan's National Center Test for University Admissions, especially in the reasoning of "falsehood" of sentences and bridging of the "granularity difference" between target sentences and knowledge resources.

1 Introduction

Automatic judgment of truth/falsehood of natural language sentences is one of the most challenging tasks for semantic processing. We face many hurdles when performing such a task; the largest hurdle is the gaps between "known" facts in knowledge resources and the semantic contents of target sentences. A promising approach to solve this issue is to enrich knowledge resources to cover many facts and linguistic variations in their descriptions. However, a resource that encompasses all commonly known facts is not realistically feasible. Furthermore, even if a resource is made as rich as possible, the following issues are still difficult to solve:

- **How to reason "falsehood" of obviously false sentences (e.g., sentences that describe nonexistent historical events).** Based on the open-world assumption, "facts" in knowledge resources are only those that are "known to be true," and we do not know the truth/falsehood of information missing from the resources.
- **How to reason truth/falsehood of natural language sentences that describe events in different granularity from the descriptions in knowledge resources**.

In this paper, we introduce a world history ontology that supports the reasoning of truth/falsehood of historical descriptions in natural languages. We also point out several properties of events that address the above issues and incorporate an event classification according to the properties in the ontology. Such properties include (1) coexistence of event participants; (2) closeness between event participants; (3) possibility of

© Springer International Publishing Switzerland 2014
Y. Nakano et al. (Eds.): JSAI-isAI 2013, LNAI 8417, pp. 42–50, 2014.
DOI: 10.1007/978-3-319-10061-6_3

recurrence; (4) sensitivity to incompatibility between participants; (5) "part-to-whole" and "whole-to-part"implication; and (6) "leader-to-organization" and "organization-to-leader" implication.

This study is a part of the Todai Robot Project (http://www.21robot.org), which aims to develop an AI that can solve problems in university entrance examinations. In Sect. 2, we outline world history problems and the challenges involved in solving them with AIs. In Sect. 3, we introduce the ontology and the event classification as its core. In Sect. 4, we discuss the relationship between our ontology and other related studies.

2 The Task: Solving World History Problems in University Entrance Examinations

The National Center Test for University Admission is a standardized test used by public universities in Japan. Approximately 70 % of the world history questions in the past examinations were true/false questions, such as the following:

Choose the one correct sentence concerning events that occurred during the 8th.

(1) *Pepin destroyed the Kingdom of the Lombards.*
(2) *The reign of Harun al-Rashid began.*
    (2009 Main Examination: World History B)

Most sentences in the true/false questions are simple sentences. They rarely include anaphoric expressions or modality and are carefully designed to avoid ambiguities.

Automatic answering to historical true/false questions is addressed by applying textual entailment recognition (Miyao 2012) and question answering (Kanayama 2012). The present best system based on textual entailment can correctly answer approximately 60 % of the questions using data sets that are manually constructed from world history textbooks and Wikipedia. However, the entailment approach makes it difficult to judge false sentences (sentences (1) and (3–5)) as false.

(3) *Francis I and Otto I maintained rivalry.*
(4) *Ottoman empire was founded in the Balkan Peninsula and expanded to Asia Minor.*
(5) *The Russian Empire claimed Pan-Germanism and advanced southward.*
    (From 2011 Main Exam: World History B and 2009 Main Exam: World History A)

Although we can obtain information on statements such as "Karl, the child of the Pepin III, destroyed the Kingdom of the Lombards," "Francis I and Charles V maintained an intense personal rivalry," "the Ottoman Empire was founded in north-western Anatolia," and "the Russian Empire claimed Pan-Slavism," they alone cannot be direct evidences of the falsehood of (1) and (3)–(5). To reason their falsehood, we need more general knowledge about events such as "destruction of a nation needs to occur during the period when its agent (destroyer) exists," "maintaining rivalry needs its participants to exist in the same period," and "foundation of a nation occurs only once." Humans often use these conditions in daily situations (e.g., proving of an alibi), but computers cannot use them unless they are provided in an explicit and usable form.

Another problematic case is the granularity mismatch between descriptions in resources and target sentences. We often see question sentences such as "Nazi occupied Paris" and find descriptions such as "Nazi occupied Northern France" in knowledge resources. The fact that the latter entails the former is obvious to humans. One might base this inference on the part–whole relationship between "Paris" and "Northern France," but the situation is not that simple. Not every event allows the same inference as occupation. For example, partitioning of an area does not necessarily imply a partitioning of all its parts.

3   World History Ontology and Its Event Classes

The world history ontology presented in this paper is constructed to compensate for the weaknesses of current approaches. We aim to enable computers to judge truth/falsehood of sentences that are obvious to humans and distinguish them from sentences whose truth/falsehood is truly unknown (e.g., "Oswald killed Kennedy"). The ontology supports such reasoning by combining parts of relatively small knowledge while adopting the open-world assumption.

3.1   Top-Level and Main Classes

The ontology is constructed under the top-level ontology of YAMATO (Mizoguchi, 2012). The reason why we adopt YAMATO as the upper ontology is that the level of abstraction and granularity of its classes is suitable for our purpose. Using YAMATO, we do not need to fill in the gap or solve the conflict between the classes in the upper ontology and the classes we need in the world history domain. YAMATO is also useful in that it has several classes which are useful to solve the ambiguity of terms and words of natural language; for example, the distinction between extrinsic accomplishment and intrinsic accomplishment.

A part of main classes and their main properties in the world history ontology are shown in Table 1.

As in Table 1, for instances of most classes, we describe time spans during which they exist (the starting time and the ending time). For organizations, including nations, their founders are given. The subregion property for "geographical object" specifies part-whole relationships between instances of the class. For the classes of "thought/theory," "religion" and "style of art," we describe incompatibility among instances.

Event classes are divided into complex event classes and simple event classes. Complex events are loosely described as those referred with proper names, for example, incidents, accidents, social movements, wars, organized events, etc. They can be viewed as complex of events, and their extent of the impact, starting times and ending times are often determined by social, political, or academic perspectives. For example, "Hundred Year's War" consists of subevents such as battles such as Battle of Crécy, the siege of Orleans, and execution of Jeanne d'Arc. For each complex event instance, we describe its main subevents, its starting and ending time, and if possible, its location. However, it is often the case that participants such as agent or theme cannot be specified.

Table 1. Main classes and properties in the world history ontology

Upper classes in YAMATO	Main classes in the world history ontology	Main properties
Weak agent	Person	nationality, time of existence, occupation
	Ethnic group (as a crowd)	
	Organization	nationality, time of existence, founder, purpose
	Nation (as a subclass of organization)	territory, time of existence, founder
Artifact	Construction	time of existence, location
	Book, painting, sculpture	time of existence, author
Event	Complex event	classes of participants, starting_time, ending_time, location, subevent
	Simple event	
Proposition	Law, social system, policy	nationality, purpose, time of existence, institutor
	Treaty and accord	participants, purpose, time of existence
	Thought/theory	time of existence, incompatibility among instances
	Statement	time of existence
Content_2	Style of art	time of existence, incompatibility among instances
	Culture	time of existence, location
Complex content	Religion	time of existence, founder, incompatibility among instances
	Academic discipline	time of existence
	Technology	time of existence, inventor
Geographical object	Region	subregion (which specifies part–whole relationships between instances)
Time	Time period	starting_time, ending_time
Role	Social role	
	Event-dependent role	

Simple events are generally those described by verbs. For each simple event instance, we describe its participants (such as agent, theme, source, and goal) in addition to its starting time, ending time and location. Simple events are classified according to the properties introduced in the following sections.

3.2 Event Properties that Serve to Reason Falsehood of False Sentences

We introduce four properties useful to reason falsehood of the obviously false sentences in world history questions. Examples of event descriptions according to the four properties are shown in Table 2.

Table 2. Example of event descriptions according to coexistence of participants, closeness, possibility of reccurrence, and sensitivity to incompatibility (*st and et indicate *starting_time* and *ending_time*, respectively.)

Event class	Participants and its classes	Coexistence of participants	Closeness	Recurrence	Sensitivity to incompatibility
Engaging in warfare	agent1: nation, agent2: nation	Required	No restriction	Possible	N/A
Founding of a nation	agent: weak agent, theme: nation	Required, st(agent) < st(theme) & st(theme) < et(agent)*	No restriction	Not possible	N/A
Killing of a person	agent: person, theme: person	Required, st (agent) < et(theme) & et(theme) < et(agent)	The theme is in a range of view of the agent	Not possible	N/A
Claiming of an ideology	agent: weak agent, theme: thought	Not required, st(theme) < et(agent)	N/A	Possible	Sensitive (theme)
Invading of an area	agent: person, theme: geographical object	Required	No restriction	Possible	N/A
Occupation of an area	agent: weak agent, theme: geographical object	Required	No restriction	Possible	N/A
Partitioning of an area	agent: weak agent, theme: geographical object	Required	No restriction	Possible	N/A

3.2.1   Coexistence of Event Participants

For many events, time spans of their participants existence must overlap, i.e., coexistence of all participants is required (e.g., maintaining rivalry, engaging in warfare, and buying and selling). In other words, for these events, if certain times of existence of their participants do not overlap, the events do not exist in the real world. This is a strong reason to judge sentences such as (3) to be false since Francis I lived in the 15th century, while Otto I lived in the 10th century. Some events are more restricted: in making events (such as founding of a nation), maker's time of existence has to include the time at which what is made starts to exist, while in annihilating events (such as destruction of a nation), annihilator's time of existence should include the time at which what is annihilated ceases to exist.

It should be noted that the coexistence of all participants is not a universal property for events, since in certain events some participants' time of existence may or may not overlap each other. For example, events such as knowing and evaluating can occur even if the agent of the events lives in a period when the theme does not exist (e.g., it is possible for us to know of a person who died much before our birth). The only requirement for these events is that the time where the theme starts to exist is earlier than the time where the agent ceases to exist, since one cannot know or evaluate something that occurs after his/her death. Although we can assume the existence of many patterns of overlapping and non-overlapping events, only a few exist (Table 2).

3.2.2 Closeness Between Event Participants

Closeness between event participants is another key to reason falsehood of sentences that describes nonexistent events. Some events require their participants to be sufficiently close so that they can touch each other (e.g., touching, hitting, or putting together). Others require the theme to be in the range of view of the agent/experiencer (e.g., meeting, reading, or killing), while many others do not have any restriction (e.g., buying, controlling, or giving).

3.2.3 Possibility of Recurrence

Many events (e.g., maintaining a rivalry, engaging in warfare) can occur more than once with the same combination of participants, possibly in different locations from the first time, but a few events do not occur with the above condition. Events that involve starting and ending an existence are categorized as the latter events, which include birth and death of humans as well as making and annihilating of things. Irreversible changes such as turning twenty years old are also included in the same category. This property is useful for the truth/falsehood judgment when combined with particular known facts in the knowledge resources. For example, the falsehood of (4) can be derived from the fact that "Ottoman Empire was founded in north-western Anatolia," "the Balkan Peninsula is an area disjoint from Anatolia," and the knowledge that founding of a nation occurs only once.

3.2.4 Sensitivity to Incompatibility Between Participants

As we have seen in Table 1, for the classes of "thought/theory," "religion," and "style of art," we describe "incompatibility" among instances (e.g., incompatibility between "Pan-Germanism" and "Pan-Slavism"). We also classify events into two types: events that are sensitive to incompatibility and those that are not. Events such as claiming an ideology and converting to a religion are included the former category, in that if an event instance exists, another event of the same type cannot exist simultaneously with an incompatible participant. This type of knowledge is necessary to reason the falsehood of (5) from the fact that "the Russian Empire claimed Pan-Slavism." On the contrary, events such as knowing or studying are not sensitive to such incompatibilities.

3.3 Event Properties for Resolving Granularity Problems

Here we introduce two properties which serve to resolve the granularity mismatch between descriptions in resources and target sentences. Examples of event descriptions according to the properties are shown in Table 3.

3.3.1 "Part-to-Whole" and "Whole-to-Part" Implication

We describe part–whole relationships between instances in the "region" class and classify events on the basis of whether the existence of an event instance involving a region instance implies that an event of the same type exists within a part or a larger area of the region. Buying, having, selling, and occupation of an area imply that the same types of events exist in any part of the area and do not necessarily have the same implication for a larger area. Invading of an area is an example of events that implies

Table 3. Example of event descriptions according to "part-to-whole" and "whole-to-part" implication and "leader-to-organization" and "organization-to-leader" implication

Event class	Part–whole implication	Leader-organization implication
Engaging in warfare	N/A	leader $\Rightarrow$ org(agent), org $\Rightarrow$ leader(agent)
Founding of a nation	N/A	leader $\Rightarrow$ org(agent), org $\Rightarrow$ leader(agent)
Killing of a person	N/A	no implication
Claiming of an ideology	N/A	leader $\Rightarrow$ org(agent), org $\Rightarrow$ leader(agent)
Invading of an area	part $\Rightarrow$ some larger area (theme)	leader $\Rightarrow$ org(agent), org $\Rightarrow$ leader(agent)
Occupation of an area	whole $\Rightarrow$ part (theme)	leader $\Rightarrow$ org(agent), org $\Rightarrow$ leader(agent)
Partitioning of an area	no implication	leader $\Rightarrow$ org(agent), org $\Rightarrow$ leader(agent)

the existence of an event of the same type in some larger area. Events such as partitioning of an area do not have either implication.

3.3.2 "Leader-to-Organization" and "Organization-to-Leader" Implication

We find the cases of granularity mismatch between world history resources and target sentences attributed to the exchangeability between an organization and its leader (representative person) in a description of an event. It is often the case that an action of an organization is described as that of its leader. In events such as declaration of a war, dominance of the territory, conclusion of a treaty or sending troops, the action of an organization can be equated with that of the leader, and vice versa. On the other hand, many events do not have such an implication. For example, a talk by national leaders does not imply the talk is held between the nations, and dissolution of an organization never entails that the same thing is happened to its leader.

4 Comparison with Other Resources

There are several studies on world history ontology such as Ide (2007), Kauppinen (2007), and Ishikawa (2008). Ontologies proposed in these studies are developed to structure historical knowledge for use in historical studies. As discussed above, our ontology is not designed for historical research, but for the general task of reasoning truth/falsehood of natural language sentences. Historical knowledge in our ontology is only high-school level; however, we aim to describe general properties of events that can serve for the truth/falsehood judgment of sentences.

The first four properties listed in Sect. 3 can be considered as necessary conditions for the existence of an event, whereas the last property can be considered as a sufficient condition. Although there are resources that include detailed descriptions about events (such as SUMO, FrameNet, and Kaneiwa (2007)'s upper event ontology), these necessary/sufficient conditions have not been studied yet. One may think the conditions would follow from the descriptions in these resources; it seems to be true for some of them. For example, the coexistence restrictions on making and annihilating events will partly follow from the characterization of the Temporal Existence Change function of events in Kaneiwa (2007). However, the derivation process is not straightforward and still needs some trivial assumptions (such as "an agent of an event has to exist during the event").

One advantage of the design of our ontology is that it can support automatic truth/ falsehood judgment of sentences without detailed description of events. For example, we can judge a sentence like "Francis I and Otto I maintained rivalry" as false even if we do not know precisely what "maintaining rivalry" is. All we need is the knowledge of the years of the participants' birth and death, and the "coexistence" property of the event. We consider that our event classification can supplement characterization of events within other resources.

5 Current Status

The ontology is still under construction. Currently it contains 420 classes (classes in YAMATO are not counted), of which 225 are event classes. 304 verbs/verbal expressions are related to the classes. This covers 50.1 % of the all occurrences of verbs in sentences in truth/false questions in world history examinations in the National Center Tests in the past 21 years. The ontology will be published online after we achieve enough coverage.

6 Conclusion

The ontology proposed in this paper is designed for the task of truth/falsehood judgment of simple historical descriptions in the National Center Test for University Admission. We introduced an event classification that supports the reasoning of existence or nonexistence of a described event from rather small, basic information.

Acknowledgments. We gratefully acknowledge the National Center for University Entrance Examination and JC Educational Institute for the Center Test data.

References

Kanayama, H., Miyao, Y., Prager, J.: Answering Yes/No questions via question inversion. In: Proceedings of COLING (2012)

Kaneiwa, K., Iwazume, M., Fukuda, K.: An upper ontology for event classifications and relations. In: Orgun, M.A., Thornton, J. (eds.) AI 2007. LNCS (LNAI), vol. 4830, pp. 394–403. Springer, Heidelberg (2007)

Ide, N., Woolner, D.: Historical ontologies. In: Ahmad, K., Brewster, C., Stevenson, M. (eds.) Words and Intelligence II: Essays in Honor of YorickWilks, pp. 137–152. Springer, The Netherlands (2007)

Mizoguchi, R.: Theory and Practice of Ontology Engineering. Ohmsha, Tokyo (in Japanese)

Ishikawa, T., Kitauchi, A., Shirotsuka, O.: Extraction of person information from historical materials for building historical ontology. J. Nat. Lang. Process. **15**(4), 3–18 (2008). The Association for Natural Language Processing. (in Japanese)

Kauppinen, T., Hyvonen, E.: Modeling and reasoning about changes in ontology time series. In: Kishore, R., Sharman, R., Ramesh, R., et al. (eds.) Ontologies: A Handbook of Principles, Concepts and Applications in Information Systems, pp. 319–338. Springer, New York (2007)

Miyao, Y., Shima, H., Kanayama, H., Mitamura, T.: Evaluating textual entailment recognition for university entrance examinations. ACM Trans. Asian Lang. Inf. Process. **11**(4), 13 (2012)

Hypersequent Calculi for Modal Logics Extending S4

Hidenori Kurokawa[✉]

Department of Information Science, Kobe University, Kobe, Japan
hidenori.kurokawa@gmail.com

Abstract. In this paper, we introduce hypersequent calculi for some modal logics extending S4 modal logic. In particular, we uniformly characterize hypersequent calculi for S4, S4.2, S4.3, S5 in terms of what are called "external modal structural rules" for hypersequent calculi. In addition to the monomodal logics, we also introduce simple bimodal logics combing S4 modality with another modality from each of the rest of logics. Using a proof-theoretic method, we prove cut-elimination for the hypersequent calculi for these logics and, as applications of it, we show soundness and faithfulness of Gödel embedding for the monomodal logics and the bimodal logic combining S4 and S5.

1 Introduction

There are not too many modal logics that traditional Gentzen-style sequent calculi can formulate in a cut-free manner. Cut-free sequent calculi for S4 and some of its sublogics, i.e., K, D, K4, T, KD4 can be formulated (e.g., [22]), but other modal logics are difficult to be naturally formulated by sequent calculi (except K45, KD45 in [20]).

On the other hand, there are several proof-theoretic frameworks which are extensions of traditional sequent calculi and in which many other modal logics can be formulated in a cut-free manner. Some of them use labels to represent possible worlds, e.g., prefixed tableau systems [9]. Other sequent calculi (say [17]) explicitly mention accessibility relations. Yet others allow the "nesting" of meta-symbols corresponding to "⇒," "," in sequents. Display calculi [21] and nested sequents [4] can be counted as these. As representative cases, we mention that until recently, only labeled sequents [17] and display calculi [12] handled S4.2 and S4.3, which are located between S4 and S5.[1]

However, hypersequent calculi, which are less radical extensions of traditional sequent calculi, can work well for these logics. In this paper, we formulate hypersequent calculi for these logics in addition to S4 and S5 so as to present these logics from a uniform perspective in which all these logics have the common

---

[1] However, see [2] for critical discussions about display calculi. Also, there exist sequent calculi for these logics, but the one for S4.2 is not cut-free [23], and the one for S4.3 does not satisfy one of the desiderata for good proof systems in [2].

© Springer International Publishing Switzerland 2014
Y. Nakano et al. (Eds.): JSAI-isAI 2013, LNAI 8417, pp. 51–68, 2014.
DOI: 10.1007/978-3-319-10061-6_4

modal rules for S4 and the differences of these logics can be formulated exactly by using different "modal (hyper)structural rules," which are modal analogues of pure (external) structural rules in hypersequent calculi. The main results of this paper is the cut-elimination theorems for these hypersequent calculi via a uniform proof-theoretic reduction method.

Although there are some very recent works independently done, such as [11], which covers S4.3, and [14], which covers more logics, our method is different from them, since they use a semantic method of proving admissibility of cut.

Our results may not be very surprising (due to their connection to super-intuitionistic logics [3,5]), but they can contribute to calibrating the expressive capacity of hypersequents as a proof-theoretic framework. To further explore the issue, we introduce bimodal logics combing S4 and another modality from the remaining logics, and we show cut-elimination for these combined logics uniformly. As applications of these, we syntactically prove soundness and faithfulness of Gödel modal embedding from the superintuitionistic counterparts into the modal logics.

The significance of our technical contributions should also be understood as a research towards a satisfactory treatment of the meaning of logical constants (in particular modal operators) from the viewpoint of proof-theoretic semantics. Proof-theoretic semantics originally started as an attempt of explaining the meaning of logical constants via introduction and elimination rules in natural deduction but has recently been extended to a view in which the meaning of logical constants can be given via combinations of operational rules for logical constants and structural rules (e.g., [19]). Our approach is along this line of research and partly motivated by "Došen's principle" [6], i.e., different logics should be understood as having common operational rules and different structural rules (i.e., as "structural variants"). This naturally leads to a view of the meaning of logical constants which divides two different elements of the meaning of logical constants, i.e., the operational meaning and the global meaning [18]. Our results may be taken to be an application of these ideas to modal operators and become a basis for applying the method of proof-theoretic semantics to yet other linguistic expressions.

2   Hilbert-Style Axiomatic Systems for Modal Logics

Let us first present the language of (bi)modal logic. For the sake of brevity, we present the language of bimodal logic, which contains two modal operators $\Box$ and $\boxdot$. The language of modal logic $\Box$. The grammar of the language looks as follows.

$$A := P_i \,|\, \bot \,|\, \neg A \,|\, A_1 \to A_2 \,|\, A_1 \wedge A_2 \,|\, A_1 \vee A_2 \,|\, \Box A \,|\, \boxdot A.$$

We now present Hilbert-style axiomatic systems for the (bi)modal logics.

I. Hilbert-style axiomatization of the single modal logics

Axioms   (0) Axioms of Propositional Logic

(1) Axioms of K4: K axiom $\Box(A \to B) \to (\Box A \to \Box B)$
    4 axiom $\Box A \to \Box\Box A$
(2) Axioms for S4 : K4 + T : $\Box A \to A$
(3) Axioms for S4.2 : S4 + .2 :[2] $\neg\Box\neg\Box A \to \Box\neg\Box\neg A$
(4) Axioms for S4.3 : S4 + .3 : $\Box(\Box A \to B) \vee \Box(\Box B \to A)$
(5) Axioms for S5 : S4 + 5 : $\neg\Box A \to \Box\neg\Box A$

Rules of Inference: Modus Ponens $\dfrac{A \to B \quad A}{B}$ Necessitation $\dfrac{A}{\Box A}$

II. Hilbert-style axiomatizations of bimodal modal logics

In addition to these monomodal logics, here we consider certain bimodal logics that combine certain pairs among those modal logics. We have considered all combinations of S4 (S4, S4.2, S4.3, S5) by using a very simple combining axiom. However, only S4 + L, where L = S4, S4.2, S4.3, or S5, can have simple formulations in hypersequent calculi.[3] Our axiomatizations of the combined logics are given by:

0. Axioms and rules for each modal logic with $\boxdot$ for L;
1.[4] $\boxdot A \to \Box A$.

3 Hypersequent Calculi for Modal Logics

Here we present hypersequent calculi for the modal logics in the previous section.

Axioms $A \Rightarrow A \quad \bot \Rightarrow$ **Structural rules** LIW $\dfrac{G|\Gamma \Rightarrow \Delta}{G|\Gamma, A \Rightarrow \Delta}$ RIW $\dfrac{G|\Gamma \Rightarrow \Delta}{G|\Gamma \Rightarrow \Delta, A}$

LIC $\dfrac{G|\Gamma, A, A \Rightarrow \Delta}{G|\Gamma, A \Rightarrow \Delta}$ RIC $\dfrac{G|\Gamma \Rightarrow \Delta, A, A}{G|\Gamma \Rightarrow \Delta, A}$ EC $\dfrac{G|H|H}{G|H}$ EW $\dfrac{G}{G|H}$

Operational rules   L$\neg$ $\dfrac{G|\Gamma \Rightarrow \Delta, A}{G|\Gamma, \neg A \Rightarrow \Delta}$ R$\neg$ $\dfrac{G|\Gamma, A \Rightarrow \Delta}{G|\Gamma \Rightarrow \Delta, \neg A}$

L$\wedge$ $\dfrac{G|A, \Gamma \Rightarrow \Delta}{G|A \wedge B, \Gamma \Rightarrow \Delta}$ $\dfrac{G|B, \Gamma \Rightarrow \Delta}{G|A \wedge B, \Gamma \Rightarrow \Delta}$ R$\wedge$ $\dfrac{G|\Gamma \Rightarrow \Delta, A \quad \Gamma \Rightarrow \Delta, B}{G|\Gamma \Rightarrow \Delta, A \wedge B}$

L$\vee$ $\dfrac{G|\Gamma, A \Rightarrow \Delta \quad B, \Gamma \Rightarrow \Delta}{G|A \vee B, \Gamma \Rightarrow \Delta}$ R$\vee$ $\dfrac{G|\Gamma \Rightarrow \Delta, A}{G|\Gamma \Rightarrow \Delta, A \vee B}$ $\dfrac{G|\Gamma \Rightarrow \Delta, B}{G|\Gamma \Rightarrow \Delta, A \vee B}$

L$\to$ $\dfrac{G|\Gamma \Rightarrow \Delta, A \quad B, \Pi \Rightarrow \Theta}{G|A \to B, \Gamma, \Pi \Rightarrow \Delta, \Theta}$ R$\to$ $\dfrac{G|\Gamma, A \Rightarrow \Delta, B}{G|\Gamma \Rightarrow \Delta, A \to B}$

Cut $\dfrac{G|\Gamma \Rightarrow \Delta, A \quad H|A, \Pi \Rightarrow \Theta}{G|H|\Gamma, \Pi \Rightarrow \Delta, \Theta}$

---

[2] $\Diamond\Box A \to \Box\Diamond A$ if we use $\Diamond$ in the language, but we do not consider $\Diamond$ in this paper.
[3] This way of combining logics is not a product or a fibring, which is known in the literature of modal logic. It is close to fusion, but with an additional combining axiom. The author does not know how to call it (since apparently there is no established technical term for this case), but thanks for an anonymous referee who suggested him to clarify this.
[4] For this axiom, see [8].

S4 Modal rules for □ $L\Box\ \dfrac{G|A,\Gamma\Rightarrow\Delta}{G|\Box A,\Gamma\Rightarrow\Delta}$ $R\Box\ \dfrac{G|\Box\Gamma\Rightarrow A}{G|\Box\Gamma\Rightarrow\Box A}$

(The ⊡ cases are similar.)

The axioms and rules presented so far constitute a hypersequent calculus for S4. We call it HS4. In addition to HS4, we consider the following modal external structural rules or modal (hyper)structural rules.

Modal Splitting (MS) $\dfrac{G|\Box\Gamma,\Delta\Rightarrow\Pi}{G|\Box\Gamma\Rightarrow|\Delta\Rightarrow\Pi}$

Restricted MS (RMS) $\dfrac{G|\Box\Gamma,\Box\Delta\Rightarrow}{G|\Box\Gamma\Rightarrow|\Box\Delta\Rightarrow}$

Modal Communication (MC) $\dfrac{G|\Sigma,\Box\Gamma\Rightarrow\Pi\quad G|\Theta,\Box\Delta\Rightarrow\Lambda}{G|\Sigma,\Box\Delta\Rightarrow\Pi|\Theta,\Box\Gamma\Rightarrow\Lambda}$

The systems obtained by adding these rules to HS4 are summarized as follows.

HS4.2 (a hypersequent calculus for S4.2) : HS4 + RMS
HS4.3 (a hypersequent calculus for S4.3) : HS4 + MC
HS5 (a hypersequent calculus for S5) : HS4 + MS

Example: S4.3⊢ $\Box(\Box(\Box A\vee\Box B))\to A)\vee\Box(\Box(\Box A\vee\Box B)\to B)$.

$$\dfrac{\dfrac{\dfrac{A\Rightarrow A}{\Box A\Rightarrow A}}{\Box A\Rightarrow A|\Box A\Rightarrow B}\quad\dfrac{\dfrac{A\Rightarrow A}{\Box A\Rightarrow A}\quad\dfrac{B\Rightarrow B}{\Box B\Rightarrow B}}{\Box B\Rightarrow A|\Box A\Rightarrow B}}{\dfrac{\Box A\vee\Box B\Rightarrow A|\Box A\Rightarrow B}{\dfrac{\Box A\vee\Box B\Rightarrow A|\Box A\vee\Box B\Rightarrow B}{\Box(\Box A\vee\Box B)\Rightarrow A|\Box(\Box A\vee\Box B)\Rightarrow B}}\quad\dfrac{\dfrac{B\Rightarrow B}{\Box B\Rightarrow B}}{\Box A\vee\Box B\Rightarrow A|\Box B\Rightarrow B}}$$

Rules for combined logics S4 + L, where L ∈ {S4, S4.2, S4.3, S5}

$L\Box\ \dfrac{G|A,\Gamma\Rightarrow\Delta}{G|\Box A,\Gamma\Rightarrow\Delta}$ $L\boxdot\ \dfrac{G|A,\Gamma\Rightarrow\Delta}{G|\boxdot A,\Gamma\Rightarrow\Delta}$ $R\Box\ \dfrac{G|\Box\Gamma,\boxdot\Delta\Rightarrow\varphi}{G|\Box\Gamma,\boxdot\Delta\Rightarrow\Box\varphi}$ $R\boxdot\ \dfrac{G|\boxdot\Gamma\Rightarrow\varphi}{G|\boxdot\Gamma\Rightarrow\boxdot\varphi}$

S4.2 $\dfrac{G|\boxdot\Gamma,\boxdot\Delta\Rightarrow}{G|\boxdot\Gamma\Rightarrow|\boxdot\Delta\Rightarrow}$ S4.3 $\dfrac{G|\Theta,\boxdot\Gamma\Rightarrow\Pi\quad G|\Xi,\boxdot\Delta\Rightarrow\Lambda}{G|\Theta,\boxdot\Delta\Rightarrow\Pi|\Xi,\boxdot\Gamma\Rightarrow\Lambda}$ S5 $\dfrac{G|\boxdot\Gamma,\Delta\Rightarrow\Pi}{G|\boxdot\Gamma\Rightarrow|\Delta\Rightarrow\Pi}$

The combined logics that are obtained by combining these rules and that we discuss here are as follows. (For the other combinations, see footnote 5). Note that S4 + S5 can be a hypersequent calculus Goranko and Passy's logic with universal modality [10].

S4+S4 = Classical rules + bimodal L□, R□, L⊡, R⊡
S4+S4.2 = S4 + S4 + S4.2 for ⊡
S4+S4.3 = S4 + S4 + S4.3 for ⊡
S4+S5 = S4 + S4 + S5 for ⊡

Example: S4+S5 ⊢ $\Box(\boxdot(\Box A\vee\Box B)\to A)\vee\Box(\boxdot(\Box A\vee\Box B)\to B)$. (Call it Φ).

$$\frac{\dfrac{\dfrac{\dfrac{A \Rightarrow A}{\Box A \Rightarrow A}}{\boxdot(\Box A \vee \Box B), \Box A \Rightarrow A}}{\Box A \Rightarrow \boxdot(\Box A \vee \Box B) \to A}}{\Box A \Rightarrow \Box(\boxdot(\Box A \vee \Box B) \to A)}$$

$$(1)\Box A \Rightarrow \varPhi$$

$$\frac{\dfrac{\dfrac{\dfrac{B \Rightarrow B}{\Box B \Rightarrow B}}{\boxdot(\Box A \vee \Box B), \Box B \Rightarrow B}}{\Box B \Rightarrow \boxdot(\Box A \vee \Box B) \to B}}{\Box B \Rightarrow \Box(\boxdot(\Box A \vee \Box B) \to B)}$$

$$(2)\Box B \Rightarrow \varPhi$$

$$\frac{\Box A \vee \Box B \Rightarrow \varPhi}{\boxdot(\Box A \vee \Box B) \Rightarrow \varPhi}$$

$$\frac{\boxdot(\Box A \vee \Box B) \Rightarrow \mid \Rightarrow \varPhi}{\Rightarrow \Box(\boxdot(\Box A \vee \Box B) \to A)\mid \Rightarrow \varPhi} \text{ S5 rule}$$

$$\frac{\Rightarrow \varPhi\mid \Rightarrow \varPhi}{\Rightarrow \varPhi}$$

4   Deductive Equivalence Between Hilbert-Style Systems and Hypersequent Calculi

By using the following translation from hypersequents to formulas, we can prove that the hypersequents and Hilbert-style axiom systems are deductively equivalent.
To show this, we use the following embeddings from hypersequents to formulas.

$$\mathcal{I}_1(\varGamma_1 \Rightarrow \varDelta_1\mid \ldots \mid \varGamma_n \Rightarrow \varDelta_n) = \Box(\bigwedge \varGamma_1 \to \bigvee \varDelta_1) \vee \cdots \vee \Box(\bigwedge \varGamma_n \to \bigvee \varDelta_n)$$
$$\mathcal{I}_2(\varGamma_1 \Rightarrow \varDelta_1\mid \ldots \mid \varGamma_n \Rightarrow \varDelta_n) = \boxdot(\bigwedge \varGamma_1 \to \bigvee \varDelta_1) \vee \cdots \vee \boxdot(\bigwedge \varGamma_n \to \bigvee \varDelta_n).$$

Note that we do not use $\Box$ but we use $\boxdot$.[5] Here L stands for one of the logics discussed above and HL stands for the hypersequent calculus for each of these.

Theorem 1 (Deductive Equivalence). *HL⊢ $\varGamma_1 \Rightarrow \varDelta_1\mid \ldots \mid \varGamma_n \Rightarrow \varDelta_n$ if and only if L⊢ $\Box(\bigwedge \varGamma_1 \to \bigvee \varDelta_1) \vee \cdots \vee \Box(\bigwedge \varGamma_n \to \bigvee \varDelta_n)$.*
In the case of combined logics, we have HL⊢ $\varGamma_1 \Rightarrow \varDelta_1\mid \ldots \mid \varGamma_n \Rightarrow \varDelta_n$ if and only if L ⊢ $\boxdot(\bigwedge \varGamma_1 \to \bigvee \varDelta_1) \vee \cdots \vee \boxdot(\bigwedge \varGamma_n \to \bigvee \varDelta_n)$.

Proof. Proof by induction on the length of the derivation for both directions.
⟹) It suffices to prove that all the axioms of Hilbert-style systems are derivable and derivability is preserved under the rules. We present the case of .2 as a representative case.

---

[5] This is not an arbitrary choice. Apparently, there is no meaningful way of defining embedding into formulas by using $\Box$. The problem consists in $R\boxdot$. (Note that $\nvdash \Box A \to \Box \boxdot A$, which we need in order to prove soundness of the rule w.r.t. Hilbert-style system under the translation using $\Box$.) Also, note that S4.2 + S4.2, S4.3 + S4.3, S5 + S5 have a problem even if we use $\boxdot$ for a translation into the object language. Proving soundness w.r.t. the Hilbert-style system for e.g., S4.2 + S5, apparently requires $\Box A \to \boxdot\Box A$, but this is not provable in the system, which can be checked by an easy model-theoretic argument.

$$\frac{\varphi \Rightarrow \varphi}{\varphi, \neg\varphi \Rightarrow}$$

$$\frac{\Box\varphi \Rightarrow, \Box\neg\varphi \Rightarrow}{\Box\varphi \Rightarrow |\Box\neg\varphi \Rightarrow}$$ restricted modal splitting

$$\frac{\Rightarrow \neg\Box\varphi| \Rightarrow \neg\Box\neg\varphi}{\Rightarrow \Box\neg\Box\varphi| \Rightarrow \Box\neg\Box\neg\varphi}$$

$$\frac{\neg\Box\neg\Box\varphi \Rightarrow |\Rightarrow \Box\neg\Box\neg\varphi}{\neg\Box\neg\Box\varphi \Rightarrow \Box\neg\Box\neg\varphi} \ EC$$

$\Longleftarrow$) Conversely, we show that axioms and all the translated forms of rules in the hypersequent calculi are derivable. We show a representative case.

(1) .2 rule: By IH, $\vdash \bigvee G \vee \Box(\bigwedge \Box\Gamma \wedge \bigwedge \Box\Delta \to \bot)$. But in S4.2 we can derive:

$\vdash (\bigwedge \Box\Gamma \wedge \bigwedge \Box\Delta \to \bot) \to (\Box \bigwedge \Gamma \wedge \Box \bigwedge \Delta \to \bot)$.

$\vdash \Box(\bigwedge \Box\Gamma \wedge \bigwedge \Box\Delta \to \bot) \to \Box(\Box \bigwedge \Delta \to \neg\Box \bigwedge \Gamma)$.

$\vdash \Box(\bigwedge \Box\Gamma \wedge \bigwedge \Box\Delta \to \bot) \to (\Box\Box \bigwedge \Delta \to \Box\neg\Box \bigwedge \Gamma)$.

$\vdash \Box(\bigwedge \Box\Gamma \wedge \bigwedge \Box\Delta \to \bot) \to (\Box \bigwedge \Delta \to \Box\neg\Box \bigwedge \Gamma)$.

$\vdash \Box(\bigwedge \Box\Gamma \wedge \bigwedge \Box\Delta \to \bot) \to (\neg\Box\neg\Box \bigwedge \Gamma \to \neg\Box \bigwedge \Delta)$.

$\vdash \Box(\bigwedge \Box\Gamma \wedge \bigwedge \Box\Delta \to \bot) \to \Box\neg\Box\neg\Box \bigwedge \Gamma \to \Box\neg\Box \bigwedge \Delta$ (∗).

On the other hand, by S4.2 axiom and taking a substitution instance in which $\Box \bigwedge \Gamma$ is plugged into A, $\vdash \neg\Box\neg\Box\Box \bigwedge \Gamma \to \Box\neg\Box\neg\Box \bigwedge \Gamma$. Also, $\vdash \Box \bigwedge \Gamma \to \Box\Box \bigwedge \Gamma$ yields $\vdash \neg\Box\neg\Box \bigwedge \Gamma \to \neg\Box\neg\Box\Box \bigwedge \Gamma$. Hence, $\vdash \neg\Box\neg\Box \bigwedge \Gamma \to \Box\neg\Box\neg\Box \bigwedge \Gamma$.

Therefore, by (∗), $\vdash \Box(\bigwedge \Box\Gamma \wedge \bigwedge \Box\Delta \to \bot) \to (\neg\Box\neg\Box \bigwedge \Gamma \to \Box\neg\Box \bigwedge \Delta)$.
This yields $\vdash \Box(\bigwedge \Box\Gamma \wedge \bigwedge \Box\Delta \to \bot) \to \Box(\bigwedge \Box\Gamma \to \bot) \vee \Box(\bigwedge \Box\Delta \to \bot)$.
Thus, $\vdash \bigvee G \vee \Box(\bigwedge \Box\Gamma \to \bot) \vee \Box(\bigwedge \Box\Delta \to \bot)$.

(2) $R\Box$ in combined logics: By IH,

S4 + L$\vdash \bigvee G \vee \Box(\bigwedge \Box\Gamma \wedge \bigwedge \Box\Delta \to \varphi)$.

S4 + L$\vdash (\bigwedge \Box\Gamma \wedge \bigwedge \Box\Delta \to \varphi) \to (\Box \bigwedge \Gamma \wedge \Box \bigwedge \Delta \to \varphi)$.

S4 + L$\vdash \Box(\bigwedge \Box\Gamma \wedge \bigwedge \Box\Delta \to \varphi) \to \Box(\Box \bigwedge \Gamma \wedge \Box \bigwedge \Delta \to \varphi)$.

S4 + L$\vdash \Box(\bigwedge \Box\Gamma \wedge \bigwedge \Box\Delta \to \varphi) \to (\Box\Box \bigwedge \Gamma \wedge \Box\Box \bigwedge \Delta \to \Box\varphi)$.

S4 + L$\vdash \Box(\bigwedge \Box\Gamma \wedge \bigwedge \Box\Delta \to \varphi) \to (\Box \bigwedge \Gamma \wedge \Box \bigwedge \Delta \to \Box\varphi)$ (S4+L $\vdash \Box A \to \Box\Box A$).

S4 + L$\vdash \Box(\bigwedge \Box\Gamma \wedge \bigwedge \Box\Delta \to \varphi) \to \Box(\Box \bigwedge \Gamma \wedge \Box \bigwedge \Delta \to \Box\varphi)$.

Hence, we have S4 + L$\vdash \bigvee G \vee \bigvee \Box(\bigwedge \Box\Gamma \wedge \bigwedge \Box\Delta \to \Box\varphi)$.

5   Cut-Elimination for the Hypersequent Calculi

Here we syntactically prove cut-elimination for the hypersequent calculi reformulated via the following multiple-cut a la Gentzen (to handle internal contraction).

Multiple-cut $\dfrac{G|\Gamma \Rightarrow \Delta, A^\kappa \qquad H|A^\nu, \Pi \Rightarrow \Theta}{G|H|\Gamma, \Pi \Rightarrow \Delta, \Theta} \quad (\kappa, \nu > 0)$

We start from some definitions, which are obtained by modifying the ones in [1].

A marked hypersequent is a hypersequent with occurrences of a formula A distinguished, written $(H|\Gamma \Rightarrow \underline{A}^\kappa, \Delta)$ or $(H|\Gamma, \underline{A}^\nu \Rightarrow \Delta)$ $(\kappa, \nu > 0)$. A marked rule instance is a rule instance with the principal formula, if there is one, marked. (Due to the presence of internal contraction, the substitutivity given below is based on a one-shot substitution, analogous to the multiple-cut, unlike [1]. Suppose that G is a (possibly marked) hypersequent and H a marked hypersequent of the forms:

$G = (\Gamma_1, [A]^{\lambda_1} \Rightarrow \Delta_1| \dots |\Gamma_n, [A]^{\lambda_n} \Rightarrow \Delta_n)$ and $H = (H'|\Pi \Rightarrow \underline{A}^\kappa, \Sigma)$ $(\kappa > 0)$, where A does not occur unmarked in $\biguplus_{i=1}^n \Gamma_i$.

Then $CUT(G, H)$ contains, for all $0 \le \mu_i \le 1$ for $i = 1 \dots n$: $H'|\Gamma_1, \Pi^{\mu_1} \Rightarrow \Sigma^{\mu_1}, \Delta_1| \dots |\Gamma_n, \Pi^{\mu_n} \Rightarrow \Sigma^{\mu_n}, \Delta_n$.

Similarly, suppose that A does not occur unmarked in $\biguplus_{i=1}^n \Delta_i$ with:

$G = (\Gamma_1 \Rightarrow [A]^{\lambda_1}, \Delta_1| \dots |\Gamma_n \Rightarrow [A]^{\lambda_n}, \Delta_n)$ and $H = (H'|\Pi, \underline{A}^\nu \Rightarrow \Sigma)$ $(\nu > 0)$.

Then $CUT(G, H)$ contains, for all $0 \le \mu_i \le 1$ for $i = 1 \dots n$: $H'|\Gamma_1, \Pi^{\mu_1} \Rightarrow \Sigma^{\mu_1}, \Delta_1| \dots |\Gamma_n, \Pi^{\mu_n} \Rightarrow \Sigma^{\mu_n}, \Delta_n$.

A rule (r) is *substitutive* if for any:

1. marked instance $\dfrac{G_1 \dots G_n}{G}$ of (r); 2. $G' \in CUT(G, H)$,

then there exist $G'_i \in CUT(G_i, H)$ for $i = 1 \dots n$ such that $\dfrac{G'_1 \dots G'_n}{G'}$ of (r)

The point of introducing the notion of substitutivity is to use a way of eliminating cut without directly using the double induction used in Gentzen's original proof of cut-elimination. (The length of proofs is used differently.) This is useful to handle external contraction.[6]

Lemma 1. *All non-modal rules in these logics (all the structural rules and operational rules for $\wedge$, $\vee$, $\rightarrow$) are substitutive.*

Proof. The proof is just checking that each such rule satisfies the property, which is a (tedious) routine.

For modal cases, there is a problem discussed in [16]. An arbitrary substitution on the antecedent disturbs $R\square$ rule, since the antecedent has to be modalized for the rule to be applied ("moralized" means that all the formulas in the context have $\square$ as the outermost logical symbol). However, it suffices to consider only modalized sequents in a hypersequent of the form $H|\square\Sigma \Rightarrow$ in order to permute applications of other rules and $R\square$. Here are a few definitions. The *length* $|d|$ of a derivation d is (the maximal number of applications of inference rules)+1 occurring on any branch of d. The complexity of a formula A is the number of occurrences of its connectives. The *cut rank* $\rho(d)$ of d is (the maximal complexity of cut formulas in d)+1. Note that $\rho(d) = 0$ if d is cut-free.

---

[6] See [3] for some problem raised to another method of avoiding this problem.

Lemma 2. *Let L be such that $L \in \{S4, S4.2, S4.3, S5\}$ or $L = S4 + L$ where $L = S4, S4.2, S4.3, S5$. Let d_l and d_r be derivations in the hypersequent calculus HL for L such that: (1) d_l is a derivation of $(G|\Gamma_1, [A]^{\lambda_1} \Rightarrow \Delta_1| \dots |\Gamma_n, [A]^{\lambda_n} \Rightarrow \Delta_n)$ (All occurrences of A are made explicit. There is no occurrence in G.); (2) d_r is a derivation of $(H|\Sigma \Rightarrow A^\kappa, \Pi)$ $(\kappa > 0)$; (3) $\rho(d_l) \leq |A|$ and $\rho(d_r) \leq |A|$; (4) A is a compound formula and d_r ends with either a right logical rule or a modal rule introducing A to $A^{\kappa-1}$ (for the modal case, $\kappa - 1 = 0$).*

Then a derivation d of $(G|H|\Gamma_1, \Sigma^{\mu_1} \Rightarrow \Delta_1, \Pi^{\mu_1}| \dots |\Gamma_n, \Sigma^{\mu_n} \Rightarrow \Delta_n, \Pi^{\mu_n})$ such that $\rho(d) \leq |A|$ and $\mu_i = 0$ or 1 can be constructed in HL.

Proof. By induction on $|d_l|$. Base case is the one in which d_l is an axiom. In this case, there is nothing to do. The conclusion immediately holds.

Inductive cases: we have different cases depending on the last rule applied to d_l. Due to the lack of space, we omit the proofs of the simple cases.[7]

Case 4. The last inference is $L\square$ ($R\square$ is a special case of the bimodal rule.)

Subcase 4.1. A is principal and $A = \square B$. The last inference of d_l looks as follows.

$$\frac{G|\Gamma_1, B, [\square B]^{\lambda_1} \Rightarrow \Delta_1| \dots |\Gamma_n, [\square B]^{\lambda_n} \Rightarrow \Delta_n}{G|\Gamma_1, [\square B]^{\lambda_1+1} \Rightarrow \Delta_1| \dots |\Gamma_n, [\square B]^{\lambda_n} \Rightarrow \Delta_n}$$

Since d_r ends as the condition (4) states, the last inference of d_r is $\dfrac{H|\square \Sigma' \Rightarrow B}{H|\square \Sigma' \Rightarrow \square B}$. Note that we have only one occurrence of $\square B$ in this case. By IH, we can get $G|H|\Gamma_1, B, (\square \Sigma')^{\mu_1} \Rightarrow \Delta_1| \dots |\Gamma_n, (\square \Sigma')^{\mu_n} \Rightarrow \Delta_n$ ($\mu_i = 0$ or 1). The derivation of it has cut rank with $\leq |A|$. By using $H|\square \Sigma' \Rightarrow B$, cut, EW, EC, and possibly IC (depending on whether $\mu_1 = 0$ or 1), we can derive the desired $G|H|\Gamma_1, \square \Sigma' \Rightarrow \Delta_1| \dots |\Gamma_n, (\square \Sigma')^{\mu_n} \Rightarrow \Delta_n$. Note that the cut rank of the entire derivation is $\leq |A|$.

Subcase 4.2. The rule of the last inference of d_l is $L\square$ and the principal formula is not A. Then the last inference of d_l looks as follows.

$$\frac{G|\Gamma_1, B, [A]^{\lambda_1} \Rightarrow \Delta_1| \dots |\Gamma_n, [A]^{\lambda_n} \Rightarrow \Delta_n}{G|\Gamma_1, \square B, [A]^{\lambda_1} \Rightarrow \Delta_1| \dots |\Gamma_n, [A]^{\lambda_n} \Rightarrow \Delta_n}$$

By IH, $G|H|\Gamma_1, B, \Sigma^{\mu_1} \Rightarrow \Delta_1, \Pi^{\mu_1}| \dots |\Gamma_n, \Sigma^{\mu_n} \Rightarrow \Delta_n, \Pi^{\mu_n}$ ($\mu_i = 0$ or 1). The derivation of it has cut rank $\leq |A|$. $L\square$ applied, the desired hypersequent $G|H|\Gamma_1, \square B, \Sigma^{\mu_1} \Rightarrow \Delta_1, \Pi^{\mu_1}| \dots |\Gamma_n, \Sigma^{\mu_n} \Rightarrow \Delta_n, \Pi^{\mu_n}$ is derived with cut rank $\leq |A|$.

Case 5. The rule of the last inference in d_l is modal splitting.

Subcase 5.1. $A = \square B$. Then the last inference of d_l looks as follows. ($\lambda_1 > 0$)

$$\frac{G|\Gamma_1', \square\Theta, [\square B]^{\lambda_1} \Rightarrow \Delta_1| \dots |\Gamma_n, [\square B]^{\lambda_n} \Rightarrow \Delta_n}{G|\square\Theta, [\square B]^\lambda \Rightarrow |\Gamma_1', [\square B]^{\lambda_1-\lambda} \Rightarrow \Delta_1| \dots |\Gamma_n, [\square B]^{\lambda_n} \Rightarrow \Delta_n}$$

---

[7] Case 1. The last rule is applied on only side sequents G. Case 2. The last rule is any non-modal rule that does not have A as the principal formula. Case 3. The last inference is an application of non-modal left introduction rule whose principal formula is A.

d_r ends with $(H|\Box\Sigma' \Rightarrow \Box B)$ via $R\Box$. By IH, $G|H|\Gamma_1', \Box\Theta, (\Box\Sigma')^{\mu_1} \Rightarrow \Delta_1|\ldots|\Gamma_n, (\Box\Sigma')^{\mu_n} \Rightarrow \Delta_n$ $(\mu_i = 0)$. The derivation of it has cut rank $\leq |A|$. Applying modal splitting and possibly IW, we get $G|H|\Box\Theta, (\Box\Sigma')^{\lambda'} \Rightarrow |\Gamma_1, (\Box\Sigma')^{\lambda''} \Rightarrow \Delta_1|\ldots|\Gamma_n, (\Box\Sigma')^{\mu_n} \Rightarrow \Delta_n$ $(\lambda' = 1$ if $\lambda > 0$; ow $\lambda' = 0$, and $\lambda'' = 1$ if $\lambda_1 - \lambda > 0$; ow, $\lambda'' = 0$.) This is the desired hypersequent and its derivation has cut rank $\leq |A|$.

Subcase 5.2. A is not of the form $\Box B$. The last inference of d_l looks as follows.

$$\frac{G|\Gamma_1', \Box\Theta, [A]^{\lambda_1} \Rightarrow \Delta_1|\ldots|\Gamma_n, [A]^{\lambda_n} \Rightarrow \Delta_n}{G|\Box\Theta \Rightarrow |\Gamma_1', [A]^{\lambda_1} \Rightarrow \Delta_1|\ldots|\Gamma_n, [A]^{\lambda_n} \Rightarrow \Delta_n}$$

Here $d_r \vdash H|\Sigma \Rightarrow A^\kappa, \Pi$ (it may be that $\kappa > 1$). By IH, $G|H|\Gamma_1, \Box\Theta, \Sigma^{\mu_1} \Rightarrow \Pi^{\mu_1}, \Delta_1|\ldots|\Gamma_n, \Sigma^{\mu_n} \Rightarrow \Pi^{\mu_n}, \Delta_n$ $(\mu_i = 0$ or $1)$. The derivation of it has cut rank $\leq |A|$. Applying MS, we get $G|H|\Box\Theta \Rightarrow |\Gamma_1, \Sigma^{\mu_1} \Rightarrow \Pi^{\mu_1}, \Delta_1|\ldots|\Gamma_n, \Sigma^{\mu_n} \Rightarrow \Pi^{\mu_n}, \Delta_n$. This is the desired hypersequent and its derivation has cut rank $\leq |A|$.

Case 5'. The rule of the last inference of d_l is restricted modal splitting. It has to be the case that $A = \Box B$ and we have only modal substitution via $H|\Box\Sigma' \Rightarrow \Box B$, due to the form of the premise of restricted modalized splitting. Hence, this is a special case of 5.1. with the empty succedent and there is no analogue of 5.2.

Case 6. The rule of the last inference of d_l is modal comm.

Subcase 6.1. $A = \Box B$. The inference is as follows.

$$\frac{G|\Theta_1, \Box\Theta_2, [\Box B]^{\lambda_1} \Rightarrow \Delta|\ldots|\Gamma_n, [\Box B]^{\lambda_n} \Rightarrow \Delta_n \quad G|\Xi_1, \Box\Xi_2, [\Box B]^{\lambda_1'} \Rightarrow \Lambda|\ldots|\Gamma_n, [\Box B]^{\lambda_n} \Rightarrow \Delta_n}{G|\Theta_1, \Box\Xi_2, [\Box B]^{\lambda_1'} \Rightarrow \Delta|\Xi_1, \Box\Theta_2, [\Box B]^{\lambda_1} \Rightarrow \Lambda|\ldots|\Box\Gamma_n, [\Box B]^{\lambda_n} \Rightarrow \Delta_n}$$

$d_r \vdash H|\Box\Sigma' \Rightarrow \Box B$. The substitution is a modal substitution.

By IH, we can derive $G|H|\Theta_1, \Box\Theta_2, (\Box\Sigma')^{\mu_1} \Rightarrow \Delta|\ldots|\Gamma_n, (\Box\Sigma')^{\mu_n} \Rightarrow \Delta_n$ and $G|H|\Xi_1, \Box\Xi_2, (\Box\Sigma')^{\mu_1'} \Rightarrow \Lambda|\ldots|\Gamma_n, (\Box\Sigma')^{\mu_n} \Rightarrow \Delta_n$ $(\mu_i = 0$ or 1, $\mu_1' = 0$ or $1)$, and the derivation of it has cut rank $\leq |A|$. Applying modal comm, we get $G|H|\Theta_1, \Xi_1, \Box\Xi_2, (\Box\Sigma')^{\mu_1'} \Rightarrow \Delta|\Theta_1, \Xi_1, \Box\Theta_2, (\Box\Sigma')^{\mu_1} \Rightarrow \Lambda|\ldots|\Box\Gamma_n, (\Box\Sigma')^{\mu_n} \Rightarrow \Delta_n$. This is the desired hypersequent, and the derivation has the cut rank $\leq |A|$.

Subcase 6.2. A is not of the form $\Box B$. The last inference is as follows.

$$\frac{G|\Theta_1, \Box\Theta_2, [A]^{\lambda_1} \Rightarrow \Delta|\ldots|\Gamma_n, [A]^{\lambda_n} \Rightarrow \Delta_n \quad G|\Xi_1, \Box\Xi_2, [A]^{\lambda_1'} \Rightarrow \Lambda|\ldots|\Gamma_n, [A]^{\lambda_n} \Rightarrow \Delta_n}{G|\Theta_1, [A]^{\lambda_1}, \Box\Xi_2 \Rightarrow \Delta|\Xi_1, [A]^{\lambda_1'}, \Box\Theta_2 \Rightarrow \Lambda|\ldots|\Box\Gamma_n, [A]^{\lambda_n} \Rightarrow \Delta_n}$$

d_r ends with $H|\Sigma \Rightarrow A^\kappa, \Pi$ (it may be $\kappa > 1$). This case is similar to that of 5.2.

Case 7. d_l ends with $R\Box$.

Subcase 7.1 $A = \Box B$ The last inference of d_l is

$$\frac{G|\Box\Gamma_1', \Box\Theta, [\Box B]^{\lambda_1} \Rightarrow C|\ldots|\Gamma_n, [\Box B]^{\lambda_n} \Rightarrow \Delta_n}{G|\Box\Gamma_1', \Box\Theta, [\Box B]^{\lambda_1} \Rightarrow \Box C|\ldots|\Gamma_n, [\Box B]^{\lambda_n} \Rightarrow \Delta_n}$$

By the condition (4), the last inference of d_r is $\dfrac{H|\Box\Sigma_1, \Box\Sigma_2 \Rightarrow B}{H|\Box\Sigma_1, \Box\Sigma_2 \Rightarrow \Box B}$, and d_r ends with $d_r \vdash H|\Box\Sigma_1, \Box\Sigma_2 \Rightarrow \Box B$. By IH, $G|H|\Box\Gamma_1', \Box\Theta, [\Box\Sigma_1, \Box\Sigma_2]^{\mu_1} \Rightarrow$

$C| \ldots |\Gamma_n, [\Box\Sigma_1, \Box\Sigma_2]^{\mu_n} \Rightarrow \Delta_n$ ($\mu_i = 0$ or 1). Applying $\Box R$, we get $G|H|\Box\Gamma_1', \Box\Theta$, $[\Box\Sigma_1, \Box\Sigma_2]^{\mu_1} \Rightarrow \Box C| \ldots |\Gamma_n, [\Box\Sigma_1, \Box\Sigma_2]^{\mu_n} \Rightarrow \Delta_n$, as is desired.

Subcase 7.2 $A = \Box B$. Similar to the above (via $\Box\Sigma \Rightarrow$ for modal substitution).

Note: For modal structural rules of the combined logics, the same argument works as above (using $\Box$ instead of $\Box$). In order to deal with modal (hyper)structural rules, it suffices to use modal substitution with respect to $\Box$ since S4 L$\Box$ rule is substitutive.

Lemma 3. *Let L be such that $L \in \{S4, S4.2, S4.3, S5\}$ or $L = S4 + L$ where $L = S4, S4.2, S4.3, S5$. Let d_l and d_r be derivations in the hypersequent calculus HL for L such that: (1) d_l is a derivation of $(G|\Gamma, A^\nu \Rightarrow \Delta)$ $(\nu \geq 1)$; (2) d_r is a derivation of $(H|\Sigma_1 \Rightarrow [A]^{\lambda_1}, \Pi_1'| \ldots |\Sigma_n \Rightarrow [A]^{\lambda_n}, \Pi_n')$; (3) $\rho(d_l) \leq |A|$ and $\rho(d_r) \leq |A|$.*

Then a derivation d of $(G|H|\Sigma_1, \Gamma^{\mu_1} \Rightarrow \Pi_1', \Delta^{\mu_1}| \ldots |\Sigma_n, \Gamma^{\mu_n} \Rightarrow \Pi_n', \Delta^{\mu_n})$ with $\rho(d) \leq |A|$ and $\mu_i = 0$ or 1 can be constructed in HL.

Proof. By induction on $|d_r|$. Base case: If the rule is the last inference is an axiom, then the statement obviously holds since there is no cut involved in d_r. (Hence the cut rank of $d =$ the cut rank of d_l. Thus, it is $\leq |A|$ by assumption.)

Inductive cases: The first three cases are similar to the previous lemma.

Case 4. The rule of the last inference of d_r is L$\Box$.

In this case, A is not a principal formula. The last inference of d_r is as follows.

$$\frac{H|\Sigma_1, B \Rightarrow [A]^{\lambda_1}, \Pi_1|\Sigma_2, \Rightarrow [A]^{\lambda_2}, \Pi_2| \ldots |\Sigma_n, \Rightarrow [A]^{\lambda_n}, \Pi_n}{H|\Sigma_1, \Box B, \Rightarrow [A]^{\lambda_1}, \Pi_1|\Sigma_2, \Rightarrow [A]^{\lambda_2}, \Pi_2| \ldots |\Sigma_n, \Rightarrow [A]^{\lambda_n}, \Pi_n}$$

d_l ends with $G|\Gamma, A^\nu \Rightarrow \Delta$. By IH, $G|H|\Sigma_1, B, \Gamma^{\mu_1} \Rightarrow \Delta^{\mu_1}, \Pi_1|\Sigma_2, \Gamma^{\mu_2} \Rightarrow \Delta^{\mu_2}, \Pi_2| \ldots |\Sigma_n, \Gamma^{\mu_n} \Rightarrow \Delta^{\lambda_n}, \Pi_n$ with the cut rank $\leq |A|$ ($\mu_i = 0$ or 1). Applying the L$\Box$ rule, we get $G|H|\Sigma_1, \Box B, \Gamma^{\mu_1} \Rightarrow \Delta^{\mu_1}, \Pi_1|\Sigma_2, \Gamma^{\mu_2} \Rightarrow \Delta^{\mu_2}, \Pi_2| \ldots |\Sigma_n, \Gamma^{\mu_n} \Rightarrow \Delta^{\mu_n}, \Pi_n$, as is desired. The cut rank of the derivation is $\leq |A|$.

Case 5. The rule of the last inference of d_r is R$\Box$. In this case, A has to be the principal formula of the form $\Box B$. The last inference of d_r is as follows.

$$\frac{H|\Box\Theta \Rightarrow B|\Sigma_2 \Rightarrow [\Box B]^{\lambda_2}, \Pi_2| \ldots |\Sigma_n \Rightarrow [\Box B]^{\lambda_n}, \Pi_n}{H|\Box\Theta \Rightarrow \Box B|\Sigma_2 \Rightarrow [\Box B]^{\lambda_2}, \Pi_2| \ldots |\Sigma_n \Rightarrow [\Box B]^{\lambda_n}, \Pi_n}$$

d_l ends with $G|\Gamma, (\Box B)^\nu \Rightarrow \Delta$ (it may be $\nu > 1$).

By IH, $G|H|\Box\Theta \Rightarrow B|\Sigma_2, \Gamma^{\mu_2} \Rightarrow \Delta^{\mu_2}, \Pi_2| \ldots |\Sigma_n, \Gamma^{\mu_n} \Rightarrow \Delta^{\mu_n}, \Pi_n$ with cut rank $\leq |A|$ ($\mu_i = 0$ or 1). Applying R$\Box$, we get $H|\Box\Theta \Rightarrow \Box B|\Sigma_2, \Gamma^{\mu_2} \Rightarrow \Delta^{\mu_2}, \Pi_2| \ldots |\Sigma_n, \Gamma^{\mu_n} \Rightarrow \Delta^{\mu_n}, \Pi_n$ with cut rank $\leq |A|$. By the previous lemma, the claim holds since this case satisfies the condition of application of the lemma.

Case 6. The rule of the last inference of d_r is modal splitting. Here we do not have to divide subcases, since A goes to only one place.

$$\frac{H|\Sigma_1', \Box\Theta \Rightarrow [A]^{\lambda_1}, \Pi_1'|\Sigma_2 \Rightarrow [A]^{\lambda_2}, \Pi_2| \ldots |\Sigma_n \Rightarrow [A]^{\lambda_n}, \Pi_n}{H|\Box\Theta \Rightarrow |\Sigma_1' \Rightarrow [A]^{\lambda_1}, \Pi_1'|\Sigma_2 \Rightarrow [A]^{\lambda_2}, \Pi_2| \ldots |\Sigma_n \Rightarrow [A]^{\lambda_n}, \Pi_n}$$

By IH, $G|H|\Sigma_1', \Box\Theta, \Gamma^{\mu_1} \Rightarrow \Delta^{\mu_1}, \Pi_1'|\Sigma_2, \Gamma^{\mu_2} \Rightarrow \Delta^{\mu_2}, \Pi_2|\ldots|\Sigma_n, \Gamma^{\mu_n} \Rightarrow \Delta^{\mu_n}, \Pi_n$ is derivable with cut rank $\leq |A|$ and $\mu_i = 0$ or 1. Applying modal splitting, we can obtain the following hypersequent $G|H|\Box\Theta \Rightarrow |\Sigma_1', \Gamma^{\mu_1} \Rightarrow \Delta^{\mu_1}, \Pi_1'|\Sigma_2, \Gamma^{\mu_2} \Rightarrow \Delta^{\mu_2}, \Pi_2|\ldots|\Sigma_n, \Gamma^{\mu_n} \Rightarrow \Delta^{\mu_n}, \Pi_n$. It is easy to check that this is the desired hypersequent, and the cut rank of the derivation is $\leq |A|$.

Case 6'. The rule of the last inference of d_r is restricted modal splitting. The succedent of the relevant sequent is empty; otherwise, this case is similar to 6.

Case 7. The rule used in the last inference in d_r is modal comm. (Since Γ and Δ in $G|\Gamma, A^\nu \Rightarrow \Delta$ are arbitrary, we do not have to divide subcases here.)

$$\frac{H|\Theta_1, \Box\Theta_2 \Rightarrow [A]^{\lambda_1}, \Phi|\ldots|\Sigma_n \Rightarrow [A]^{\lambda_n} \Rightarrow \Pi_n \qquad H|\Xi_1, \Box\Xi_2 \Rightarrow [A]^{\lambda_1'}, \Psi|\ldots|\Sigma_n \Rightarrow [A]^{\lambda_n} \Rightarrow \Pi_n}{H|\Theta_1, \Box\Xi_2 \Rightarrow [A]^{\lambda_1}, \Phi|\Xi_1, \Box\Theta_2 \Rightarrow [A]^{\lambda_1'}, \Psi|\ldots|\Box\Sigma_n \Rightarrow [A]^{\lambda_n}, \Pi_n}$$

By IH, we can derive $G|H|\Theta_1, \Box\Theta_2, \Gamma^{\mu_1} \Rightarrow \Delta^{\mu_1}, \Phi|\ldots|\Sigma_n, \Gamma^{\mu_n} \Rightarrow \Delta^{\mu_n}, \Pi_n$ and $G|H|\Xi_1, \Box\Xi_2, \Gamma^{\mu_1'} \Rightarrow \Delta^{\mu_1'}, \Psi|\ldots|\Sigma_n, \Gamma^{\mu_n} \Rightarrow \Delta^{\mu_n}, \Pi_n$ ($\mu_i = 0$ or 1 and $\mu_1' = 0$ or 1). Both derivations have cut rank $\leq |A|$. Applying modal comm, we get $G|H|\Theta_1, \Box\Xi_2, \Gamma^{\mu_1} \Rightarrow \Delta^{\mu_1}, \Phi|\Xi_1, \Box\Theta_2, \Gamma^{\mu_1'} \Rightarrow \Delta^{\mu_1'}, \Psi|\ldots|\Box\Sigma_n, \Gamma^{\mu_n} \Rightarrow \Delta^{\mu_n}, \Pi_n$. This is the desired hypersequent. The derivation of it has cut rank $\leq |A|$.

Case 8. d_r ends with $R\Box$. The only case is $A = \Box B$.

$$\frac{H|\Box\Sigma_1, \Box\Theta \Rightarrow B|\Sigma_2 \Rightarrow [\Box B]^{\lambda_2}, \Pi_2|\ldots|\Sigma_n \Rightarrow [\Box B]^{\lambda_n}, \Pi_n}{H|\Box\Sigma_1, \Box\Theta \Rightarrow \Box B|\Sigma_2 \Rightarrow [\Box B]^{\lambda_2}, \Pi_2|\ldots|\Sigma_n \Rightarrow [\Box B]^{\lambda_n}, \Pi_n}$$

$d_l \vdash \Gamma, (\Box B)^\nu \Rightarrow \Delta$ (it may be $\nu > 1$).

By IH, $G|H|\Box\Sigma_1, \Box\Theta \Rightarrow B|\Sigma_2, \Gamma^{\mu_2} \Rightarrow \Delta^{\mu_2}, \Pi_2|\ldots|\Sigma_n, \Gamma^{\mu_n} \Rightarrow \Delta^{\mu_n}, \Pi_n$ ($\mu_i = 0$ or 1). Applying $R\Box$, we get $G|H|\Box\Sigma_1, \Box\Theta \Rightarrow \Box B|\Sigma_2, \Gamma^{\mu_2} \Rightarrow \Delta^{\mu_2}, \Pi_2|\ldots|\Sigma_n, \Gamma^{\mu_n} \Rightarrow \Delta^{\mu_n}, \Pi_n$. But as in the case of $R\Box$ for monomodal logic, the condition of the previous lemma is satisfied. Hence, we can apply the above Lemma to show that we can further transform the derivation so that the cut rank of the derivation is $|A|$.

Theorem 2 (Cut-elimination). *For any modal logic L that in the set given above, cut-elimination holds for HL.*

Proof. Let d be a derivation in a hypersequent calculus for a logic in L, with $\rho(d) > 0$. The proof proceeds by the double induction on $(\rho(d), n\rho(d))$, where $n\rho(d)$ is the number of application of cut in d with cut rank $\rho(d)$. Consider an uppermost application of cut in d with cut rank $\rho(d)$. By applying the last lemma to its premises. Either $\rho(d)$ or $n\rho(d)$ decreases. Then we can apply IH.

6   Gödel Modal Embedding

Here we prove soundness and faithfulness of superintuitionistic variants of Gödel modal embedding (in [15]), which map L ={intuitionistic logic, logic of weak excluded middle (LWEM), Gödel-Dummett logic, classical logic, intuitionistic logic with classical atoms} into modal logics discussed above (we call them ML (in $\{S4, S4.2, S4.3, S5\}$). Intuitionistic logic with classical atoms (IPC$_{CA}$) is a logic discussed in [13]. The idea is to introduce a variant of intuitionistic logic in a two-sorted language of propositional logic with two different kinds of propositional variables, only in the classical one of which an atomic version of the law of the excluded middle is postulated (letting X be a classical variable, this can be put as $X \vee \neg X$). Cut-free single-conclusion and multiple-conclusion hypersequent calculi are presented in [13].

Definition 1. *The embedding is defined by induction as follows.*

1. $p^\square := \square p$; *2.* $\perp^\square := \perp$; *2.* $(A \wedge B)^\square := \square(A^\square \wedge B^\square)$; *3.* $(A \vee B)^\square := \square(A^\square \vee B^\square)$; *4.* $(A \to B)^\square := \square(A^\square \to B^\square)$; *5. for a classical atom* X, $X^\square = \boxdot X$.

Let us make two remarks about our translation here. First, we use the notation for mutisets of formulas $\Gamma^\square := B_1^\square, \ldots, B_n^\square$ and multisets of sequents $G^\square := \Gamma_1^\square \Rightarrow \Delta_1^\square | \ldots | \Gamma_n^\square \Rightarrow \Delta_n^\square$. Note that all the formulas except $\perp$ in the images of the mapping are modalized. Thus, it would make things simpler if we could treat all the formulas in the images of the formulas as modalized. This is not quite the case, since $\perp^\square = \perp$. However, if $\perp^\square(= \perp)$ is in $\Gamma_i^\square$, the statement of soundness and faithfulness of the embedding obviously holds. Hence, in our argument, we assume that $\perp \notin \Gamma_i^\square$, unless $\perp$ is explicitly mentioned.

Second, to make our argument simpler, we use both single-conclusion hypersequent calculi (we call them sHL) and multiple-conclusion hypersequent calculi (we call them mHL) for logics in L. The latter are analogous to Maehara's multiple-conclusion sequent calculus for intuitionistic logic [15]. Such multiple-conclusion hypersequent calculi are obtained as follows: 1. keep intuitionistic R→; 2. replace all other rules by multiple-conclusion versions. The equivalence of the two calculi can be proven straightforwardly. (We use cut-free single-conclusion hypersequent calculi, but our proof does not require that the multiple-conclusion ones be cut-free. In the following, let HML$^-$ and sHL$^-$ mean cut-free HML and sHL, respectively.)

Proposition 1 (Soundness). *If* $sHL^- \vdash G|\Gamma_1 \Rightarrow \varphi_1| \ldots |\Gamma_n \Rightarrow \varphi_n$, *then* HML$\vdash$ $G^\square|\Gamma_1^\square \Rightarrow \varphi_1^\square| \ldots |\Gamma_n^\square \Rightarrow \varphi_n^\square$.

Proof. Proof is by induction on the length of cut-free proofs of sHL.

Case 1. Axioms: In the case of axiom of the form sHL$\vdash p \Rightarrow p$, the claim holds as follows. By the mapping, we get $p^\square \Rightarrow p^\square$, i.e. $\square p \Rightarrow \square p$. This is certainly provable.

For the other axiom sHL$\vdash \perp \Rightarrow$, since $\perp^\square = \perp$, $\perp^\square \Rightarrow$ immediately follows.

Case 2. Operational rules: These are straightforward. As a representative case, we show the case of $R \to$ $\dfrac{G|\Gamma, A \Rightarrow B}{G|\Gamma \Rightarrow A \to B}$ (This is an intuitionistic $R \to$ rule.)

By IH, the translation of the premise of the rule $R \to$ in sHL, i.e. HML$\vdash$ $G^\square|\Gamma^\square, A^\square \Rightarrow B^\square$. Then HML$\vdash G^\square|\Gamma^\square \Rightarrow A^\square \to B^\square$ is derived via R→. And HML$\vdash G^\square|\Gamma^\square \Rightarrow \square(A^\square \to B^\square)$ is derived via R$\square$. This is in the range of the mapping ()$^\square$.

Case 3. Structural rules: The standard structural rules are all straightforward. Hence, we check only external structural rules characterizing logics.

Subcase 3.1. Splitting (L = classical logic)

$$\frac{G|\Gamma_1, \Gamma_2 \Rightarrow \varphi_1|\ldots|\Gamma_n \Rightarrow \varphi_n}{G|\Gamma_1 \Rightarrow \varphi_1|\Gamma_2 \Rightarrow |\Gamma_n \Rightarrow \varphi_n}$$

By IH, HS5$\vdash$ $G^\square|\Gamma_1^\square, \Gamma_2^\square \Rightarrow \Delta_1^\square|\ldots|\Gamma_n^\square \Rightarrow \Delta_n^\square$. Note that $\Gamma_2^\square$ are already modalized. By applying modal splitting and cuts, we get the following inference.

$$\frac{G^\square|\Gamma_1^\square, \Gamma_2^\square \Rightarrow \varphi_1^\square|\ldots|\Gamma_n^\square \Rightarrow \varphi_n^\square}{G^\square|\Gamma_1^\square \Rightarrow \varphi_1^\square|\Gamma_2^\square \Rightarrow |\Gamma_n^\square \Rightarrow \varphi_n^\square}$$

But the lower hypersequent is the image of the mapping $()^\square$ of the lower hypesequent of classical splitting. So, the inference is preserved under this mapping. (In the case of S4+S5, our single-conclusion hypersequent calculus for IPC$_{CA}$ has a special rule of splitting, applied to formulas consisting of classical propositional variables only. Due to the definition of our mapping, the only relevant case is an atomic case $X^\square = \square X$, which allows applications of modal splitting for S5 $\square$.)

Subcase 3.2. Restricted splitting (L is LWEM): $\dfrac{G|\Gamma_1, \Gamma_2 \Rightarrow |\ldots|\Gamma_n \Rightarrow \varphi_n}{G|\Gamma_1 \Rightarrow |\Gamma_2 \Rightarrow |\Gamma_n \Rightarrow \varphi_n}$

This is simply a special case of the case of splitting with the empty succedent.

Subcase 3.3. Communication (L is Gödel-Dummett logic)

$$\frac{G|\Gamma_1, \Pi \Rightarrow \varphi_1|\ldots|\Gamma_n \Rightarrow \varphi_n \qquad G|\Gamma_2, \Sigma \Rightarrow \varphi_2|\ldots|\Gamma_n \Rightarrow \varphi_n}{G|\Gamma_1, \Sigma \Rightarrow \varphi_1|\Gamma_2, \Pi \Rightarrow \varphi_2|\ldots|\Gamma_n \Rightarrow \varphi_n}$$

By IH, we have as follows HS4.3$\vdash$ $G^\square|\Gamma_1^\square, \Pi^\square \Rightarrow \varphi_1^\square|\ldots|\Gamma_n^\square \Rightarrow \varphi_n^\square$ and HS4.3$\vdash$ $G^\square|\Gamma_2^\square, \Sigma^\square \Rightarrow \varphi_2^\square|\ldots|\Gamma_n^\square \Rightarrow \varphi_n^\square$. Note again that $\Pi^\square$ and $\Sigma^\square$ are modalized.

$$\frac{G^\square|\Gamma_1^\square, \Pi^\square \Rightarrow \varphi_1^\square|\ldots|\Gamma_n^\square \Rightarrow \varphi_n^\square \qquad G^\square|\Gamma_2^\square, \Sigma^\square \Rightarrow \varphi_2^\square|\ldots|\Gamma_n^\square \Rightarrow \varphi_n^\square}{G^\square|\Gamma_1^\square, \Sigma^\square \Rightarrow \varphi_1^\square|\Gamma_2^\square, \Pi^\square \Rightarrow \varphi_2^\square|\ldots|\Gamma_n^\square \Rightarrow \varphi_n^\square}$$

It is easy to check that the lower hypersequent is the image of the mapping of the lowersequent of the original case of communication. Hence, the derivability is preserved under this mapping. This completes the proof of the proposition. ⊠

Theorem 3 (Faithfulness). *If* HML$^-\vdash$ $G^\square|\Gamma_1^\square \Rightarrow \Delta_1^\square|\ldots$ $|\Gamma_n^\square \Rightarrow \Delta_n^\square$, *then* mHL$\vdash$ $G|\Gamma_1 \Rightarrow \Delta_1|\ldots|\Gamma_n \Rightarrow \Delta_n$.

Proof. Faithfulness of the embedding $()^\square$ is proven as follows. The idea is simply to strip off all the modal operators from a given cut-free proof of HML in the assumption.

Case 1. HML = HS5 and HL = classical logic. Stripping off all modal operators from the given proof in HS5 simply gives a proof in a multiple-conclusion hypersequent calculus with intuitionistic R$\rightarrow$ and with the additional rule of splitting, which is a classical system [2]. Since the underlying logic of HS5 is classical, the embedding is faithful. (In IPC$_{CA}$, stripping off $\boxdot$ from $\boxdot X$ gives a case of classical splitting.)

Case 2. HML = HS4.2 or HS4.3 and HL = LWEM or Gödel-Dummett logic (respectively). Modal structural rules and R→ of these cases require some additional care to prove the faithfulness of the embedding, other cases being straightforward.

Case 2.1. Modal structural rules: We check that after stripping off $\Box$, applying external structural rules gives valid inferences in the desired logics.

2.1.1. Restricted modal splitting $$\dfrac{G^\Box|\Gamma_1^\Box, \Gamma_2^\Box \Rightarrow |\dots|\Gamma_n^\Box \Rightarrow \Delta_n^\Box}{G^\Box|\Gamma_1^\Box \Rightarrow |\Gamma_2^\Box \Rightarrow |\Gamma_n^\Box \Rightarrow \Delta_n^\Box}$$

If we strip off all the modal operators, then we have the following inference.

$$\dfrac{G|\Gamma_1, \Gamma_2 \Rightarrow |\dots|\Gamma_n \Rightarrow \Delta_n}{G|\Gamma_1 \Rightarrow |\Gamma_2 \Rightarrow |\Gamma_n \Rightarrow \Delta_n}$$

This is a case of restrict splitting valid in LWEM.

2.1.2. Modal communication.

$$\dfrac{G^\Box|\Gamma_1^\Box, \Pi^\Box \Rightarrow \Delta_1^\Box|\dots|\Gamma_n^\Box \Rightarrow \Delta_n^\Box \qquad G^\Box|\Gamma_2^\Box, \Sigma^\Box \Rightarrow \Delta_2^\Box|\dots|\Gamma_n^\Box \Rightarrow \Delta_n^\Box}{G^\Box|\Gamma_1^\Box, \Sigma^\Box \Rightarrow \Delta_1^\Box|\Gamma_2^\Box, \Pi^\Box \Rightarrow \Delta_2^\Box|\dots|\Gamma_n^\Box \Rightarrow \Delta_n^\Box}$$

If we strip off all the modal operators, then we have the following inference.

$$\dfrac{G|\Gamma_1, \Pi \Rightarrow \Delta_1|\dots|\Gamma_n \Rightarrow \Delta_n \qquad G|\Gamma_2, \Sigma \Rightarrow \Delta_2|\dots|\Gamma_n \Rightarrow \Delta_n}{G|\Gamma_1, \Sigma \Rightarrow \Delta_1|\Gamma_2, \Pi \Rightarrow \Delta_2|\dots|\Gamma_n \Rightarrow \Delta_n}$$

Case 2.2. R→. We show that the applications R→ can be restricted to a single-conclusion version; then, all $\Box$'s being removed, all R →'s are intuitionistically valid. This shows faithfulness, since multiple-conclusion R→ is the only classical rule here.

First, observe that, in the range of the mapping $()^\Box$, all the occurrences of → statements are in the scope of $\Box$. Also, all the positive occurrences of formulas whose outermost logical symbol is $\Box$ in a cut-free proof of $G^\Box|\Gamma^\Box \Rightarrow \Delta^\Box$ can be introduced only in one of the following three ways: 1. Axiom; 2. Weakening; 3. R$\Box$. However, in our case, 1 is out, since in our systems only atomic formulas are used in axioms. In case 2, modal operators stripped off, weakening is still (intuitionistically) valid. The only remaining case is that $\Box(A^\Box \to B^\Box)$ introduced via R$\Box$ from $A^\Box \to B^\Box$.

Now positive occurrences of $A \to B$ can be introduced via $R \to$ or weakening (IW or EW). But the only case that matters here is again the case where this is introduced by $R \to$. Then, in order to show that all the applications of R→ in HML$^-$ can be restricted to a single-conclusion case, it suffices to show that all the cases of positive occurrences of a formula of the form $\Box(A \to B)$ are introduced in such a way that the introduction of $A \to B$ in HML$^-$ can be restricted to the single-conclusion version of R→ and then R$\Box$ is applied. This is because only cases in the range of the mapping $()^\Box$ matter here, and there is no way of applying R□ to a sequent in which we have more than one formula on the succedent.[8] We now claim the following.

Claim. If the only case of introducing $A^\square \to B^\square$ on the succedent via $R \to$ are the ones in which R□ is later applied to $A^\square \to B^\square$ as the only formula on the succedent of a hypersequent occurring lower part of a cut-free proof, then we can replace all of these applications of R→ by R→ to single-conclusion sequents.

Proof. (claim) By induction on the number of the applications of R→ whose uppersequent has more than one formula on the succedent in a proof of HML⁻.

Base case: the number of R→ whose uppersequent have more than one formula on the succedent is 0. In this case, every positive occurrence of a formula of the form $\square(A^\square \to B^\square)$ can be derived from $A^\square \to B^\square$ by R→ as follows.

$$
\frac{\dfrac{\dfrac{G^\square|\Gamma', A^\square \Rightarrow B^\square}{G^\square|\Gamma' \Rightarrow A^\square \to B^\square}}{G^\square|\square\Gamma \Rightarrow A^\square \to B^\square}}{G^\square|\square\Gamma \Rightarrow \square(A^\square \to B^\square)}\ R\square
$$

Inductive case: Suppose that we have n applications of R→ in a cut-free proof that has more than one formula on the succedent of the relevant sequent. Pick up the lowermost among such application of $R \to$, which looks as follows. (In the following, "$\square\Psi(C^\square \to D^\square)$" means a schematic expression of a formula possibly derivable in this form of derivation.)

$$
\frac{\dfrac{\dfrac{\dfrac{G^\square|\Gamma'', A^\square \Rightarrow B^\square, C^\square, \Delta''}{G^\square|\Gamma'' \Rightarrow A^\square \to B^\square, C^\square, \Delta''}\ (1)\ R\to*}{G^\square|\Gamma' \Rightarrow A^\square \to B^\square, C^\square, \Delta'} \qquad G^\square|\Theta', D^\square \Rightarrow}{G^\square|\Gamma', \Theta', C^\square \to D^\square \Rightarrow A^\square \to B^\square, \Delta'}\ (2)\ L\to}{\dfrac{G^\square|\square\Gamma, \square\Delta, \square\Psi(C^\square \to D^\square) \Rightarrow A^\square \to B^\square}{G^\square|\square\Gamma, \square\Delta, \square\Psi(C^\square \to D^\square) \Rightarrow \square(A^\square \to B^\square)}\ (3)\ R\square**}
$$

We divide the case into two subcases.

Case 1. $A^\square \to B^\square$ does not occur in Δ'' or any other succedent of a sequent on a branch of a proof tree leading to the application of $R\square**$. Observe the following.

(1) Two sequential applications of rules (except R□) in a cut-free proof whose principal formulas are two different ones can be permuted.

(2) Also, by the condition that all the formulas eventually derived in the proof must be in the range of the mapping $()^\square$, we have some further conditions on the proof. Namely, until (3), we cannot have any other application of R□

---

[8] Note that L→ is the only rule that lowers the number of formulas on the succedent in a cut-free proof, except contraction, since we use only context-sharing rules for ∧ and ∨.

or R→ in the proof, since R□ requires that we have only one formula on the succedent of the premise.

(3) By permuting down $R \to$ to $L \to$ (and the other rules), we can reduce the number of the formulas on the succedent when $R \to$ rule is applied as follows.

$$\cfrac{\cfrac{\cfrac{\cfrac{\cfrac{G^\square | \Gamma'', A^\square \Rightarrow B^\square, C^\square, \Delta''}{G^\square | \Gamma', A^\square \Rightarrow B^\square, C^\square, \Delta'} \qquad G^\square | \Theta', D^\square \Rightarrow}{G^\square | \Gamma', \Theta', C^\square \to D^\square, A^\square \Rightarrow B^\square, \Delta'} \text{(2) L→}}{G^\square | \square\Gamma, \square\Theta, \square\Psi(C^\square \to D^\square), A^\square \Rightarrow B^\square}}{G^\square | \square\Gamma, \square\Theta, \square\Psi(C^\square \to D^\square) \Rightarrow A^\square \to B^\square} \text{(1) R→*}}{G^\square | \square\Gamma, \square\Theta, \square\Psi(C^\square \to D^\square) \Rightarrow \square(A^\square \to B^\square)} \text{(3) R□**}$$

Note that R→* is now applied to a sequent whose succedent has only one formula. Hence, permutations of applications guarantee that this case can be reduced to a single-conclusion $R \to$, which is intuitionistically valid.

The idea of permuting rules can be made precise by using an inductive argument based on the sum of the number of hypersequents starting from the introduction of $A^\square \to B^\square$ via R→ to the relevant application of R□**. Since any application of a rule immediately below the application of R→ whose auxiliary formula is not this $A^\square \to B^\square$ can be permuted with R→, we can apply induction hypothesis.

Case 2. $A^\square \to B^\square$ occurs in Δ'' or the succedent of other premise(s) leading to the application of R□** (these are the cases where contraction is used in the proof).

To accommodate the cases, we consider all the chains of sequents that contain $A^\square \to B^\square$ on the succedent whose conclusion lead to R□**, i.e. the part of the branches of a proof tree containing occurrences of the formula on the succedent.

We apply the same permutation argument that we gave above to all of these occurrences of $A^\square \to B^\square$. We change all the relevant hypersequents of the form $G^\square | \Gamma \Rightarrow A^\square \to B^\square, \Delta$ to $G^\square | \Gamma, A^\square \Rightarrow B^\square, \Delta$ (by permutating rules) so that we can reduce applications of R→ to the only one immediately above R□**. Such permutations are possible since in the relevant parts of the original proof tree we can never apply R□.

This idea of permuting rules can be made precise by an inductive argument based on the sum of the number of hypersequents starting from the introduction of $A^\square \to B^\square$ to the succedent of a sequent to the application of R□**. By the argument given in the previous paragraph, we can reduce the number of applications of R→ where the premise has more than one formula on the succedent to a number strictly smaller than it. By IH, all such applications of R→ can be eliminated from the given cut-free proof.[9] ⊠(claim)

By this claim, we can ensure that in a derivation of formulas that are in the range of the mapping $()^\square$, we can dispense with the applications of R→ to

---

[9] This turns out to be a hypersequent variant of the arguments in [15] and [7].

sequents that have more than one formula on the succedent. Therefore, if we strip off all the modal operators from an entire proof figure of a hypersequent whose final conclusion is in the range of $()^\square$, then every application of a rule in the proof figure is a valid inference in each of the superintuitionistic (or intuitionistic) logics at issue here. $\boxtimes$ (theorem)

7 Discussions

We have shown cut-elimination for hypersequent calculi for logics extending S4 by .2 axiom, .3 axiom, and S5 axiom, and simple combinations of these. It is well-known that modal logics obtained by adding slight variants of .2 and .3 axioms to K4, KD4, GL, Grz can also be formulated. Which of these cases can be handled by hypersequents and by a proof method similar to the one used here appears to be an interesting topic for future research.

References

1. Metcalfe, G., Ciabattoni, A., Montagna, F.: Algebraic and proof-theoretic characterizations of truth stressers for MTL and its extensions. Fuzzy Sets Syst. **161**(3), 369–389 (2010)
2. Avron, A.: The method of hypersequents in the proof theory of propositional non-classical logics. In: Hodges, W., Hyland, M., Steinhorn, C., Truss, J. (eds.) Proceedings of Logic Colloquium, Logic: From Foundations to Applications, Keele, UK, 1993, pp. 1–32. Oxford University Press, New York (1996)
3. Baaz, M., Ciabattoni, A., Fermüller, C.G.: Hypersequent calculi for Gödel logics - a survey. J. Logic Comput. **13**, 1–27 (2003)
4. K. Brünnler. Nested sequents (habilitationsschrift) (2010)
5. Ciabattoni, A., Gabbay, D.M., Olivetti, N.: Cut-free proof systems for logics of weak excluded middle. Soft Comput. **2**(4), 147–156 (1998)
6. Došen, K.: Logical constants as punctuation marks. Notre Dame J. Formal Logic **30**(3), 362–381 (1989)
7. Došen, K.: Modal translations in substructural logics. J. Philos. Logic **21**(3), 283–336 (1992)
8. Fitting, M.: Logics with several modal operators. Theoria **35**, 259–266 (1969)
9. Fitting, M.: Proof Methods for Modal and Intuitionistic Logic. Reidel Publishing Company, Dordrecht (1983)
10. Goranko, V., Passy, S.: Using the universal modality: gain and questions. J. Logic Comput. **2**(1), 5–30 (1992)
11. Indrzejczak, A.: Cut-free hypersequent calculus for S4.3. Bull. Sect. Logic **41**(1), 89–104 (2012)
12. Kracht, M.: Power and weakness of the modal display calculus. Proof Theory of Modal Logic, pp. 93–121. Kluwer Academic Publishers, Dordrecht (1996)
13. Kurokawa, H.: Hypersequent calculi for intuitionistic logic with classical atoms. Ann. Pure Appl. Logic **161**(3), 427–446 (2009)
14. Lahav, O.: From frame properties to hypersequent rules in modal logics. In: LICS, pp. 408–417 (2013)
15. Maehara, S.: Eine darstellung der intuitionistischen logik in der klassischen. Nagoya Math. J. **7**, 45–64 (1954)

16. Mints, G.: Cut elimination for S4C: a case study. Stud. Logica. **82**(1), 121–132 (2006)
17. Negri, N.: Proof analysis in modal logic. J. Philos. Logic **34**(5–6), 507–544 (2005)
18. Paoli, F.: Quine and Slater on paraconsistency and deviance. J. Philos. Logic **32**(2), 531–548 (2003)
19. Sambin, G., Battilotti, G., Faggian, C.: Basic logic: reflection, symmetry, visibility. J. Symbolic Logic **65**(3), 979–1013 (2000)
20. Shvarts, G.F.: Gentzen style systems for K45 and K45D. In: Meyer, A.R., Taitslin, M.A. (eds.) Logic at Botik '89. LNCS, vol. 363, pp. 245–256. Springer, Heidelberg (1989)
21. Wansing, H.: Displaying Modal Logic. Kluwer Academic Publishers, Dordrecht (1998)
22. Wansing, H.: Sequent systems for modal logics. In: Gabbay, D.M., Guenthner, F. (eds.) Handbook of Philosophical Logic, vol. 8, 2nd edn, pp. 61–146. Kluwer Academic Publishers, Amsterdam (2002)
23. Zeman, J.J.: Modal Logic/The Lewis-Modal System, 1st edn. Oxford University Press, Oxford (1973)

Discourse-Level Politeness and Implicature

Elin McCready[1](✉) and Nicholas Asher[2]

[1] Aoyama Gakuin University, Tokyo, Japan
mccready@cl.aoyama.ac.jp
[2] CNRS-IRIT, Toulouse, France
Nicholas.Asher@irit.fr

Abstract. This paper considers politeness at the discourse level in terms of strategic choice. We begin with a discussion of the nature and levels of linguistic politeness from semantic and pragmatic perspectives, then turning to the way in which such strategies can be realized in natural language. A distinction is drawn between *formal* and *polite* linguistic behavior. We then provide a formal analysis in terms of the topological analysis of game strategies in an infinitely repeated game. This analysis extends that of [2]. It improves on that earlier work in three ways: (i) by considering a wider range of player 'types', (ii) by implementing the distinction between formality and politeness, and (iii) by analyzing a much wider range of kinds of politeness strategies, together with their positions in the Borel hierarchy [8].

1 Introduction

Politeness is an ubiquitous phenomenon, yet one which is, from the perspective of formal linguistics, nonetheless not well-understood. The recent semantics and pragmatics literature contains a good deal of work on the topic, but most of it concentrates on a subset of the available phenomena, namely the basic motivations for politeness, and the semantic implementation of phenomena like honorification. In this paper we consider a different domain, that of discourse-level politeness. This domain is dependent on the existence of the means of expressing politeness and on having motivations for doing so, but remains distinct. We take politeness at the discourse level to be the set of facts relating to how speaker choices about politeness and formality evolve over the course of interactions, how speakers can choose between the possibilities for 'politeness choice' at the discourse level, and how this linguistic domain should be given a proper formal modeling which allows researchers to state hypotheses about politeness behavior.

The present paper extends the earlier work on these issues of the authors [2]. There, an analysis was given of discourse-level politeness phenomena in terms of

---

We would like to thank the audience of LENLS 13 for helpful comments and discussion. The first author would also like to acknowledge the support of JSPS Kiban C Grant #25370441, which partially supported this research.

© Springer International Publishing Switzerland 2014
Y. Nakano et al. (Eds.): JSAI-isAI 2013, LNAI 8417, pp. 69–81, 2014.
DOI: 10.1007/978-3-319-10061-6_5

Gale-Stewart games [7]. Section 4 of this paper outlines this framework and its application to politeness phenomena, together with some of the analysis provided by [2]. Before turning to these theoretical issues, though, we will present the general picture of a theory of politeness we are working with, and summarize briefly some of the existing formal work on the issue of politeness. We then turn to a distinction which has not so far been considered in such formal work, the distinction between formality and politeness: these two concepts are often equated in the literature, but we will show both that distinguishing them is possible and that it is useful. Here we also present the basic set of data which we will analyze. Section 5 provides that analysis; the strategy is to complicate the models used by [2], use the resulting expressivity to characterize a broader set of possible strategies, and analyze these strategies for complexity of execution. Section 6 concludes.

2   Levels of Politeness

The analysis of politeness seems to involve three elements. The first is lexical politeness: the content of honorifics, how composition works in this domain, the contribution of lexical hedges, and so on. The second is utterance choice: in short, whether to be polite or not. How is one to make this decision? Presumably the answer involves strategic considerations related to maximizing one's benefit in social and linguistic interaction. Finally, the third element involves sequences of utterances, politeness in discourse: the evolution of utterance choice over time, and politeness strategies at the discourse level.

The first of these areas has gotten considerable attention in the recent literature, mostly by Potts (2005) and people working in his framework ([10, 14, 17], McCready et al. 2013, i.a.). This work is mostly concerned with honorifics and how their semantic composition functions, a set of issues which are far from settled at present. Researchers in game-theoretic pragmatics have given a lot of attention to the second area, yielding a variety of diverse analyses, e.g. van Rooij and Sevenster (2006) who treat politeness in terms of handicaps [20], and Asher and Quinley [4, 15] who analyze politeness using trust games. A key insight here is that politeness crucially can be taken to involve reputation effects, which in turn can arise only in the context of repeated games; this means that the notion of repeated interaction is a key one for understanding polite behavior. Intuitively, this is sensible. The function of politeness (it has been postulated) is to minimize face threats [5]; why would one want to minimize threats to another's face if not with an eye toward future interactions? At minimum, politeness can help to achieve one's objectives in communication, which already counts as weighting future interactions.

The final element of a theory of politeness is the least-studied and the one we will focus on in this paper. This is the discourse level. As far as we know, the only formal work in this area so far is our own, though a good deal of related research can be found in sociolinguistics and anthropological linguistics (cf. [18]). But the path to such a theory is clear. As the previous paragraph pointed out, the theory of repeated games is highly useful in the analysis of politeness, just as with other kinds of social behavior [1,11]. But a consideration of repeated games makes it clear that agents are free to change their behavior at any point; there is thus a need to examine the discourse level to make clear what is optimal behavior at the level of individual interactions. It turns out that, given the right descriptive formalism, the results of analysis can straightforwardly be examined for properties such as complexity and safety, a point which will be detailed further in Sect. 4.

3 Formality vs Politeness

This section will consider some (relatively simple) aspects of the linguistic expression of politeness: the tools available to speakers at the lexical level. The facts here can be quite complex. Asher and Paul [2] discussed essentially the simplest possible case, that of European languages such as French, Spanish or German which deploy two types of second person pronouns, one formal and one informal. Here we will take up the more difficult case of Japanese, which has an array of distinct pronominals marked for politeness in both first and second person. For example, first person pronouns include, in descending order of formality, *watakushi*, *watashi*, *boku*, and *ore*; second person pronouns are of a similar level of complexity, including, again in descending order of formality, *anata*, *anta*, *kimi*, *omae*, *teme*, and *kisama*.[1] The last two second person pronouns are distinguished in not being only informal but also genuinely insulting; there is no situation in which they could be used in a manner not violating politeness norms, unlike even the least formal of the other first and second person pronouns.

Because of this range of choices in both first and second person, a wide variety of combinations of formality is available for the speaker. It is often the case that first and second person pronominals 'match' in register, but they can differ, yielding complex pragmatic effects. For instance, one could use a highly formal first person pronoun such as *watakushi* or *watashi* and an informal second person pronoun such as *omae*, which gives an impression of coldness or even anger; one might call it the 'VIP dialect' ('Very Important Person', or 'Very Important Professor' in a few notorious cases). One could also use an informal first person pronoun with a formal second person pronoun, which gives a feeling of someone who feels relaxed yet wishes to be perceived as being polite. The typical case here is *ore* together with *anata*; one might call this the 'Courtship Dialect' after its connotation of lovestruck youth. The reasons these impressions arise have to do with interaction between positive and negative face [5]. We will return to this in a moment, after making a proposal about what the formality of each pronoun amounts to.

It is usual in the literature on honorification not to use complex denotations tailored especially to honorifics. For example, [14] analyze subject honorification by means of 'expressive indices': triples consisting of a pair of individuals and a subinterval of $[-1, 1]$ corresponding to an emotive attitude, which, when

---

[1] Actually *anta* and *kimi* seem to be relatively close in level of formality.

processed, modify an initially set, more underspecified attitude stated as a set of such triples called an expressive index. This approach faces the obvious problem that honorification does not correspond in any direct way to emotive attitude. Reference [13], working in a similar compositional framework, takes formal and familiar pronouns to again modify expressive indices, but now analyzes them as introducing primitive and incompatible relations. This is a move forward, but since the content of these relations is never spelled out, it is hard to be fully satisfied with the analysis. These two works are, to our knowledge, the most clear formal statements of the semantic function of honorification and formal/familiar pronouns; we think it is possible to do better.

It seems to us that it is necessary to dissociate speaker attitudes from the 'honoring' function of formal pronouns (and, of course, honorifics, which we take to do similar things in at least some cases). One major reason is that it is certainly possible to despise someone but still 'honor' them by using formal speech, which means that speaker attitudes cannot be directly tied to honorification. The function of using honorifics, or formal pronouns, is, we think, to indicate a social attitude: a public commitment to a certain kind of behavior. The way in which such expressions are used then plays a role in constructing social roles and 'social reality' in general [16]. We think this 'externalist' picture of the meaning of expressions like these is appealing.

We will now try to spell it out for the pronominal case, afterward turning to the empirical payoff of the 'face implicatures' arising from mixed uses of pronouns. We will assume the following denotations for first and second person pronouns. Here we only make a binary distinction; a more fine-grained analysis which accounts for all the Japanese pronouns just amounts to adding intermediate stances with respect to social norms.[2]

(1) First person pronouns:
 a. If F used, then $1(c)$ makes public an expectation that $2(c)$ will behave formally toward her.
 b. If I used, then $1(c)$ makes public that $2(c)$ is not obliged to behave toward her in any particular way.
(2) Second person pronouns:
 a. If F used, then $1(c)$ makes public that she will behave formally with respect to $2(c)$.
 b. If I used, then $1(c)$ makes public that she will not follow any particular codes of behavior with respect to $2(c)$.

The core idea is that use of a particular pronoun commits the speaker to engaging in a particular kind of social interaction with respect to the hearer. The lexical entries vary along two dimensions. First, the person feature of the pronoun determines what individual is constrained in behavior by the pronoun use. For first person pronouns, it is the hearer; for second person pronouns, it is the hearer. This may strike the reader as counterintuitive, but we believe that the pronouns convey intentions of the speaker about the future discourse: the

---

[2] Thanks to Daisuke Bekki for extensive and useful discussion here.

second person shows how the speaker intends to behave toward the hearer, and the first person shows how the speaker expects to interacted with.[3] Second, the formal feature determines whether the behavior indicated is to be formal or not; the exact meaning of 'behave formally' in these definitions is something we must leave underspecified, but presumably it depends on various social and cultural factors. Here, we have let the informal pronoun indicate nothing more than a lack of expectation of formal behavior, so it just means 'not necessarily formal.' The behavior ultimately adopted might be formal; it is just not constrained to be so.

Consider now the various permutations of these pronominals. If both speakers use formal 1P and 2P pronouns, they agree to treat each other formally; if neither uses either formal 1P or 2P, they agree to allow each other to treat them as they like, and to do the same themselves. But what about the mixed cases? In the VIP dialect, the speaker requires formal treatment for himself,[4] but does not agree to abide by the constraints formal behavior requires with respect to his interlocutor. This is obviously rude; it does not respect the face of the addressee in any way. We thus derive the cold and arrogant impression such pronominal use induces. For the courtship dialect, the opposite situation arises: the speaker makes known his intention to treat his interlocutor with respect, but does not ask anything in return; this sounds humble, an effect due to its 'ultra-respect' for the interlocutor's face. The particular lovelorn impression may arise from this (possibly excessive) respect and treading-on-eggshells quality.[5] We take the possibility of deriving these pragmatic facts to be an indication that our lexical entries are on the right track.

One important point to note is that we have been careful to use the term 'formal' and not 'polite,' unlike our earlier work [2,12]. The reason is similar to our motivation for rejecting the analysis of [14]: one can easily be formal yet rude, or informal yet, intuitively, polite. This distinction will play a role in our formal analysis in Sect. 5, where we take a formal, rude discourse move to be a genuine game-changer.

4   Gale-Stewart Games and Politeness

Asher, McCready, and Paul [2,12] analyze discourse-level politeness in terms of Gale-Stewart games [7]; these games involve two players, 1 and 2, who each play one element of a given set A in their turn, where winning (for player 1) requires that the sequence corresponding to the sequentially concatenated moves of both players be in a given set. The sequences corresponding to the games are potentially infinite in length. (If player 1 loses, player 2 wins.) AMP took A to consist of possible moves P(olite) and I(nformal), which are meant to correspond roughly

---

[3] As usual, whether this commitment is accepted by the hearer is independent of its introduction; cf. e.g. [6].

[4] Surely most of these cases involve 'himself' rather than the other option(s).

[5] Presumably this dialect also has a strategic use, by those who wish to appear lovelorn. We will disregard this complexity here.

to formal and informal pronouns in many European languages. It turns out that the winning strategies of these 'politeness games' fall into different complexity classes corresponding to levels of the Borel hierarchy [8]. In this previous work we discussed strategies in classes Σ_1, Π_1 and Σ_2 corresponding to different sorts of coordination (and non-coordination) on P or I. In fact, Gale-Stewart showed that Gale-Stewart games with Borel Σ_1 and Π_1 determinable winning conditions are determinate, which means that there is a winning strategy for player 1 or player 2, and [9] showed this was true of all Gale-Stewart games with Borel determinable winning conditions. Determinacy of closed sets of A^ω for arbitrary A is equivalent to the axiom of choice over ZF [8], but the determinacy of all subsets of ω^ω is inconsistent with the axiom of choice. The latter result is extremely interesting when one takes into account the full contents of conversational moves in a game (players are then playing with a countably infinite collection of vertices or moves), but we leave that for future work. For full motivations and development of the theory, see the above references (and also [3]); here we only define Gale-Stewart games, give a brief overview of this previous work, and discuss some of its limitations which we aim to rectify in the remainder of the paper.

Gale-Stewart games, in the present setting, consist of two players and a finite vocabulary of politeness expressions V, which we can partition into 2 sets V_1 and V_2 (the politeness expressions for players 1 and 2). To illustrate with a simple example, we set $V_1 = V_2 = \{P, I\}$. (This is the vocabulary used in [2, 12], which one aim of this paper is to extend.) The game describes a graph $< V, E >$, where V is the set of vertices, and E is a set of edges. A strategy for player p is a function which maps each play prefix $v_0...v$ ending in a vertex $v \in V_p$ to a suitable 'next vertex', i.e. some v_α with $(v, v_\alpha) \in E$. Conversations are large dynamic games with no necessary stopping points and so can be seen as infinite; the game plays are thus strings in $\{P, I\}^\omega$. We impose exogenously imposed winning conditions and say that player 1 wins if she is able to achieve an infinite string of a certain definable type Win. Otherwise, player 2 wins. In particular, player 1 wins if the infinite sequence w resulting from game play is such that $w \in Win$, and otherwise player 2 does. The winning conditions originate in pragmatic aspects of interaction, much of which can presumably be modeled within game theory. In our previous work, we simply took them to be given (to the extent that they could be intuitively motivated), a practice we will continue in our extensions here.

Define a topological space on X^ω, the set of infinite strings over X, such that the basic open sets are sets of the form xX^ω, where $x \in X^*$ is a finite string over X. Thus a basic open set is a finite string x over X followed by all its possible continuations. We denote the open set xX^ω by $O(x)$. As usual, the set of open sets, denoted $O(X)$, is the closure of the basic open sets under countable unions and finite intersections. A closed set is a complement of an open set. A Gale-Stewart game is said to be determined if one of the players always has a winning strategy. Gale and Stewart [7] showed that all GS games where win is either an open or a closed set are determined. Determinacy of the second level of the Borel hierarchy for these games was shown by Wolfe in 1955 [19]. Finally Martin [9] in

1975 proved that all GS games where the winning set is Borel are determined. This result carries over directly to GS games as set up to analyze politeness: it means that, given a Borel-definable winning condition, it is possible to achieve an interaction that fulfills the politeness-related aims of the participants.

In our earlier work, we showed that politeness strategies at the discourse level can be characterized using the Borel hierarchy over sets, on the basis of the winning conditions they place on the infinite sequences resulting from game play. The Borel hierarchy is defined over the complexity of sets with respect to a topological space. Suppose that we have a winning condition characterizable by basic open sets or unions of such sets, or which corresponds to a number of possible such sets. Such a winning condition is Σ_1, which amounts to a kind of reachability condition.

(3) Suppose R is a subset of X. Then $Reach(R) = \{x \in X^\omega \mid R \subseteq occ(x)\}$ is the set of all strings which contain at least one element of R.

In our previous work, we showed that, given a vocabulary $V = \{P, I\}$ (interpretable as (P)olite and (I)nformal), a number of natural winning conditions could be characterized as Σ_1. For instance, consider an interaction which starts out formally and remains formal indefinitely, one standard pattern among languages which utilize formal and informal pronouns. This amounts to a requirement that the sequence w consist of (some subsequence of) an infinite sequence of Ps; this is an open set, and hence Σ_1.

Along with the Borel hierarchy, the complexity of winning conditions can rise. The next level of complexity is Π_1, which corresponds to winning conditions which are complements of Σ_1 conditions and equivalent to *safety*, as defined in (4). Intuitively, Π_1 conditions correspond to conditions which *avoid* conditions of rank Σ_1.

(4) Suppose S is a subset of X (the 'safe' set). Then $Safe(S) = \{x \in X^\omega \mid occ(x) = S\}$ is the set of all strings which contains elements from S alone. That is, the strings remain in the safe set and do not move out of it.

One example of such a condition is a sequence $\{P, I, P, I, \ldots\}$, so where player 1 always plays P and player 2 always plays I. This corresponds, for instance, to a conversation between an employee and a boss in a company of a certain type. Such situations are also common.

Next are Σ_2 conditions, which correspond to countable unions of Π_1 sets. An example here is a conversation which begins in a formal mode and switches to informal at some point, a very common (in fact canonical) kind of situation; as the conversational participants become 'closer,' their formality decreases. This corresponds to a union of sets of the form $P^i \cup I^j$ for some $i, j \prec w$; since P^i and I^j are Π_1 as we have just seen, the winning condition corresponding to this kind of interaction is Σ_2. As we will see, such conditions are very common in politeness interactions, and most of the cases we will analyze in the next section are either of this type or are Π_1.

There are still more complex winning conditions for politeness coordination, as the hierarchy continues. Our previous work considered a case in which the conversation moves from a sequence of Ps then to a sequence of I, and then back to P and then eventually back to I. Supposing that a win for player 1 requires the play to remain longer in I than in P, for instance, we have a set of plays characterized by the intersection of a set of complements of Σ_2 sets. This characterizes a Π_2 winning condition. Intuitively, we might see this sort of case for someone who occasionally affronts his conversational partner 1, leading 1 to use the formal register, but then gets 1 back eventually to a longer play in the I register. We will not have much to say about conditions of this and higher levels of complexity in the present paper.

5   Politeness, Formality, and Complexity

This section extends the model of [2,12], considering a wider range of strategies and a more complex class of possible moves. Doing so requires a more articulated set of types A from which the players can draw; after performing this extension, we consider the effects it has on the complexity of strategies. We begin with the extended vocabulary we will assume, which works along two dimensions: distinguishing formal moves from polite ones, and distinguishing first and second person polite pronouns. We then examine some strategies which can be used with this enriched vocabulary, presenting them in order of complexity as understood in terms of the Borel hierarchy for strategies outlined in the last section.

5.1   Vocabulary and Motivation

Recall that our original work made use of two types of move, P and I, so that the GS games considered there involved only combinations of these two moves. Here we will extend this strategy in two directions, both motivated already in Sect. 3.

The first distinction is between polite and formal moves. As we showed earlier, it is possible to find people being formal yet impolite, and equally possible to find discourse moves which are informal but not in any sense rude. This motivates a distinction between two kinds of moves, formal ones and polite ones. We model this observation by using two binary 'features': F(ormal) and N(ot formal) for formality, and P(olite) and I(mpolite) for politeness. The first type of feature will be used here basically to model lexical content: our primary focus being pronouns, most of our examples will be drawn from this domain, but the same distinction should be used for morphological honorifics as well. The second feature involves content, which can be genuinely rude, or not, irrespective of the particular language chosen. We leave the characterization of politeness and rudeness vague for the present, but most likely it simply involves some level of face threat.

The result of this discussion is that each speaker move is some combination of formal/informal and polite/impolite. This means that there are the four possible moves shown in Fig. 1.

FP	formal and polite
FI	formal and impolite
NP	informal and polite
NI	informal and impolite

Fig. 1. Formality versus politeness

The second distinction we draw is between the use of formal and informal first and second person pronouns. As we also saw in Sect. 3, the possible combinations also induce different politeness strategies, and 'politeness profiles' like the VIP dialect. We will also use a pair of binary features to model this behavior, here distinguishing formal and informal varieties of first and second person, so giving $1F/1N$ and $2F/2N$ and their possible combinations, shown in Fig. 2.

$1F2F$	formal speech
$1F2I$	VIP dialect
$1I2F$	courtship dialect
$1I2I$	informal speech

Fig. 2. First and second person pronominals and formality

We will use these possible moves to state winning conditions in the following sections. In this paper, we will restrict attention to sequences of the moves shown in Fig. 1, and sequences of those in Fig. 2, without attempting to combine the two. We think that the choices made between the options in Fig. 2 constrain those in Fig. 1; in particular, it is difficult to make moves of type FP or NP while using the VIP dialect, as it has impolite connotations. But these interactions are beyond the scope of the present paper.

We should say something about the notation we will use to describe the strategies. We are interested in winning conditions, and so in descriptions of sets of sequences of moves of the general form $\{g|P\}$, where g is a string and P some description that must be satisfied as usual. Because the game structure is that of the natural numbers – each player moves, followed by the next, meaning that we have an infinite sequence of moves with all odd-numbered ones those of player 1 and all even-numbered ones those of player 2 – we'll also sometimes use the notation $g(n)$ to describe the nth element of the string g. Where the formal realization is obvious, we'll sometimes also resort to simple descriptions.

5.2 Σ_1 Conditions

We start with strategies of the Σ_1 complexity class. Recall that such strategies only require the play of a certain move at some point in the game. There seem not to be so many of these in the domain of politeness, compared to the case of discourse with communicative intent, where there seem to be relatively many [3]. We mention only two types of case.

The first involves status marking. Suppose that someone wants to indicate their status relative to the other person, as higher or lower (in some domain). One natural way to do so is to use a formal first-person pronoun together with an informal second-person pronoun, i.e. to use the VIP dialect: this is the move $1F2I$ in our framework. Arguably, consistently sticking to this strategy isn't required. In many cases at least, seeing only one instance of this move is enough to categorize the player as of the VIP type; supposing that one wants to be so categorized, a single play of $1F2I$ is sufficient. Formally, this corresponds for player 1 to the condition $win = \{g | \exists n[g(n) = 1F2I \wedge odd(n)]\}$. It is easy to imagine corresponding strategies for the courtship dialect: if one wants to communicate that one is placing oneself metaphorically lower than the other player, a single instance of $1N2F$ will be enough, a winning condition where $win = \{g | \exists n[g(n) = 1N2F \wedge odd(n)]\}$. Other possible realizations of this sort of Σ_1 strategy are cases where a speaker would like to indicate that she is being formal, at least once as a formality, or being informal, to demonstrate independence. We find these slightly less plausible as distinct strategies though, as speakers most often remain consistent in their use of formal and informal pronouns across discourse stretches, which corresponds to the more complex kinds of strategies we will consider below. Still, it may be that there are strategies of the form $\{g | \exists n[odd(n) \wedge g(n) = NX]\}$, which yield a win for player 1 if she is the first to make a move with an informal pronoun, or even $\{g | \exists n[odd(n) \wedge g(n) = XI]\}$ (where $X \in \{F, N\}$), where player 1 wins if she manages to be impolite, something which might be desirable in certain kinds of social situations.

5.3 Π_1 Conditions

There are many winning conditions of the Π_1 class in the domain of politeness. This is to be expected; in this domain, Π_1 merely amounts to the requirement that a speaker exhibit consistent politeness behavior. Such strategies are expected to be quite common.

Thus, consider the following strategy: $win = \{g | \forall n[g(n) = FP]\}$, so each element of the sequence is a move which is both formal and polite. This condition characterizes a great deal of polite behavior in the standard sense. A related condition is $\{g | \forall n[g(n) = NP]\}$, where we have an unchanging sequence of moves which are polite yet informal, like much friendly discourse. In general, $\{g | \forall n[g(n) = XP]\}$ will be a winning condition as well, so either formal or informal language is available as long as the interaction remains polite.

We can find something very similar in the domain of first and second person pronominals. Suppose that consistent use of pronominals results in a winning strategy. Then we arrive at the following set of winning conditions:

$$\{g | \forall n[g(n) = 1F2F]\} \quad \sim \quad \text{formal speech}$$
$$\{g | \forall n[g(n) = 1F2N]\} \quad \sim \quad \text{VIP dialect}$$

$$\{g|\forall n[g(n) = 1N2F]\} \quad \sim \quad \text{courtship dialect}$$
$$\{g|\forall n[g(n) = 1N2N]\} \quad \sim \quad \text{informal speech}$$

The condition requiring sequences to be of the form $\{P, I, P, I, \ldots\}$ from Sect. 4 is also of this type, of course, though here we must replace P-moves with some variant of F-moves, and I-moves with some variant of N-moves. All these will remain winning strategies in politeness. We take even the sequence of the form $\{1F2F, 1F2N, 1F2F, 1F2N, \ldots\}$, where player 1 is unfailingly polite and player 2 uses the VIP dialect, to be a winning condition for player 1 in some circumstances, for example when he needs something from player 2 badly enough to be willing to grovel for it. Ultimately, regardless of the 'justice' or general desirability of the strategy, it will count as winning if it assists the player to achieve his objective; in this sense, even humiliation can yield victory.

5.4 Σ_2 Conditions

Finally, we turn to Σ_2 conditions. Recall our example of a Σ_2 condition from Sect. 4: a conversation that moves from a sequence of P-moves to a sequence of I-moves at some point. As we saw there, a Σ_2 condition is essentially just one which involves one or more such transitions, since such conditions consist of unions of Π_1-characterizable winning conditions.

In the present extension of the vocabulary used in [2,12], the above condition amounts to a winning condition of the form $\{g|\exists n\forall m[(m < n \rightarrow g(m) = FX) \land (m \geq n \rightarrow g(m) = IX)]\}$, where $X \in \{P, N\}$, as expected. We can also have less specified conditions of the form $\{g|\exists n\forall m[m \geq n \rightarrow g(m) = IX\}$, where no requirement is placed on the sequence preceding n in the game. Nonetheless, such sequences are still Σ_2, because a transition between sequence types is induced by the winning condition.

Finally, we can find winning conditions that we might characterize as recovery strategies. Consider a winning condition of the form $\{g|\exists n[g(n) = XI \land \forall m[m \geq n \rightarrow g(m) = XP]\}$. Here, an impolite discourse move is made, and then all is polite thereafter; In some sense, this amounts to the speaker acknowledging having made an error; such strategies are natural, for there is always a risk of error and misjudgment in conversation, since speakers are fallible in various respects. A similar kind of case can be seen for the case of formality: it is easy to get the timing wrong for the shift to the use of informal speech, and speakers often must hastily revert to a more formal speech pattern. This corresponds to the winning condition $\{g|\exists n[g(n) = NX \land \forall m[m \geq n \rightarrow g(m) = FX]\}$.

5.5 Admissible Continuations

Finally, we want to briefly mention one additional thing we take to be needed for a theory of discourse politeness. It seems to us that the set of moves available to a player alters with what has come before. For instance, if player 1 plays an impolite move in every turn, it eventually becomes bad strategy for player 2 to keep being polite. Indeed, we think that each move in some sense literally changes the game, by constraining the set of moves subsequently available.

A means of formalizing this notion is already available in our definition of strategies as functions from initial sequences to continuations, in that the particular sequence can condition the continuation; a notion of admissible continuation just amounts to constraining what the strategies can be.

We cannot go into details for reasons of space, but we want to consider one particular special case, a kind of game-ender. We take utterances which are both impolite and formal to, in some sense, take the game 'out of bounds'; there is no 'best response' to such moves, and they cannot be interpreted as intended to indicate friendship or closeness, as impolite and informal moves can sometimes be. We therefore take a play of FI to indicate a loss for the other player, and, consequently, a win for the player who uses FI, given the zero-sum nature of GS games.

This move can play into a number of strategy types at different levels of the complexity hierarchy. A first Σ_1 strategy involves just playing FI; this is an automatic win for the user of this move. Still we would not want to say that it's an optimal move in most circumstances. Here the notion of an admissible continuation comes into play; only certain kinds of discourses will allow a player to use FI, for instance one where the other player is excessively impolite. A real-life example might involve a drunken and rude student whose supervisor finally loses patience and plays FI, leaving the student without a best response. For this situation to be avoided, the student ought to play (have played) the Π_1 strategy whose winning condition disallows use of FI by the other player. A proper formalization requires a spelled-out analysis of admissible continuations, though, so we will not take it up here.

6   Conclusion

In this paper we have extended our previous analysis to a more complex vocabulary, which we have used to discuss a much larger class of strategies than in our initial work on the topic. In particular, we are now able to handle differences in first and second person pronoun usage and the distinction between politeness and formality. Of course, what we have done is far from exhaustive. There are many more strategies of the Σ_1, Π_1 and Σ_2 complexity classes than we have been able to address here, and we have not even touched on higher complexity classes such as Σ_2, which has already been shown to have applications to the theory of communicative interaction [3]. A full picture of politeness also requires the development of the idea of admissible continuations which we briefly discussed in Sect. 5.5. We will leave these issues for future work.

References

1. Alexander, J.: The Structural Evolution of Morality. Cambridge University Press, Cambridge (2007)
2. McCready, E., Asher, N., Paul, S.: Winning strategies in politeness. In: Motomura, Y., Butler, A., Bekki, D. (eds.) JSAI-isAI 2012. LNCS, vol. 7856, pp. 87–95. Springer, Heidelberg (2013)

3. Asher, N., Paul, S.: Conversations as Banach-Mazur games (2012). Submitted to Discourse and Dialogue
4. Asher, N., Quinley, J.: Begging questions, their answers and basic cooperativity. In: Okumura, M., Bekki, D., Satoh, K. (eds.) JSAI-isAI 2012. LNCS, vol. 7258, pp. 3–12. Springer, Heidelberg (2012)
5. Brown, P., Levinson, S.: Politeness: Some Universals in Language Usage. Cambridge University Press, Cambridge (1987)
6. Davis, C.: Constraining interpretation: sentence final particles in Japanese. Ph.D. thesis, University of Massachusetts at Amherst (2010)
7. Gale, D., Stewart, F.M.: Infinite games with perfect information. Ann. Math. Stud. **28**, 245–266 (1953)
8. Kechris, A.: Descriptive Set Theory. Springer, New York (1995)
9. Martin, D.A.: Borel determinacy. Ann. Math. **102**(2), 363–371 (1975)
10. McCready, E.: Varieties of conventional implicature. Semant. Pragm. **3**, 1–57 (2010)
11. McCready, E.: Reliability in pragmatics. Oxford University Press (2013) (in press)
12. McCready, E., Asher, N., Paul, S.: Winning strategies in politeness. In: Motomura, Y., Butler, A., Bekki, D. (eds.) JSAI-isAI 2012. LNCS, vol. 7856, pp. 87–95. Springer, Heidelberg (2013)
13. Potts, C.: The expressive dimension. Theor. Linguist. **33**, 165–198 (2007)
14. Potts, C., Kawahara, S.: Japanese honorifics as emotive definite descriptions. In: Proceedings of SALT XIV (2004)
15. Quinley, J.: Trust games as a model for requests. In: Lassiter, D., Slavkovik, M. (eds.) New Directions in Logic, Language, and Computation. LNCS, vol. 7415, pp. 221–233. Springer, Heidelberg (2012)
16. Searle, J.: The Construction of Social Reality. Free Press, New York (1997)
17. Sells, P., Kim, J.B.: Korean honorification: a kind of expressive meaning. J. East Asian Linguist. **16**, 303–336 (2007)
18. Watts, R.: Politeness. Cambridge University Press, Cambridge (2003)
19. Wolfe, P.: The strict determinateness of certain infinite games. Pac. J. Math. **5**(1), 841–847 (1955)
20. Zahavi, A., Zahavi, A.: The Handicap Principle. Oxford University Press, New York (1997)

Bare Plurals in the Left Periphery
in German and Italian

Yoshiki Mori[✉] and Hitomi Hirayama

University of Tokyo, Tokyo, Japan
mori@boz.c.u-tokyo.ac.jp, hirayama@phiz.c.u-tokyo.ac.jp

Abstract. Chierchia's comparative analysis of nominals based on the two features [± arg] and [± pred] has lead to many discussions on the semantics of nominals in argument positions and predicate positions. On the other hand, many syntactic results about left dislocation and topicalization have been accumulated. In this paper, we will try to elucidate possibilities of bare plurals in the left periphery and differentiate their readings. By examining data from Italian and German we claim that different demands on foregoing contexts are organized as constructions, as far as the left periphery is concerned. In addition, those constructions also reflect an organization of discourse.

Keywords: Bare plural · Kind-denoting nouns · Left periphery · Left dislocation

1 Introduction

In this research, we will see whether bare plurals can appear in the left dislocated position in German and Italian. We also compare bare plurals with definite plurals in different left-dislocated construction. From the data we obtained, we claim the followings: First, there is a difference in the interpretation of dislocated nominals between in Italian and in German. Second, Italian bare plurals can be licensed in the left periphery in a certain discourse condition.

In all, the aim of this research is to show that how constructions and contexts interact with each other. In addition, we claim that nominals located in the left periphery are interpreted differently from those located in argument positions or predicate positions. This is because using a certain type of construction affects nominal interpretation and the left periphery imposes a sort of specificity on nominals located there.

This paper is organized as follows. First, we introduce some general description of German and Italian NPs. Then previous literature on nominal interpretation is reviewed in Sect. 2. In Sect. 3, we illustrate interpretation of nouns in the left periphery, which cannot be anticipated from the previous research on nominals in argument positions and predicate positions. Next, we show how Italian bare plurals in the left periphery, which are problematic for the literature, can be explained in Sect. 4. Finally we conclude in Sect. 5.

© Springer International Publishing Switzerland 2014
Y. Nakano et al. (Eds.): JSAI-isAI 2013, LNAI 8417, pp. 82–97, 2014.
DOI: 10.1007/978-3-319-10061-6_6

1.1   German and Italian NPs

The main difference between German and Italian NPs is that Italian bare plurals have restricted distribution whereas German NPs occur much more freely. Moreover, although German bare plurals can be interpreted as kinds like those in English, Italian bare plurals, which must be licensed by lexical or functional heads, cannot denote kinds but they have only existential readings (Chierchia 1998).

(1)   a. * Studenti hanno telefonato.
         students have   phoned

      b. * Hanno telefonato studenti.
         have   phoned   students

      c. Luigi odia i   gatti.
         Luigi hates the cats
         'Luigi hates (the) cats.'

      d. Luigi odia gatti.
         Luigi hates cats.

(1a) and (1b) show that both preverbal and postverbal bare plural subjects are disallowed in Italian. In addition, kind denoting nouns must accompany the definite article at least in argument positions unlike English. This means that Italian definite plurals are ambiguous between kind readings and definite ones. For example, in (1c), a plural *gatti* 'cats' occurs with the definite article *i*. (1c) is ambiguous between the reading that 'Luigi hates cats in general' and the reading that 'there are contextually salient cats which Luigi hates.' Although bare plurals cannot occur in subject positions, Italian bare plurals can be licensed in object positions. However, they are always interpreted existentially unlike English. So, (1d) has only the reading that 'there are some cats such that Luigi hates.' Here *gatti* cannot refer to kinds as the English counterpart *cats* can.

We have seen that Italian bare plurals cannot occur freely in argument positions. According to Zamparelli (1995), even in the left periphery, bare plurals are not good, either.

(2) ?Ragazzi, li   ho   visti, (*ragazzi).
     boys   cl.3pl have.1sg seen boys

(2) is an example of Italian CLLD (CLitic Left Dislocation). Zamparelli judges left-dislocated bare plurals are marginal. On the other hand, right-dislocated bare plurals are considered as clearly ungrammatical. However, bare plurals become grammatical when left-dislocated elements are stressed and bear focus (3a) or when left dislocation is accompanied by a clitic *ne* (3b) (Chierchia 1998).

(3)   a. POLLO io voglio, non pesce.
         chicken I want, not fish

      b. Soldi, non ne ho.
         money not NE have

(2) and (3) show that on some conditions bare plurals can be licensed. However, as seen in (2), the instances of bare plurals discussed in the literature on the left periphery (see Sect. 4) are not kind-denoting ones[1]. Whether kind denoting bare plurals can survive in the left periphery is our main concern in this paper.

2   Previous Research on Interpretation of NPs

Before discussing nominals in the left periphery, let us review the previous research on nominal semantics. There are now two main approaches to interpretation of NPs. One comprises so-called ambiguity approaches, which consider bare nouns ambiguous between kind readings and existential readings (Gerstner and Krifka 1987). In this framework, the predicates determine which of the two readings is available. The meaning of a sentence is calculated by a generic operator or existential closure.

On the other hand, what we call neo-Carlsonian approaches consider that bare nominals unambiguously refer to kinds following Carlson (1977). For this type of approach, it is necessary to define how to derive existential readings from kind readings when a predicate is not suitable for kind denoting nouns.

In this paper, the neo-Carlsonian approach is adopted to discuss nominal interpretation since we are interested in how and in what environment a noun phrase can get a kind reading.

Three theories on the nominal semantics are introduced hereafter. One is Chierchia (1998), which is the seminal paper of this approach. We adopt his idea and some basic operations to discuss the problems in this paper. Then we review Dayal (2004), which adds some important modifications to Chierchia's theory, to discuss some problems which include German kind-denoting nouns. Finally, Zamparelli's analyses on Italian nominals will be introduced since he has investigated a variety of Italian noun phrases, and being somewhat for Chierchia and in some respects against him, his research sheds interesting insight on the research of this area.

2.1   Chierchia (1998)

As mentioned above, Chierchia assumes that bare plurals refer to kinds based on Carlson's idea. Kinds are considered as 'nominalizations' of predicative common nouns, namely, properties of them. Conversely, properties are regarded as 'predicativization' of kinds. That is, there is a correspondence between kinds and properties.

For the relationship between kinds and properties, Chierchia defines a function that maps properties into kinds and one that maps kinds into properties: ∩ and ∪, respectively. Let DOG be the property of being a dog, and d is the

---

[1] *Pollo* and *soldi* in (3) seem to refer to kinds. However, *pollo* is a mass noun. Moreover, although *soldi* is a plural form of *soldo*, this word is always used in a plural form in this meaning. Therefore, it is not clear whether bare plurals can denote kinds in the left periphery as well.

dog-kind, then $^\cap$DOG = d and $^\cup$d = DOG. More precisely, these functions are defined as follows (Chierchia 1998:350–351):

(4) Let d be a kind. Then for any world/situation s,
$^\cup$d= λx [x $\leq$ d$_s$], if d$_s$ is defined
= λx [FALSE], otherwise
where d$_s$ is the plural individual that comprises all of the atomic members of the kind

(5) For any property P and world/situation s,
$^\cap$P= λs ι P$_s$, if λs ι P$_s$ is in K,
undefined, otherwise
where P$_s$ is the extension of P in s

Meaning of nouns is calculated by these two operations and other type shifts proposed by Partee (1987).

Besides, Chierchia introduces nominal mapping parameters to describe how nominals are interpreted in a certain language. The two parameters are [± arg], and [± pred]. Here [+arg] and [+pred] indicate that the category N can be mapped onto the type of arguments and predicates, respectively.

For example, Italian is a [−arg, +pred] language. This type of language does not allow N to be mapped onto arguments. That is, every noun is a predicate. Therefore, a kind-denoting noun cannot occur in an argument position without projecting D. As a result, kind referring nouns always occur with the definite article.

On the other hand, English belongs to the [+arg, +pred] languages. In this kind of language, bare plurals can freely occur in argument positions. Therefore, in a sentence 'Dogs bark.', dogs can refer to the dog-kind without the definite article. On the other hand, in English, the dogs cannot refer to the dog-kind unlike Italian. An English definite plural only refers to a salient group of entities: in this case, the dogs is interpreted as denoting a salient group of dogs. To explain this difference in the meaning of the definite article between two languages, Chierchia proposes a restriction called Blocking Principle.

(6) Blocking Principle ('Type Shifting as Last Resort')
For any type shifting operation τ and any X:*τ(X)
if there is a determiner D such that for any set X in its domain,
D(X) = τ(X)

This principle defines covert type shifting as last resort. That is, covert type shifting is blocked when there are other overt tools for type shifting like the definite article. If the English definite article could function ι and $\cap$, dogs could not refer to kinds since covert $\cap$ would not be available. Therefore, it can be concluded that the English definite article can function only as ι, and not as $\cap$. As a result, the dogs can be interpreted only as a contextually salient group of dogs and not as the dog-kind. Conversely, a bare plural cannot be interpreted as a salient group of dogs.

Although the use of the definite article and the interpretation of nouns in a language is largely explained by these parameters and the Blocking Principle, we should find out how to derive existential readings from kind readings in this approach, since bare plurals can have existential readings as well as kind readings in English.

(7)   a. Dogs love to play.

     b. Dogs are ruining my garden.

Dogs in (7a) refers to the dog-kind whereas the one in (7b) refers to indefinite dogs. For Ambiguity Approaches, there is no problem to get these two readings since they allow bare plurals to be ambiguous between kinds and indefinite plurals, and predicates determine in which of these readings bare nominals should be interpreted.

    Following Carlson's approach, however, we need to find some way to derive the existential readings from kind readings when there is a mismatch between nouns and the predicate. To derive existential readings from kind readings, Chierchia proposes an operation called DKP (Derived Kind Predication) defined as follows:

(8)   Derived Kind Predication:
    If P applies to objects and k denotes a kind, then
    $P(k) = \exists \; [^{\cup}k(x) \wedge P(x)]$

By using this operation, we can get an appropriate meaning of *dogs* in (7b). Moreover, this operation enables us not only to get the readings we want, but also to explain the special behavior of scope regarding to bare plurals. It is well known that bare plurals always take narrow scope in relation to other operators like negations.

(9)   a. Bob is not looking for ship parts.

     b. Bob is not looking for ship parts of the "Titanic".

     c. Bob is looking for ship parts of the "Titanic", and John is looking for them, too.

Negation in (9a) always takes wide scope over the bare nominal, *ship parts*. Therefore (9a) is unambiguously interpreted that 'it is not the case that there exist ship parts such that Bob is looking for them.'

    Chierchia argues that this scope limitation is due to DKP. Applying DKP, nominal interpretation ends up with taking narrower scope than other operators. This is manifested by the fact that if there is no kind reading available, a bare plural can take wide scope, too. In (9b–c), bare nominals even can take wide scope although they usually cannot. This is because bare nominals are modified with rigid modifiers, namely *of the "Titanic"* in (9b–c). In this case, modification makes it impossible for bare plurals to get kind readings since modified nominals are no longer well-established kinds. Therefore, it can be concluded that whether nominals can be interpreted as kinds determines the scopal behavior of bare plurals. This assumption also explains the fact that this scopal characteristic of bare plurals is shared by both English and Italian.

We have seen how Chierchia deals with nominal interpretation on the basis of the assumption that bare plurals basically denote kinds. Whether bare plurals can refer to kinds in argument positions and predicate positions depends on language-dependent parameters. To derive existential readings we can utilize DKP. Although his approach seems to treat nominal interpretation very well so far, Chierchia's argument has some problems as shown in the next section introducing subsequent work by Dayal, who is also following the neo-Carlsonian view.

2.2 Dayal (2004)

Dayal (2004) expands Chierchia's theory to Hindi and Russian in order to study whether it is cross-linguistically valid. Moreover, she looks into German, which looks like a counterexample to Chierchia's framework, because it seems to violate the Blocking Principle. After she challenges these problems, she concludes that Chierchia's theory and neo-Carlsonian approaches are on the right track.

Dayal investigates two languages, Hindi and Russian, where the definite article is missing. In these two languages, bare nominals are ambiguous between kind readings and definite readings. However, they sometimes can have existential readings as well. In those cases, their distribution is restricted and the existential operator always takes narrow scope. That is, this bare nominal's scopal behavior is the same as that in Italian and English. Considering all these characteristics, Dayal concludes that in these two languages, too, existential readings are derived from kind readings by DKP and this explains the scope limitation of bare nouns.

The other problem which Dayal tries to deal with in her paper is German kind-denoting nouns.

(10)   a. (Die) Pandabären sind vom Aussterben bedroht.
          '(The) pandas are facing extinction.'
       b. (#The) pandas are facing extinction.

German is a [+arg, +pred] language like English. In English, bare plurals refer to kinds and definite plurals do not as illustrated by (10b). By contrast, both bare plurals and definite plurals have kind readings in German (10a). This looks like a violation of the Blocking Principle, which prohibits the covert type shifts when there are overt type shifts available.

Dayal proposes that the Blocking Principle applies only to a canonical meaning of the definite article. That is, canonical meaning of the German definite article is ι, and not $\cap$. As a result, German bare plurals have existential readings by covert $\exists$ and kind readings by covert $\cap$, but they do not have definite readings since covert ι is not available. On the other hand, definite plurals have definite readings by overt ι and kind readings by overt $\cap$. This means that the Blocking Principle is actually working in German, too. It is the canonical meaning of the definite article that decides to which meaning Blocking Principle might be applied.

As far as we have seen, Chierchia's theory combined with Dayal's modification looks very attractive to explain nominal interpretation since it almost correctly predicts possible nominal interpretation. Moreover, Chierchia's analysis now seems to be valid cross-linguistically with the support by Dayal, who has shown that Chierchia's approach works in other languages than English or Italian, which Chierchia mainly investigated. However, Roberto Zamparelli, a linguist who has been concentrating on Italian nominal phrases, casts some doubt on Chierchia's analysis.

2.3   Zamparelli

Zamparelli is not against all of Chierchia's arguments. In Zamparelli (2002), he indicates that DKP is actually necessary to analyze English. However, he denies the idea that Italian bare plurals refer to kinds. First, as a matter of fact, Italian bare plurals cannot denote kinds in argument positions as mentioned in the preceding parts. To refer to kinds, the definite article is obligatory with both singulars and plurals. This means that there is no evidence which shows that Italian bare plurals refer to kinds.

In addition, Zamparelli argues that DKP, which is an operation which assumes that bare plurals are kinds, is not necessary to give an explanation to nominal interpretation. For example, Chierchia explains scopelessness of bare plurals by using DKP. Once they are modified with other rigid modifiers, wide scope readings become available. This is because nominals cannot be interpreted as kinds any more, and DKP becomes unavailable in such a situation.

According to Zamparelli's idea, it is some kind of projection in D that is required to have wide scope readings. Bare plurals do not have any projection in D. Therefore, bare plurals cannot take wide scope. On the other hand, when they are modified with a rigid modifier like *of the "Titanic"*, this modifier works as D. In his framework, *ship parts of the "Titanic"* is assumed to have a structure like: [$_{DP}$[$_D$ *of the "Titanic"*]$_i$ [$_{NP}$ *ship parts* t$_i$]]. Rigid modifiers play the same role as articles, and this enables bare nominals to have both narrow and wide scope readings. His arguments mean that DKP is no longer necessary for interpreting bare plurals in Italian. Together with the fact that bare nominals cannot appear as kinds in Italian, he concludes that bare nominals never denote kinds in Italian, opposed to the traditional neo-Carlsonian approach.

3   Kind-Denoting Nouns in the Left Periphery

The papers cited in the previous chapter mainly discuss nominals in argument positions. What we are interested in here is what happens to bare plurals in the left periphery.

Before discussing NPs dislocated to the left periphery, it is important to first clarify what we are interested in this research. The focus of this paper is how bare plurals are interpreted in the left periphery. That is, what a dislocated noun refers to as a topic or focus. Let us give some examples.

(11)   a.   Monkeys, I saw in the Ueno Zoo.
            Reading: That kind of animal, I saw in the Ueno Zoo.

b. The monkeys, I saw in the Ueno Zoo.
Reading: # That kind of animal, I saw in the Ueno Zoo.

In example (11a), *monkeys* refers to a kind. In this case, *monkeys* can be replaced with *that kind of animal* under an appropriate context. It should be noted that this kind interpretation is valid only in the topic position. After the bare plural is reconstructed in the object position, it must be subject to type-shifting from a predicate interpretation by the covert ∪-operator or some similar operation. In contrast, the definite plural *the monkeys* in (11b) cannot be replaced with *that kind of animal* without changing meaning. That is, the topic of this sentence should be a set of contextually salient monkeys. In this case, the interpretation in the topic position corresponds to that in the object position.

It should be noted that nouns can refer to kinds in topic position independently of the predicate in the main clause. It is possible that the discourse topic is a certain kind in general, and the main clause conveys more specific information. So, predicates in the main clause need not to be so-called kind predicates.

If what the literature on nominals in argument positions predicts remains the same in the left periphery, in German both definite plurals and bare plurals should be able to refer to kinds, whereas in Italian only definite plurals should. This prediction seems not to be borne out according to our informants.

(12)   a. Mozartopern, die habe ich gesehen.

b. Die Mozartopern, die habe ich gesehen.
'Operas of Mozart/The operas of Mozart, I saw'

To begin with, in German, when definite plurals are dislocated, kind readings become unavailable. The definite plural *die Mozartopern* 'The operas of Mozart' in (12b) only refers to a contextually salient group of operas by Mozart. By contrast, in (12a), a bare plural *Mozartopern* can refer to a kind. That is, operas by Mozart in general. Therefore, in the left periphery, German is similar to English in that definite plurals cannot refer to kinds but to some definite plural individuals. In other words, the definite article does not work as ∩ in the left periphery.

In Italian, on the other hand, the situation seems more complex.

(13)   Le rivoluzioni , per fortuna, non le fanno i giudici.
'Fortunately, revolutions , judges do not cause.'

(14)   ... contar palle è una cosa che non mi viene bene, ... e io che dovrei fare?
Bugie , non mi va di raccontar<u>ne</u>, non ci si può più prendere in giro.
'...I do not feel like telling a lie, ... and what should I do? Lies , I do not feel like telling about them.'

(15)   Bambini , ne volete? Sono in programma. Abbiamo cominciato a provarci un paio di mesi fa.
' Children , do you want? I am planning it. We began to try to do it a few months ago.'

According to the data from the Italian written corpus CORIS (CORpus di Italiano Scritto), it seems possible that definite plurals can refer to kinds as they can in argument positions as shown in (13). Moreover, when bare plurals are dislocated with the clitic *ne* (*ne*-topicalization hereafter), they also refer to lies or children in general as shown in (14) and (15). This is a violation of the Blocking Principle, which says, 'Do not type-shift covertly when there is an overt tool available.'

Based on this comparison, we have made a descriptive generalization that definite and bare plurals are distributed differently in Italian and German topic positions: In German, the bare plural is used for the kind reading, whereas the definite plural prevailingly refers to individuals (anaphorically or situation bound). By contrast, in Italian, definite plurals can be a kind expression. In addition, bare plurals can also denote kinds[2], even though they cannot do so in argument positions. The permission of Italian bare plurals in the topic position is problematic for the previous account for the nominal semantics we introduced earlier.

We claim that this obvious difference between German and Italian in the naturalness of bare plural topics is construction-dependent. For example in German, as we have seen, bare plurals as well as definite plurals are compatible with the kind-reading when it comes to the normal prefield topicalization as shown by example (10). As for this sort of German dislocated nominals, it is possible to explain this difference by the canonical meaning of the definite article as Dayal (2004). That is, in the left periphery the only available meaning of the definite article is ι, which is the canonical meaning of the German definite article. As a result, definite plurals cannot be interpreted as denoting kinds in the left periphery in German.

What is problematic is Italian bare plurals in the left periphery. We claim that this unique behavior of Italian bare plurals can be explained by the discourse-sensitive properties of the left periphery. Next, we would like to show that there are two types of discourse structures where bare plurals are licensed in Italian left periphery.

4 Bare Plurals and Nominal Interpretation

4.1 Two Types of Licensing

Our data suggest that two conditions seem to govern the use of kind-denoting bare plurals in the left periphery. In one case, bare plurals are introduced as

---

[2] However, it is a little difficult to show that these bare plurals are kinds. We cannot check whether bare plurals really denote kinds with kind predicates because the predicate in the main clause does not have to be a kind-predicate. Moreover, in the first place, bare plurals licensed by *ne* cannot occur with kind-predicates (Zamparelli p.c.). For example *Ratti, ne stermino* 'Rats, ne (I) exterminate.' is ungrammatical. We assume this ungrammaticality is due to the meaning of *ne*, which existentially quantifies entities rather than denying that bare plurals are kinds. From contexts, bare plurals seem to denote neither specific indefinite entities nor contextually salient entities.

topics when their superkinds have been introduced as a discourse topic in the context as illustrated by example (16).

(16)   a.   Quale animale hai   visto?
            which animal   have.2sg seen
            'Which animal did you see?'

       b.   (Di) cani, ne ho   visti.
            of   dogs NE have.1sg seen
            'I saw dogs.'

This example shows that bare plurals in the left periphery become felicitous when the preceding context provides a topic related to kinds. In this case, the preceding context semantically licenses the bare plural.

In the second case, bare plurals are licensed as frame-setting topics, because of accommodation between speech participants. This point is shown by the example given in (17).

(17)   a.   *Quando litighiamo mi viene sempre da pensare che l'unico motivo
            per cui mi ha sposato sia che voleva la cittadinanza americana. Ma
            non capita spesso. Nove giorni su dieci andiamo d'accordo. Sul serio.*
            'When we argue, I always feel like thinking that the only motivation
            for which she married me might have been that she wanted American
            citizenship. But it does not happen often. Nine days out of ten days,
            we reach agreement. Seriously.'

       b.   Bambini, ne volete?
            boys   NE want.2pl
            'Do you want children?'

In this sort of licensing, crucially, the preceding context is not related to a kind; rather we argue that this irrelevance is the factor which licenses the bare plural. In (17), an utterance (17b) suddenly changes the discourse topic and introduces the bare plural *bambini* 'children' as the frame-setting topic.

In these two cases, we claim that bare plurals are licensed in different manners. In the first case, bare plurals are licensed by the preceding context. On the other hand, in the second case, the specificity attributed to the topic position in the left periphery licenses kind-denoting bare plurals. In what follows, both licensing mechanisms are explained in detail.

4.2   Question Under Discussion and Nominal Interpretation

In the first case, we claim that Question under Discussion (QuD) proposed by Roberts (1996) can affect nominal interpretation. That is, the alternative answers to a question license some sort of nominal interpretation: in this case, a reference to kinds.

Here is an example to show what is QuD and how it is represented in the construction.

(18)   a. Mario, lo   rivedrò.
          Mario him I will see again.

       b. MARIO, rivedrò   (non Luigi).
          Mario   I will see again, not   Luigi

In (18), it should be the case that the addressee already knows the fact that the speaker will see someone again, but does not know who they are. That is, *Mario* is new information to the hearer in terms of the hearer's knowledge about who the speaker will meet again.

From another point of view, answering by (18a) and (18b) respectively assume different types of questions. For example, (18a) is an appropriate answer to a question *'What will happen to Mario?'*. However, (18b) seems not to be so. Putting the focus on *Mario* and preposing it to the beginning of the sentence indicates that the presupposed question is only *'Who will you see again?'*. The fact that (18b) is not compatible with more general question indicates that non-focused parts must be known to he hearer.

We assume a kind-denoting bare plural in (16) is licensed in the similar way. In this case, the question is 'Which animal did you see?'. The question denotes the set of possible answers. More generally, this question denotes a set of answers $A=\{x \in Q|$ I saw $x\}$, where the variable x determines each possible answer in a given domain Q, which corresponds to animal kinds in (1a). We claim that as suggested by this example, kind-denoting bare plurals are licensed whenever the domain of possible answers is composed of kinds.

To sum up, bare plurals can be semantically licensed when the QuD has established a discourse topic related to kinds before bare plurals are introduced. Adopting Roberts' definition of information structure, we propose the following.

(19)   *Licensing Condition 1*[3]
       Bare plurals in the topic position at m_k can be licensed as kind-denoting nominals when m_i $(n \leq i \leq k-1)$ provides a question α where m_n is the move which starts a new domain goal of conversation, and D, which is the domain of the model, is made up of by kinds.

This licensing condition states that appropriate kind-related topics must be introduced so that bare plurals can denote kinds.

4.3   Frame Setting Topics and Bare Plurals

Next, let us see how bare plurals are licensed in the second case. In this case, we claim that bare plurals are licensed to denote kinds due to the specificity imposed by the topic position.

---

[3] In representing discourse structure where bare plurals are licensed, we adopt the definition of the information structure by Roberts (1996: 10). M is the set of moves in the discourse. That is, M can be regarded as the set of utterances in the discourse. The number attached to m indicates the precedence relation. The smaller the number is, the earlier the m is uttered in the discourse.

Although it is commonly assumed that a topic should be a discourse-old entity, some linguists argue that this is not necessarily the case. For example, Benincà (1988) and Brunetti (2009) argue that an aboutness topic does not need to be discourse-old, even though their ideas are different from each other. Benincà claims that the speaker must assume that the entity to which the topic refers is known to the addressee. On the other hand, Brunetti claims that such an assumption is not necessary, providing the example below (Brunetti 2009: 760).

(20)   Sai?   Dante lo   hanno   bocciato all'esame   di chimica.
       know.2sg Dante cl.3sg have.3pl failed   at the exam of chemistry

       Do you know? Dante failed the chemistry exam.

In Italian, 'Sai' introduces an out-of-the-blue context. Therefore, in the example (20), Dante is regarded as a discourse-new element. In addition, Dante is dislocated by CLLD, so it is interpreted as an aboutness topic. Contrary to the assumption that an aboutness topic should be discourse-old, to Brunetti and Benincà, this sentence does not sound unusual at all. Therefore, they agree to the idea that an aboutness topic does not need to be discourse-old. However, they have different conditions which make this sentence felicitous in their minds. According to Benincà, Dante should be known to the hearer even though it is not necessary for him to have been introduced in the previous discourse. Conversely, according to Brunetti, such an assumption is not required. Therefore, for Brunetti, example (20) is felicitous even when a hearer does not know Dante.

Although both arguments make sense, in this research we adopt Brunetti's account. For the utterance to be regarded felicitous, discourse-oldness is an adequate condition although it is not a necessary condition. Moreover, the referent that is situated as a topic does not have to be known to the hearer so that the sentence itself sounds felicitous. However, at least it must be the case that the speaker presupposes that the addressee knows the referent.

By contrast, from a viewpoint of the hearer, when an unknown entity is introduced as a topic he has at least two strategies in the discourse. One is to accommodate the information at stake into his knowledge without mentioning that he does not know the referent, and the other is to interrupt the conversation to clarify the topic. In both cases the hearer should modify his knowledge to repair inconsistencies between actual common ground and the one assumed by the speaker. In this accommodation, the addressee modifies his knowledge on the assumption that the referent in the topic is specific.

There is another example which shows that the topic position imposes specificity. This characteristic of a topic is also manifested by the indefinite dislocated as an external topic (Ebert et al. 2008). Let us examine the German example.

(21)   a. [Einen   Song von Bob Dylan]T, den   kennt jeder.
           Some-ACC song of   Bob Dylan,   RP-ACC knows everybody.

           ' Everybody knows some song of Bob Dylan. '

       b. Einen   Song von Bob Dylan kennt jeder.
           Some-ACC song of   Bob Dylan knows everybody.

' Everybody knows a song of Bob Dylan. '

In (21a) and (21b), the indefinite noun phrase *einen song von Bob Dylan* is located at the beginning of the sentence. However, there is difference in their interpretation. In (21a), the indefinite is left dislocated by German Left Dislocation (GLD). Therefore, this indefinite is interpreted as a topic of this sentence. This dislocated indefinite has only a wide scope reading in relation to *jeder* 'everybody'. It should be noted that this topic is located external to the sentence. On the other hand, in (21b), the indefinite is simply located at the beginning of the sentence without dislocation to an external topic position. That is, the indefinite in (21b) cannot be interpreted as an external topic of this sentence. This indefinite is ambiguous between a wide scope reading and a narrow scope one unlike the indefinite (21a).

In addition, it should be noted that the topic in the second condition is what is called frame-setting topic, a new topic used to mark the beginning of a discourse.

It is important to bear in mind that the frame-setting topic should be differentiated from the aboutness topic as the German example from Frey (2005) suggests below.

(22)   Ich habe etwas   in der Zeitung   über   Hans gelesen.
       I   have something in the newspaper about Hans read

       'I read something about Hans in the news paper.'

(23)   a. Den   Hans, den   will   der Minister zum Botschafter
           the-ACC Hans RP-ACC wants the minister to   ambassador
           ernennen.   (GLD)
           (to) appoint

       b. #Hans ↓ , der Minister will   ihn zum Botschafter ernennen.
           Hans   the minister wants him to   ambassador (to) appoint
           (HTLD)

Let us assume that each sentence in example (23) is uttered after the sentence shown by (22), which explicitly introduces *Hans* as a discourse topic.

In this context, (23a) is natural while (23b) is not. Let us see what is a difference. In each sentence, *Hans* is dislocated by different type of dislocation. In (23a), *Hans* is left-dislocated by GLD, which marks an aboutness topic as we have seen earlier. On the other hand, *Hans* is topicalized by HTLD (Hanging Topic Left Dislocation) in (23b). The reason (23b) sounds unnatural is that HTLD generally introduces a new discourse topic as a frame-setting topic. The reason (23b) sounds inappropriate is that the topic is not changed between the two utterances (22) and (23b). This example clearly illustrates that frame-setting topics and aboutness topics are not the same, and some languages like German have strategies to differentiate these topics.

We reviewed the specificity imposed by the topic position and the fact that frame-setting topics should be differentiated. Based on these properties of topics, we claim that bare plurals can be licensed as kinds when it is dislocated as a frame-setting topic due to the accommodation by the hearer. When a discourse topic is suddenly changed and a bare plural is located at the topic position, it is a kind that occurs to hearer's mind as an interpretation of the bare plural. In other words, kind names can always be called upon if necessary since they are considered as a part of the common ground. We call this type of specificity as *kind-specificity* distinguishing so-called "specificity" regarding individual entities from one we discuss here.

In sum, we propose the following generalization, again adopting the formalism in Roberts' (1996):

(24) *Licensing Condition 2*
      Bare plurals at m_k can be licensed as kind denoting nouns when a domain goal is achieved at m_i ($i \leq k - 1$), and they are left-dislocated as a topic with *ne*.

This says that bare plurals can be interpreted as kinds when they are frame-setting topics. It should be noted that frame-setting topics cannot be used anytime. For the topic to be changed, it is necessary that speech participants have achieved a domain goal, which is a part of the biggest goal of conversation, what the world is like. To change the topic before achieving the domain goal would violate Grice's maxim of relation (Grice 1975). Therefore, a topic shift should be made only after the speaker judges that he and the addressee should commit to another domain goal.

4.4 The Difference Between Two Conditions

We would like to point out that these licensing conditions are different. First, bare plurals are licensed through different processes. *Licensing Condition 2* requires that speech participants have achieved their domain goal before a topic bare plural is introduced. In other words, a bare plural licensed by *Licensing Condition 2* is a part of a new domain goal whereas *Licensing Condition 1* indicates that a bare plural is licensed by a domain goal which has already been introduced in the context.

In addition, in each condition, dislocated nominals are different kinds of topics. As Frey (2005) suggest, a topic cannot be considered as a unitary function. Although the same construction, namely *ne*-topicalization is used in both licensing conditions, we suggest that each topic works differently. On one hand, as we have seen in the examples in Sect. 3, bare plurals licensed by *Licensing Condition 2* serve as frame-setting topics, which introduce a new discourse topic.

By contrast, bare plurals in *Licensing Condition 1* restrict a domain given by the previous context like some adverbs which can serve as topics as shown the example below from Ernst (2004: 104, underlines ours).

(25) What have they done in their last two years in office?
      - Well, economically, they have passed new tax legislation; politically, they have raised far more money for the party than was expected.

In (25), the first sentence sets up a topic. Then the adverbs serve as restricting the domain of the events. Similarly, we claim that bare plurals in *Licensing Condition 1* can restrict the domain which has already been introduced in the previous context.

5   Conclusion

In this paper, we argued that both in German and in Italian, the nominal semantics in the left periphery is different from that in argument positions and predicative positions. Italian bare plurals can be a topic in the left periphery and they can refer to kinds. To the contrary, German definite plurals cannot refer to kinds in the left periphery unlike ones in argument positions.

We claim that this difference is due to the property of the left periphery, which is very sensitive to information packaging. That is, different demands on foregoing contexts are organized as constructions. From this viewpoint, the information packaging is not only packaging of the own sentence, but also one of the foregoing context. As a result, some constructions affect semantic interpretation of nominals. This mechanism is responsible for why nominals in the left periphery are interpreted differently from those in argument positions or predicate positions, where Chierchia's parameters are viable.

References

Benincà, P.: L'ordine delle parole e le costruzioni marcate. In: Renzi, L., Salvi, G., Cardinaletti, A. (eds.) Grande Grammatica Italiana di Consultazione, vol. 1, pp. 129–239. Il Mulino, Bologna (1988)

Brunetti, L.: On links and tails in Italian. Lingua **119**, 756–781 (2009)

Carlson, G. N.: Reference to Kinds in English. Ph.D. Thesis, University of Massachusetts at Amherst (1977)

Chierchia, G.: Reference to kinds across languages. Nat. Lang. Seman. **6**, 339–405 (1998)

Dayal, V.: Number marking and (In)definiteness in kind terms. Linguist. Philos. **27**(4), 393–450 (2004)

Ebert, C., Endriss, C., Hinterwimmer, S.: A unified analysis of conditionals as topics. In: Proceedings of SALT, 18, Amherst (2008)

Ernst, T.: Domain adverbs and the syntax of adjuncts. In: Austin, J.R., Engelberg, S., Rauh, G. (eds.) Adverbials, pp. 103–129. John Benjamins, Amsterdam (2004)

Frey, W.: Pragmatic properties of certain german and english left peripheral constructions. Linguistics **43**(1), 89–129 (2005)

Gerstner, C., Krifka, M.: Genericity, an introduction. Ms., Universität Tübingen (1987)

Grice, P.: Logic and conversation. In: Cole, P., Morgan, J. (eds.) Syntax and Semantics: Speech Acts, vol. 3, pp. 41–58. Academic Press, New York (1975)

Partee, B.H.: Noun phrase interpretation and type-shifting principles. In: Groenendijk, J., et al. (eds.) Studies in Discourse Representation Theory and the Theory of Generalized Quantifiers, pp. 115–144. Foris, Dordrecht (1987)

Roberts, C.: Information structure in discourse: towards an integrated formal theory of pragmatics. In: Yoon, J.-H., Kathol, A. (eds.) OSU Working Papers in Linguistics, 49: Papers in Semantics, The Ohio State University (1996)

Zamparelli, R.: Layers in the Determiner Phrase. Ph.D. Thesis, University of Rochester (1995)

Zamparelli, R.: Definite and bare kind-denoting noun phrases. In: Beyssade, C. Bok-Bennema, R., Drijkoningen, F., Monachesi, P. (eds.) Selected papers from Going Romance 2000, pp. 305–342. John Benjamins, Amsterdam (2002)

CORpus di Italiano Scritto, Università di Bologna. http://corpora.ficlit.unibo.it/

Analyzing Speech Acts Based on Dynamic Normative Logic

Yasuo Nakayama[✉]

Graduate School of Human Sciences, Osaka University, Suita, Japan
nakayama@hus.osaka-u.ac.jp

Abstract. In a conversation, different kinds of speech acts are performed. Logic for communication has to deal with these various kinds of speech acts ([5]: 52). Additionally, for interpretation of conversations, it will be appropriate to take shared beliefs among communication partners into consideration. In this paper, we show that this problem can be dealt with in a framework that is a dynamic extension of the logic for normative systems.

Keywords: Speech acts · Dynamic normative logic · Logic for normative systems · Logic for communication · Shared attitudes · Common belief

1 Introduction

It is an aim of this paper to propose a logical framework for communication. The framework is based on *dynamic normative logic* (DNL) proposed in [19]. We interpret a communication as a game played by communication partners. A communication game consists of verbal and physical actions. Verbal actions can be interpreted as speech acts. As physical actions change physical sates in the world, successful performances of speech acts change normative states that are shared by communication partners. It may update shared beliefs and shared norms.

2 Logic for Normative Systems

In [12], I proposed a new logical framework that can be used to describe and analyze normative phenomena in general. I called this framework *Logic for Normative Systems* (LNS).[1] LNS takes not only assertive sentences but also normative sentences into consideration. In other words, LNS distinguishes two kinds of information, namely propositional and normative information. Assertive sentences, which express propositional information, are true or false. They describe

---

[1] A characteristic of LNS is its dynamic behaviors. LNS is quite flexible, so that LNS can be applied to describe complex normative problems including ethical problems. See [12,13].

© Springer International Publishing Switzerland 2014
Y. Nakano et al. (Eds.): JSAI-isAI 2013, LNAI 8417, pp. 98–114, 2014.
DOI: 10.1007/978-3-319-10061-6_7

physical facts and other kinds of facts such as social facts. If they properly describe the corresponding facts, then they are true. In contrast, normative sentences seem to have no truth value. They are related with social norms. They are accepted or rejected by a certain group and they influence the decision making of agents.

LNS is a quite flexible formal framework and can explicitly express both propositional and normative constraints. In LNS, the validity of assertive sentences remains independent of normative requirements, while that of normative sentences depends on the presupposed set of assertive sentences. The explicitness of LNS makes it possible to apply it to analysis of legal systems, paradoxes in deontic logics, and ethical problems.

In this paper, we modify the previous version of LNS, so that we can express mental states of agents.[2] Here, we mainly deal with belief states and normative states of agents. For the sake of simplicity, we use &, $\Rightarrow$, and $\Leftrightarrow$ as meta-logical abbreviations for *and*, *if ... then*, and *if and only if*.

Definition 1. *Suppose that each of T and OB be a set of sentences in First-Order Logic (FOL), more precisely a set of sentences in Many-sorted Logic.*[3]

(1a) *A pair $\langle T, OB \rangle$ consisting of belief base T and obligation base OB is called a normative system (NS = $\langle T, OB \rangle$).*

(1b) *A sentence q belongs to the belief set of normative system NS (abbreviated as $B_{NS}q$) $\Leftrightarrow$ q follows from T.*

(1c) *A sentence q belongs to the obligation set of NS (abbreviated as $O_{NS}q$) $\Leftrightarrow$ T$\cup$OB is consistent & q follows from T$\cup$OB & q does not follow from T.*[4]

(1d) *A sentence q belongs to the prohibition set of NS (abbreviated as $F_{NS}q$) $\Leftrightarrow$ $O_{NS}\neg q$.*

(1e) *A sentence q belongs to the permission set of NS (abbreviated as $P_{NS}q$) $\Leftrightarrow$ T$\cup$OB$\cup\{q\}$ is consistent & q does not follow from T.*

(1f) *A normative system $\langle T, OB \rangle$ is consistent $\Leftrightarrow$ T$\cup$OB is consistent.*

(1g) *In this paper, we interpret that NS represents a normative system accepted by a person or by a group in a particular time interval. Thus, we insert what a person (or a group) believes to be true into the belief base and what he believes that it ought to be done into the obligation base.*

We read formulas of LNS as follows:

$B_{NS}q$: "It is believed in NS that q."
$O_{NS}q$: "It is obligatory in NS that q."

---

[2] The main difference between two versions of LNS consists in the use of some notions in (1a). We use now the notion *belief base* instead of *propositional system*. Some effects of this change will become visible, when we start to analyze interactions among normative systems of different agents.

[3] The many-sorted logic is reducible to FOL. Thus, this difference is not essential.

[4] In this paper, we require the consistency of T$\cup$OB from two reasons, namely to justify the claim that obligation implies permission and to smoothly describe rule-following behaviors.

$\mathbf{F}_{NS}q$: "It is forbidden in NS that q."
$\mathbf{P}_{NS}q$: "It is permitted in NS that q."

Based on Definition 1, we can easily prove the following main theorems that characterize LNS.

Theorem 2. *The following sentences are meta-logical theorems of LNS. Here, we assume* $NS = \langle T, OB \rangle$.

(2a1) $(\mathbf{B}_{NS}(p \to q) \ \& \ \mathbf{B}_{NS}p) \Rightarrow \mathbf{B}_{NS}q$.
(2a2) $\mathbf{B}_{NS}p \Leftrightarrow T \vdash p$.
(2b1) $(\mathbf{O}_{NS}(p \to q) \ \& \ \mathbf{O}_{NS}p) \Rightarrow \mathbf{O}_{NS}q$.
(2b2) $\mathbf{F}_{NS}p \Leftrightarrow \mathbf{O}_{NS}\neg p$.
(2b3) $\mathbf{O}_{NS}p \Rightarrow \mathbf{P}_{NS}p$.
(2b4) $\mathbf{F}_{NS}p \Rightarrow not \ \mathbf{P}_{NS}p$.
(2c) $\mathbf{B}_{NS}p \Rightarrow (not \ \mathbf{O}_{NS}p \ \& \ not \ \mathbf{F}_{NS}p \ \& \ not \ \mathbf{P}_{NS}p)$.
(2d1) $(\mathbf{O}_{NS}(p \to q) \ \& \ \mathbf{B}_{NS}p) \Rightarrow \mathbf{O}_{NS}q$.
(2d2) $(\mathbf{O}_{NS}(p \wedge q) \ \& \ not \ \mathbf{B}_{NS}p) \Rightarrow \mathbf{O}_{NS}p$.
(2d3) $(\mathbf{O}_{NS}(p \wedge q) \ \& \ \mathbf{B}_{NS}p) \Rightarrow \mathbf{O}_{NS}q$.
(2d4) $(\mathbf{O}_{NS}(p \vee q) \ \& \ \mathbf{B}_{NS}\neg p) \Rightarrow \mathbf{O}_{NS}q$.
(2d5) $(\mathbf{O}_{NS}(p \wedge q) \ \& \ \mathbf{F}_{NS}p) \Rightarrow \mathbf{O}_{NS}q$.
(2e1) $(\mathbf{O}_{NS}\forall x_1, \ldots, \forall x_n(P(x_1, \ldots, x_n) \to Q(x_1, \ldots, x_n)) \ \& \ \mathbf{B}_{NS}P(a_1, \ldots, a_n)$
$\& \ not \ \mathbf{B}_{NS}Q(a_1, \ldots, a_n)) \Rightarrow \mathbf{O}_{NS}Q(a_1, \ldots, a_n)$.
(2e2) $(\mathbf{F}_{NS}\exists x_1, \ldots, \exists x_n(P(x_1, \ldots, x_n) \wedge Q(x_1, \ldots, x_n)) \ \& \ \mathbf{B}_{NS}P(a_1, \ldots, a_n) \ \&$
$not \ \mathbf{B}_{NS}\neg Q(a_1, \ldots, a_n)) \Rightarrow \mathbf{F}_{NS}Q(a_1, \ldots, a_n)$.

Proof. (2a1) is obvious, because modus ponens holds in FOL. To prove (2b1), suppose that $\mathbf{O}_{NS}(p \to q) \ \& \ \mathbf{O}_{NS}p$. Thus, $p \to q$ does not follow from T. Now, it is sufficient to show that q does not follow from T, because *modus ponens* shows that q follows from $T \cup OB$. However, since $q \to (p \to q)$ is a theorem of FOL, q does not follow from T. In a similar way, other theorems can be easily proved. □

In LNS, we have belief operator $\mathbf{B}_{NS}$ and normative operators $\mathbf{O}_{NS}$, $\mathbf{F}_{NS}$, and $\mathbf{P}_{NS}$, where all of these operators are relativized by the given normative system NS. This relativization is a main difference of LNS to modal logics. The theorem (2a2) shows that the belief set is not influenced by normative requirements. From Theorem 2, we can see how LNS differs from the doxastic logic. For example, LNS presupposes a particular theory T, so that the belief ascription becomes dependent on T, while doxastic logic characterizes belief in a more abstract manner.

Within LNS, iterations of operators are forbidden, while modal logics usually allow any iteration of modal operators. This is a limitation of LNS. However, we can imitate iterations of operators, as we show this in Sect. 2 (See (3b) and Table 1).

We will often use inference rule (2e1) in this paper. The content of (2e1) can be paraphrased as follows: If $\forall x_1, \ldots, \forall x_n(P(x_1, \ldots, x_n) \to Q(x_1, \ldots, x_n))$ is an obligation and you believe $P(a_1, \ldots, a_n)$, then $Q(a_1, \ldots, a_n)$ is an obligation, unless you believe that it was already done. The last condition is reasonable, because you need not do again what is already done.

3   Dynamic Normative Logic

In a previous work [19], I extended LNS and proposed *Dynamic Normative Logic* (DNL). DNL is a LNS complemented with information update device. Recently, the *Dynamic Epistemic Logic* (DEL) has been established as a framework for logical description of social interactions.[5] There are many extensions of DEL and some of them deal with communication problems and change of common beliefs [23]. DNL can be considered as an alternative framework for the same purpose. We can update normative system $\langle T, OB \rangle$ by extending the belief base T or obligation base OB with new information p (i.e. $T_{new} = T \cup \{p\}$ or $OB_{new} = OB \cup \{p\}$). In [19], based on DNL, I gave a full description of social interactions in a restaurant scene discussed by van Benthem ([2]: 4).

In this paper, we sometimes say *normative state of an agent* instead of *normative system*, when this agent has a particular normative system in certain time interval.[6] When NS is the normative system that A has in t, we say "A believes in time interval t that q" instead of "It is believed in NS that q" and "A believes in time interval t that it is obligatory that q" instead of "It is obligatory in NS that q". In general, we express ascriptions of beliefs and normative states as follows.

(3a) We use $ns(X, t)$ $(ns(X, t) = \langle bel(X, t), ob(X, t) \rangle)$ to refer to the normative system that person X has in time t, where $bel(X, t)$ is the belief base of X in t and $ob(X, t)$ is the obligation base of X in t.

(3b) We use $ns(X > Y, t)$ to refer to the normative system that X ascribes to Y in time t. Furthermore, we use $ns(X > Y > Z, t)$ to refer to the normative system that X identifies in time t as the normative system that Y ascribes to Z. In this way, you may construct more complex ascriptions of normative systems.[7]

(3c) We require for any $ns(X > X, t)$ the following three conditions:
   1. $bel(X > X, t)$ is consistent.
   2. If $bel(X, t)$ is consistent, then $bel(X > X, t) = bel(X, t)$.
   3. $ob(X > X, t) = ob(X, t)$.

$bel(X > X, t)$ is a normative system for an agent who has the ability of the complete introspection. As a matter of fact, based on (3c), we can easily prove the following fact: If $bel(X, t)$ is consistent, then for any formula q, ($\mathbf{B}_{ns(X,t)}q \Leftrightarrow \mathbf{B}_{ns(X>X,t)}q$). Thus, the belief part of this kind of agents roughly corresponds to the belief representation within the doxastic logic.[8]

---

[5] For the development of the dynamic epistemic logic, you nay consult [2]. There, van Benthem characterizes the epistemic logic as logic of semantic information ([2]: 21). Compared to DEL, our approach in this paper is more syntactically orientated.

[6] As we see in the next section, a normative state of a person can be influenced by that of other persons.

[7] Note that all of theorems in Theorem 2 are applicable to $ns(X > Y, t)$ and $ns(X > Y > Z, t)$ as well, because they are all normative systems that satisfy all conditions in Definition 1.

[8] Note that $\mathbf{B}q \leftrightarrow \mathbf{BB}q$ is a theorem of the doxastic logic **D45**.

In (multiple) doxastic logic, it is possible to have a belief about other person's belief state. For example, $\mathbf{B}_i\mathbf{B}_j\mathbf{B}_kq$ means that i believes that j believes that k believes that q. The corresponding content can be expressed within LNS and DNL as $\mathbf{B}_{ns(i>j>k,t)}q$. Table 1 shows some examples of complex attitude ascriptions.

Table 1. Examples of complex attitude ascriptions

DNL-formulas	Reading of DNL-formulas
$\mathbf{B}_{ns(A,t)}q$	A believes in time t that q
$\mathbf{B}_{ns(A>B,t)}q$	A believes in time t that B believes that q
$\mathbf{B}_{ns(A>B>C,t)}q$	A believes in time t that B believes that C believes that q
$\mathbf{O}_{ns(A,t)}q$	A believes in time t that it is obligatory that q
$\mathbf{F}_{ns(A>B,t)}q$	A believes in time t that B believes that it is forbidden that q
$\mathbf{P}_{ns(A>B>C,t)}q$	A believes in time t that B believes that C believes that it is permitted that q

In order to take intentional attitudes into consideration, we interpret intention as a self-obligation. In DNL, a self-obligation has the form $\mathbf{O}_{ns(A,t)}$ $\exists t_1(do(A, action_k, t_1) \wedge t \leq t_1)$. This formula means: A believes in t that he himself is obligated to perform $action_k$. This self-obligation may cause A's performance of $action_k$, while A's knowledge of obligations of others, such as $\mathbf{O}_{ns(A,t)}$ $\exists t_1(do(B, action_k, t_1) \wedge t \leq t_1)$, has no such motivational power over A.

Now, we consider how to ascribe normative states to collective agents. We assume thereby a mereological ontology and interpret a collective agent as a mereological sum of atomic agents, because we want to avoid the use of set conception. Note that a set is an abstract entity, while a mereological sum of physical entities remains as physical.[9]

(4a) Let $BEL(X, t)$ be the belief set of (possibly collective) agent X in t. That means, any formula p that follows from $BEL(X, t)$ is already included in $Bel(X, t)$ as its element.

(4b) Let $A_1, \ldots, A_n$ be atomic agents. We construct the collective agent G as the mereological sum of $A_1, \ldots, A_n$ i.e. $G = A_1 + \cdots + A_n$. We assign G the normative system $ns(G, t)$, only if for any $A_k(1 \leq k \leq n)$ and any time t, $(BEL(G, t) \subseteq BEL(A_k, t)$ and $ob(G, t) \subseteq ob(A_k, t))$.

---

[9] In fact the notion of collective agent should be more carefully defined. See discussions about extended agents in [17,18]. For mereology, you may consult [25]. For four-dimensionalism, see [24]. For four-dimensional mereology, see [8,11].

(4c) To refer to groups of people, we accept the axiom system for *General Exten-sional Mereology* (GEM)[10]. We claim: For any time t, GEM is included in $bel(G, t)$.

(4d) From (4b) follows the following fact: For any atomic agent A who is a member of the collective agent G and for any formula p, $(\mathbf{B}_{ns(G,t)}p \Rightarrow \mathbf{B}_{ns(A,t)}p)$. This means that we may consider that $\mathbf{B}_{ns(G,t)}p$ expresses a shared belief of group G in time t.

(4e) We require for any $ns(G > G, t)$ the following three conditions:

1. $bel(G > G, t)$ is consistent.
2. If $bel(G, t)$ is consistent, then $bel(G > G, t) = bel(G, t)$.
3. $ob(G > G, t) = ob(G, t)$.

When normative system $ns(G > G, t)$ satisfies all of these conditions, we call it *common normative system for group G* and its belief *common belief*. In this case, it follows that $ns(G > G, t) = ns(G, t)$.

To express simple anaphoric relations, we use Skolem-symbols in this paper. We interpret (demonstrative) pronouns as a kind of Skolem symbols. So, not only d_k but also he_k, it_k, $this_k$, $that_k$, and the_k are used as Skolem-symbols.[11]

Definition 3. *Let $M = \langle U, V \rangle$, S be a set of formulas in which some elements of S contain Skolem-symbols, and μ be a variable assignment.*

(5a) *M^* is a Skolem expansion of M with respect to S iff (if and oly if)*
$M^* = \langle U, V^* \rangle$ & $V \subseteq V^*$ &
For all Skolem constant symbols d_k, $V^*(d_k) \in U$ &
For all n-ary Skolem function symbols d_k, $V^*(d_k)$ is a function from U^n into U.

(5b) *S is true according to M and μ iff*
There is M^* (M^* is a Skolem expansion of M with respect to S & S is true according to M and μ).

(5c) *S is true according to M iff*
S is true according to M and μ for all assignments μ.

Thus, Skolem-symbols are interpreted as constant symbols (or function symbols) whose referents can be determined from the viewpoint of the interpreter.

4 Describing Speech Acts in Dynamic Normative Logic

In this paper, we accept, by and large, Searle's analysis of speech acts [20] and his classification of illocutionary forces [21]. In Chap. 3 of [20], he distinguished *preparatory*, *sincerity*, and *essential conditions* for illocutionary acts. Prepara-tory conditions formulate indispensable conditions for the success of a speech act. The sincerity condition describes what kind of intentionality the speaker must have in order to sincerely perform certain kind of speech act. We call these

---

[10] GEM is the strongest mereological system. For GEM, you may consult [3] and [27].
[11] I did this kind of proposal in [9].

two kinds of conditions *pre+sin-conditions*. The essential condition expresses the essential feature of a speech act. For example, the undertaking of an obligation to perform a certain act is the essential condition of a promise ([20]: 60). This essential condition is the one we try to analyze in this section.

In [21], Searle distinguished five illocutionary forces, namely *Assertives*, *Directives*, *Commissives*, *Expressives*, and *Declarations*. He explains these classes of illocutionary forces in terms of a characterization of their illocutionary points.

(6a) [**Assertives**]. "The point or purpose of the members of the assertive class is to commit the speaker (in varying degrees) to something's being the case, to the truth of the expressed proposition." ([21]: 12)

(6b) [**Directives**]. "The illocutionary point of theses consists in the fact that they are attempts (or varying degrees, and hence, more precisely, they are determinates of the determinable which includes attempting) by the speaker to get the hearer to do something." ([21]: 13)

(6c) [**Commissives**]. "Commissives ... are those illocutionary acts whose point is to commit the speaker (again in varying degrees) to some future course of action." ([21]: 14)

(6d) [**Expressives**]. "The illocutionary point of this class is to express the psychological state specified in the sincerity condition about a state of affairs specified in the propositional content." ([21]: 15)

(6e) [**Declarations**]. "It is the defining characteristic of this class that the successful performance of one of its members brings about the correspondence between the propositional content and reality, successful performance guarantees that the propositional content corresponds to the world:" ([21]: 16–17)

Right now, there exist several formal approaches to speech act theory [26, 29]. However, most of them failed to work as logic for communication. For example, we should deal with anaphoric relations that keep referents over performances of different types of speech acts, while most of existing frameworks failed to solve this problem.

In this paper, we express a successful performance of a speech act through an update of a normative system among communication partners. We also use Skolem symbols in order to deal with anaphoric relations.

At first, we introduce a function t_f that maps a time stamp to a time interval (see (7)).

(7) $\forall m \forall n (t_f(m) \leq t_f(n) \leftrightarrow m \leq n)$.

In this paper, we represent a speech act through the following representation schema:

[*Speech act type, Speaker, Hearer, Time stamp, ok* (or *fail*)] (*Proposition*).

Speech act types roughly correspond to Searle's classification of illocutionary forces in [21], while our classification is a little bit more detailed than

Searle's. The fifth element of the head of the schema expresses whether the pre+sin-conditions for this type of speech act are fulfilled or not. For example, '$[assertive, s, h, n, ok]$ (p)' means that the speaker s performs an assertive speech act to h in time $t_f(n)$ with the content that p, when the pre+sin-conditions for this speech act are fulfilled. When the pre+sin-conditions are not fulfilled, we write '$[sa\text{-}type, s, h, n, fail](p)$'. It is worth to note that this representation schema also contains information about the context of the utterance which can be used for interpreting the proposition stated in the same context. In other words, a representation schema for a speech act contains information about the speaker, the hearer, and the utterance time; this information can be used to interpret demonstratives and indexicals.[12]

It is our first assumption that a collective observation creates a common belief.

Definition 4. *Presupposition and Observation as Common belief*

(8a) *[**Common belief**]. Let G be a group of agents. Then, [common-belief, $G, n](p) \Leftrightarrow B_{ns(G>G,t)}p$.*

(8b) *[**Presupposition**]. We stipulate: [presupposition, G, n] = [common-belief, G, n]. This means that what is presupposed among G is also a common belief of G.*

(8c) *[**Observation**]. We stipulate: [observation, G, n] = [common-belief, G, n]. This means that what is observed among G becomes a common belief of G.*

Now, we express a performance of speech acts as (local) information update of the given normative systems. The requirements in Definitions 5 and 6 express the update rules that describe the effects of observations and performed speech acts.

Definition 5. *Update of normative states*
Let $p(*_s/s, *_h/h, *_t/t_f(n))$ be the formula that can be obtained from p by replacing all of $*_s$, $*_h$, and $*_t$ by $s, h, t_f(n)$ respectively.[13] In the following description, $s + h$ refers to the communication partners interpreted as the mereological sum of the speaker and the hearer. Let $ns(s + h, t_f(n)) = \langle bel(s + h, t_f(n)), ob(s + h, t_f(n))\rangle$.

(9a) *[**Update of belief base**]. The belief update of a normative state in context of (s, h, n) is defined as follows: [[belief-update, s, h, n] $(p(*_s, *_h, *_t))$]/ $(ns(s + h, t_f(n))) = ns(s + h, t_f(n + 1))$, where $bel(s + h, t_f(n + 1)) = bel(s + h, t_f(n)) \cup \{p(*_s/s, *_h/h, *_t/t_f(n))\}$ and $ob(s + h, t_f(n + 1)) = ob(s + h, t_f(n))$. This means that a belief update changes only the belief base of both communication partners and their obligation base remains unchanged.*

---

[12] The classical work for semantics of demonstratives is formalized by D. Kaplan [7]. I proposed some improvements of Kaplan's framework [13,15,16].

[13] This replacement of *-terms by singular terms creates interpretations of demonstratives and indexicals.

(9b) *[Update of obligation base].* *The obligation update of a normative state in context of (s, h, n) is defined as follows: [[obligation-update, s, h, n] $(p(*_s, *_h, *_t))$] $(ns(s+h, t_f(n))) = ns(s+h, t_f(n+1))$, where $bel(s+h, t_f(n+1)) = bel(s+h, t_f(n))$ and $ob(s+h, t_f(n+1)) = ob(s+h, t_f(n)) \cup \{p(*_s/s, *_h/h, *_t/t_f(n))\}$. This means that an obligation update changes only the obligation base of both communication partners and their belief base remains unchanged.*

Definition 6. *Interpretations of simple speech acts*

(0a) *[Assertives].* *We stipulate: [assertive, s, h, n, ok] = [belief-update, s, h, n]. This means that a successful performance of an assertive speech act can be interpreted as a belief update among communication partners.*

(10b) *[Expressives].* *We stipulate: [expressive, s, h, n, ok] = [belief-update, s, h, n]. This means that a successful performance of an expressive speech act can be interpreted as a belief update among communication partners. It is a characteristic of expressive speech acts that their propositional content expresses a mental state of the speaker.*

(10c) *[Directives].* *We stipulate: [directive, s, h, n, ok] = [obligation-update, s, h, n]. This means that a successful performance of a directive speech act can be interpreted as an obligation update among communication partners. It is a characteristic of directive speech acts that their propositional content expresses an action of the hearer.*

(10d) *[Commissives].* *We stipulate: [commissive, s, h, n, ok] = [obligation-update, s, h, n]. This means that a successful performance of a commissive speech act can be interpreted as an obligation update among communication partners. It is a characteristic of commissive speech acts that their propositional content expresses an action of the speaker.*

The above update rules show that we always update the normative state of communication partners after a successful performance of a speech act. For example, suppose that A said to B that B should immediately go to the school. In this case, it is clear who is obliged, namely B. For the further inference, it is important to confirm that the information of this obligation is shared by both of the communication partners. This is the reason why we should update the normative state of $A + B$. Now, based on (4b), we can justify that after A's performance of the directive speech act, both A and B know that B should immediately go to the school.[14]

Questions can be interpreted as a kind of directive speech acts; they require certain responses from the hearer.

----

[14] From (4b) follows: $\mathbf{O}_{ns(A+B,t(n))} \exists t(\textit{go-to-school}(B,t) \wedge t(n) \leq t) \Rightarrow \mathbf{O}_{ns(A,t(n))} \exists t(\textit{go-to-school}(B,t) \wedge t_f(n) \leq t)$ & $\mathbf{O}_{ns(B,t(n))} \exists t(\textit{go-to-school}(B,t) \wedge t_f(n) \leq t)$.

Definition 7. *Speech acts for communication*

(11a) *[Yes/No question].* A *Yes/No-question is a directive speech act with a requirement of a Yes/No answer. We can, therefore, replace [interrogative-yn, s, h, n, ok] $p(*_s, *_h, *_t)$ with [directive, s, h, n, ok] (answer-yes$_{(*_s, *_h, *_t)}$(p) $\vee$ answer-no$_{(*_s, *_h, *_t)}$(p)), where* answer-yes$_{(*_s, *_h, *_t)}$(p) := $(p(*_s, *_h, *_t) \rightarrow \exists t\ (say(*_h, *_s,$ *'Yes', t)* $\wedge *_t < t))$ *and* answer-no$_{(*_s, *_h, *_t)}$(p) := $(\neg p(*_s, *_h, *_t) \rightarrow \exists t(say(*_h, *_s,$ *'No', t)* $\wedge *_t < t))$. *This means: Asked 'p?', the hearer should say 'Yes', when she (or he) believes that p, and she (or he) should say 'No', when she (or he) believes that $\neg p$.*

(11b) *[Which question].* [interrogative-which, s, h, n, ok] $(p_1 \vee \cdots \vee p_m)$ *is defined as the following obligation update:* $ob(s + h, t_f(n + 1)) = ob(s + h, t_f(n)) \cup \{(p_k \rightarrow \exists t(say(h, s, \lceil p_k \rceil, t) \wedge t_f(n) < t)) : 1 \leq k \leq m\}$, *where* $\lceil p_k \rceil$ *denotes an English sentence for proposition p_k! This means the following: Asked 'which of $\{p_1, \ldots, p_m\}$?', the hearer should say $\lceil p_k \rceil$, when she (or he) believes that p_k.*

(11c) *[Wh-question].* [interrogative-wh, $s, h, n, ok](p(x))$ *is defined as the following obligation update:* $ob(s + h, t_f(n + 1)) = ob(s + h, t_f(n)) \cup \{(p(x/c) \rightarrow \exists t(say(h, s, \lceil p(c) \rceil, t) \wedge t_f(n) < t)): c$ *is a singular term}. This means the following: Asked 'which x is $p(x)$?', the hearer should say $\lceil p(c) \rceil$, when she (or he) believes that $p(c)$.*

These proposals show that we interpret here questions as a kind of conditional obligations. In other words, a question creates a certain kind of conditional obligation to the hearer.

5   Describing Declarations in Dynamic Normative Logic

Searle interpreted a declaration as a speech act that creates facts. However, some declarations, such as a declaration of a new law, create norms; a normative requirement that is stated by an authority can be accepted by people as their *new norms*. So, we distinguish, in this paper, two types of declarations.

Definition 8. *Declarations*
Let G be a group of people.

(12a) *[Declaration of facts].* We stipulate: [declaration-fact, s, G, n, ok] = [belief-update, s, G, n]. This means that a declaration of a fact can be interpreted as a belief update for a group. Usually, a declaration of a fact expresses a social fact and not a physical fact, because a physical fact holds independent of any collective belief.[15]

---

[15] This claim implies a rejection of social constructivism of physical facts. For discussions about this topic, see [10, 22].

(12b) *[Declaration of obligations].* *We stipulate:* *[declaration-obligation, $s, G,$* *$n, ok]$ = [obligation-update, s, G, n].* *This means that a declaration of an* *obligation can be interpreted as an obligation update for a group. The* *amendment of laws deals with this kind of declaration of obligations. By* *the way, in such a case, we usually have to contract some of old articles,* *before we add the new articles, so that the law system remains consistent.*[16]

Declarations are speech acts that are addressed to a group and used to create social agreements and social norms. They are effective, only if the majority of the group members accept the corresponding authority of the speaker.

6   DNL-Analysis of Simple and Complex Speech Acts

In English, there are two different uses of conjunction, namely static and dynamic one. We can express this distinction in DNL. The statement 'p_1 and_{static} p_2' can be represented as $[assertive, s, h, n, ok](p_1 \wedge p_2)$. In this case, we obtain: $bel(s+h, t_f(n+1)) = bel(s+h, t_f(n)) \cup \{p_1(*_s/s, *_h/h, *_t/t_f(n)) \wedge p_2(*_s/s, *_h/h, *_t/t_f(n))\}$. The statement '$p_1$ $and_{dynamic}$ p_2' can be represented as $[assertive, s, h, n, ok](p_1)$ & $[assertive, s, h, n+1, ok](p_2)$. Then, we obtain $bel(s+h, t_f(n+2)) = bel(s+h, t_f(n)) \cup \{p_1(*_s/s, *_h/h, *_t/t_f(n))\} \cup \{p_2(*_s/s, *_h/h, *_t/t_f(n+1))\}$. In conclusion, the static conjunction is commutative, while the dynamic one is not.

N. Asher pointed out that speech acts, such as directives and questions, embed under some of natural language sentential connectives ([1]: 211). To examine Asher's claim, let us consider some of his examples:

(13a) Whoever stole this television bring it back.

(13b) Nobody move a muscle.

(13c) Get out of here or I'll call the police.

(13d) Go to the office and there you'll find the files I told you about.

DNL provides a straightforward analysis of these speech acts. Let $stole(x, *_h, *_t) := (television(this_1) \wedge atomic\text{-}part_{human}(x, *_h) \wedge \exists t(steal(x, this_1, t) \wedge t < *_t))$ and $bring(x, *_t) := (it_1 = this_1 \wedge \exists t \ (bring\text{-}back(x, it_1, t) \wedge *_t \leq t))$. Then, (13a) can be translated as $[directive, s, G, n, ok]$ $(\forall x \ (stole(x, *_h, *_t) \rightarrow bring(x, *_t)))$. When this normative requirement is consistent with the previous normative state, according to (9b) and (10c), we obtain : $\mathbf{O}_{ns(s+G, t_f(n+1))}$ $(\forall x$ $(stole(x, G, t_f(n)) \rightarrow bring(x, t_f(n))))$. Now, suppose that John is the person who stole the television. Then, according to (2e1), when John believes that he believes in $t_f(n)$ that he stole the television, he is obligated to bring it back, unless he has already done it. Formally:

$\mathbf{B}_{ns(John, t_f(n+1))}$ $stole(John, G, t_f(n))$ &

not $\mathbf{B}_{ns(John, t_f(n+1))}$ $bring(John, t_f(n)) \Rightarrow$

$\mathbf{O}_{ns(John, t_f(n+1))}$ $bring(John, t_f(n))$.

---

[16] The AGM theory is an established formal framework for belief revision [4]. However, the revision of normative systems is quite complex and difficult to deal with the AGM theory.

(13b) can be analyzed in the same way as (13a). Let $move(x, *_t, t_1) :=$ $(move\text{-}muscle(x, t_1) \wedge *_t < t_1)$, where t_1 is a future reference time. Then, (13b) can be interpreted as $[directive, s, G, n, ok]$ $(\forall x \, (atomic\text{-}part_{human}(x, *_h) \rightarrow \neg move(x, *_t, t_1)))$. When this normative requirement is consistent with the previous normative state, according to (9b) and (10c), we obtain : $\mathbf{O}_{ns(s+G, t_f(n+1))}$ $\forall x \, (atomic\text{-}part_{human}(x, G) \rightarrow \neg move(x, t_f(n), t_1))$. Thus, because of (2e1), for any person a who knows that he himself is a member of G: $\mathbf{O}_{ns(a, t_f(n+1))}$ $\neg move(a, t_f(n), t_1)$, which is equivalent to $\mathbf{F}_{ns(a, t_f(n+1))} \, move(a, t_f(n), t_1)$.

According to Searle's taxonomy of illocutionary forces, (13c) should be interpreted as a disjunction of a directive and a commissive speech act. However, as we saw in (10c) and (10d), we can interpret the both speech acts as two kinds of obligation update: $[obligation\text{-}update, \, s, h, n, ok]$ $(\exists t \, (get\text{-}out(*_h, here(*_s, *_t), t), t)$ $\wedge *_t \leq t) \vee \exists t \, (call\text{-}police(*_s, t) \wedge *_t \leq t))$. In this formula, $here(*_s, *_t)$ refers to the place in which the speaker is located in $*_t$. When this normative requirement is consistent with the previous normative state, we obtain: $\mathbf{O}_{ns(s+h, \, t_f(n+1))}$ $(\exists t \, (get\text{-}out(h, here(s, t_f(n)), t) \wedge t_f(n) \leq t) \vee \exists t \, (call\text{-}police(s, t) \wedge t_f(n) \leq t))$. Based on (2d4), we can show that this formula implies a conditional meaning: 'If you do not get out of here, I'll call the police'. To show this, suppose that both s and h recognize that h does not get out of here: $\mathbf{B}_{ns(s+h, t_f(n+1))} \, \neg \exists t$ $(get\text{-}out(h, here(s, t_f(n)), t) \wedge t_f(n) \leq t))$. Then, because of (2d4), we obtain $\mathbf{O}_{ns(s+h, t_f(n+1))} \, \exists t \, (call\text{-}police(s, t) \wedge t_f(n) \leq t)$, which means that the speaker is self-obligated to call the police. In conclusion, if both s and h know that h will not go out, then s is self-obligated to call the police.

In all of these three examples, I have shown that each of them can be interpreted as a *single speech act with a complex content*. Asher claims to interpret them as embedded speech acts. For example, he interprets (13b) in the following manner ([1]: 214): *for all $x \in G$, Imperative $(\phi(x))$*. Contrarily, we interpreted (13b) as $[directive, \, s, G, n, ok]$ $(\forall x (x \in G \rightarrow \phi(x)))$, and we concluded from this formal representation that *every a in G is obligated to perform an action such that $\phi(a)$*. These results suggest that the interpretation of speech acts is quite complex, because we have to consider the effect of normative inferences.

We interpret (13d) as a combination of a directive and an assertive speech act. To show this, we introduce two abbreviations: $go\text{-}office(*_h, *_t) := (office(the_1)$ $\wedge \exists t \, (go\text{-}to(*_h, the_1, t) \wedge *_t < t))$ and $find\text{-}file(*_h, *_s, *_t) := (there_1 = the_1 \wedge$ $file(the_2) \wedge \exists t \, (tell\text{-}about(*_s, *_h, the_2, t) \wedge t < *_t) \wedge \exists t \, (find(*_h, the_2, t) \wedge *_t < t))$. Now, we can interpret (13d) as follows: $[directive, s, h, n, ok]$ $(go\text{-}office(*_h, *_t))$ & $[assertive, s, h, n+1, ok]$ $(in(*_h, there_1, *_t) \rightarrow find\text{-}file(*_h, *_s, *_t))$. Thus, we obtain: $\mathbf{O}_{ns(s+h, t_f(n+1))} \, go\text{-}office(h, t_f(n))$ & $\mathbf{B}_{ns(s+h, t_f(n+2))} \, (in(h, there_1, t_f(n+1)) \rightarrow find\text{-}file(h, s, t_f(n+1)))$. This means that we read (13b) as an abbreviation of the following sentence: 'Go to the office! And if you are there, you'll find the files I told you about.'

7 DNL-Analysis of Speech Acts: Description of a Dialog

Our interpretation of speech acts can be applied to an analysis of dialogs. To demonstrate this, let us consider the following conversation.

A says, 'Give me the book over there.'
   B picks up a book and shows it to A. B asks, 'This one?'
A answers, 'No, it isn't. The book behind it.'
   B picks up another book and shows it to A. B asks, 'This one?'
A answers, 'Yes, that one.'
   B brings this book to A.

This conversational scene contains different types of sentences, such as *declaratives*, *imperatives*, and *interrogatives*. It also contains actions and their observations. Collective observations play an important role for our representation of the scene, because they bring change of belief states of the communication partners.

We assume, at first, the following meaning presupposition for a linguistic community LC that includes A and B as its members.

Elementary Theory for $LC : ET^{LC} = \{(4c), (7), (14)\}$.
(14) $\forall x \forall y \forall z \forall n (bring(x, y, z, t(n)) \rightarrow give(x, y, z, t_f(n)))$.

We use Skolem constant symbols d_k, the_k, $this_k$, and $that_k$ to express anaphoric relations. For the sake of simplicity, we consider here only a case in which *pre+sin-conditions* of all intended speech acts in the conversation are satisfied. Furthermore, we assume: $ET^{LC} \subseteq bel(A + B, t_f(0))$.

Here, we interpret imperative sentence 'Give me the book over there' as a conditional obligation 'If the book exists over there, then you ought to bring it to me'. This conditional obligation can be performed, only if the hearer understands which book the speaker means. This is the reason why the hearer asks several questions in order to identify which book the speaker means. TS in the left top cell in Table 2 expresses the time stamp. When an action of the participants takes place, the time stamp n is replaced by $n + 1$.

In $t_f(1)$, A orders B to bring him book d_1 that is located in some place far from A but in the vicinity of B. Through this order of A, B comes to be obligated to bring book d_1 to A. However, in $t_f(1)$, B is not sure which book is meant. So, B makes some trials of identifying the book which A actually meant. Finally in $t_f(5)$, B realizes which book he ought to bring to A. Thus, immediately after $t_f(5)$, B brings the book to A, so that B fulfills the original requirement of A.

To describe the development of a conversation, we introduce a *representation structure for normative systems* (RSNS). A RSNS is a n-tuple composed of the sets in form $bel(X, t_f(n))$, and $ob(X, t_f(n))$. Here, we take $\langle ob(A + B, t_f(1)), bel(A + B, t_f(6)), ob(A, t_f(4)) \rangle$ as a RSNS of the above conversation. We can directly obtain the content of this RSNS by applying local update rules, namely (9a), (9b), (10a) $\sim$ (10d), and (11a) $\sim$ (11c), to the DNL-representation-schemata described in the right column of Table 2.

Now, you can easily prove: $\mathbf{B}_{ns(B,t_f(5))}(book(the_1) \wedge over\text{-}there(the_1, A, t_f(1)) \wedge this_3 = the_1)$. Because of the description of this situation, we may assume: *not* $\mathbf{B}_{ns(B,t_f(5))}\exists t(give(B, A, the_1, t) \wedge t_f(1) \leq t \leq t_f(5))$. Thus, $\mathbf{O}_{ns(B,t_f(5))}\exists t(give(B, A, this_3, t) \wedge t_f(1) \leq t)$. So, B tries to fulfill this obligation and

Table 2. DNL-analysis of a conversation

TS	Conversation	DNL-Analysis of speech acts and observations
1	A : 'Give me the book over there'	$[directive, A, B, 1, ok]$ $(book(the_1)$ $\wedge$ $over\text{-}there(the_1, *_s, *_t)) \rightarrow$ $\exists t(give(*_h, *_s, the_1, t) \wedge *_t \le t))$ $[presupposition, A + B, 1]$ $(book(the_1)$ $\wedge$ $over\text{-}there(the_1, *_s, *_t))$
2	$(B$ picks up a book and shows it to A.) B: 'This one?'	$[observation, A + B, 2]$ $(pick\text{-}up$ $(B, d_1, *_t)$ $\wedge$ $book(d_1))$ $[presupposition, A + B, 2]$ $(this_1 = d_1)$ $[interrogative\text{-}yn, B, A, 2, ok]$ $(this_1 = the_1)$
3	A : 'No, it isn't. The book behind it'	$[assertive, A, B, 3, ok]$ $(\neg\, this_1 = the_1)$ $[assertive, A, B, 3, ok]$ $(book(the_2)$ $\wedge$ $it_1 = this_1 \wedge behind(the_2, it_1, *_t))$
4	$(B$ picks up another book and shows it to $A)$ B: 'This one?'	$[observation, A + B, 4]$ $(pick\text{-}up$ $(B, d_2, *_t)$ $\wedge$ $book(d_2))$ $[presupposition, A + B, 4]$ $(this_2 = d_2)$ $[interrogative\text{-}yn, B, A, 4, ok]$ $(this_2 = the_1)$
5	A : 'Yes, that one'	$[presupposition, A + B, 5]$ $(that_1 = this_2)$ $[assertive, A, B, 5, ok]$ $(that_1 = the_1)$
6	B brings this book to A	$[presupposition, A + B, 6]$ $(this_3 = that_1)$ $[observation, A + B, 6]$ $(book(this_3) \wedge$ $bring(B, A, this_3, *_t))$

Table 3. RSNS for the conversation described in Table 2

$ob(A + B, t_f(1)) = \{\ldots, (book(the_1) \wedge\ over\text{-}there(the_1, A, t_f(1)))$
$\rightarrow \exists t(give(B, A, the_1, t) \wedge t_f(1) \le t)\}$

$bel(A + B, t_f(6))$	$ob(A, t_f(4))$
$\ldots$	$\ldots$
$book(the_1) \wedge over\text{-}there(the_1, A, t_f(1)),$	$answer\text{-}yes_{(A,B,t_f(2))}$
$pick\text{-}up(B, d_1, t_f(2)) \wedge book(d_1),$	$(this_1 = the_1) \vee$
$this_1 = d_1,$	$answer\text{-}no_{(A,B,t_f(2))}$
$say(A, B, \text{'No'}, t_f(3)), \neg this_1 = the_1,$	$(this_1 = the_1)$
$book(the_2) \wedge behind(the_2, it_1, t_f(3)),$	
$pick\text{-}up(B, d_2, t_f(4)) \wedge book(d_2),$	$answer\text{-}yes_{(A,B,t_f(4))}$
$this_2 = d_2, that_1 = this_2,$	$(that_1 = the_1) \vee$
$say(A, B, \text{'Yes'}, t_f(5)), that_1 = the_1,$	$answer\text{-}no_{(A,B,t_f(4))}$
$this_3 = that_1,$	$(that_1 = the_1)$
$book(this_3) \wedge bring(B, A, this_3, t_f(6))$	

brings the book to A in $t_f(6)$. Additionally, because of (2c), after the performance of this obligation, bringing the book to A ceases to be an obligation for B: $\mathbf{B}_{ns(B,t_f(6))}give(B,A,the_1,t_f(6))$ & $not\mathbf{O}_{ns(B,t(6))}\exists t(give(B,A,the_1,t) \wedge t_f(1) \leq t)$.

The construction of RSNS is cumulative, which is similar to the construction principle of *Discourse Representation Structure* (DRS) [6]. Thus, a RSNS can be considered as a background context for interpretation of new speech acts and new observations.

The original speech act theory of Austin and Searle analyzed utterances of simple sentences. In this section, we analyzed speech acts in context of a conversation. Speech acts are actions among many other kinds of actions and these actions and obligations change beliefs and normative states of the communication partners. As [19] points out, these kinds of social interactions can be interpreted as a cooperative game for achieving a collective goal. In our conversation example, getting a book that A wants to have is a shared goal for A and B. They try to find the most efficient way to achieve this goal.

8   Concluding Remarks

Recently, the dynamic epistemic logic (DEL) has been established as a framework for logical description of social interactions [2]. DNL can be considered as an alternative framework for the same purpose.[17] DNL can explicitly express conditions for social behaviors and describe interactions between social actions and normative inferences in detail. In this paper, we have shown how to describe and analyze a conversational development within DNL. As [19] suggests, DNL can be also used to describe games and simple language games described in [28].[18]

References

1. Asher, N.: Dynamic discourse semantics for embedded speech acts. In: Tsohatzidis, S.L. (ed.) John Searle's Philosophy of Language: Force, Meaning and Mind, pp. 211–243. Cambridge University Press, Cambridge (2007)
2. van Benthem, J.: Logical Dynamics of Information and Interaction. Cambridge University Press, Cambridge (2011)
3. Casati, R., Varzi, A.C.: Parts and Places: The Structures of Spatial Representation. MIT Press, Cambridge (MA) (1999)

---

[17] Discussions in [2] are restricted on various kinds of extension of proposi-tional modal logics, while DNL is a framework based on FOL.

[18] This research was supported by Global COE Program Center of Human-Friendly Robotics Based on Cognitive Neuroscience of the Ministry of Education, Culture, Sports, Science and Technology, Japan, and by Grant-in-for Scientific Research, Scientific Research C (24520014): The Construction of Philosophy of Science based on the Theory of Multiple Languages. Finally, I would like to thank two reviewers for useful comments.

4. Gärdenfors, P.: Knowledge in Flux: Modelling the Dynamics of Epistemic States. MIT Press, Cambridge (MA) (1988)
5. Harrah, D.: The logic of questions. In: Gabay, D., Guenthner, F. (eds.) Handbook of Philosophical Logic, vol. 8, pp. 1–60. Kluwer Academic, Dordrecht (2002)
6. Kamp, H., Reyle, U.: From Discourse to Logic. Kluwer Academic, Dordrecht (1993)
7. Kaplan, D.: Demonstratives. In: Almog, J., Perry, J., Wettstein, H. (eds.) Themes from Kaplan, pp. 481–563. Oxford University Press, Oxford (1989)
8. Nakayama, Y.: Four-dimenaional extensional mereology with sortal predicates. In: Meixner, U., Simons, P. (eds.) Metaphysics in the Post-Metaphysical-age: Papers of the 22nd International Wittgenstein Symposium, pp. 81–87. The Austrian Ludwig Wittgenstein Society, Kirchberg am Wechsel (1999)
9. Nakayama, Y.: Dynamic Interpretations and Interpretation Structures. In: Sakurai, A., Hasida, K., Nitta, K. (eds.) JSAI 2003. LNCS (LNAI), vol. 3609, pp. 394–404. Springer, Heidelberg (2007)
10. Nakayama, Y.: An Introduction to the Philosophy of Science: Metaphysics of Knowledge, in Japanese. Keiso shobo, Tokyo (2008)
11. Nakayama, Y.: The Construction of Contemporary Nominalism: Its Application to Philosophy of Nominalism, in Japanese. Syunjyu-sha, Tokyo (2009)
12. Nakayama, Y.: Logical framework for normative systems. In: SOCREAL 2010: Proceedings of the 2nd International Workshop on Philosophy and Ethics of Social Reality, pp. 19–24. Hokkaido University, Sapporo (2010).
13. Nakayama, Y.: Norms and Games: An Introduction to the Philosophy of Society, in Japanese. Keiso shobo, Tokyo (2011)
14. Nakayama, Y.: Philosophical problems in robotics: Contexts as mutual cognitive environments and questions about the self. Kagaku tetusaku **44**(2), 1–16 (2011). In Japanese
15. Nakayama, Y.: The Self as a Shown Object: Philosophical Investigations on the Self. Syunjyu-sha, Tokyo (2012). In Japanese
16. Nakayama, Y.: Ontology and epistemology for four-dimensional mereology. In: Greek Philosophical Society & Fisp (eds.) Abstracts of the 23rd World Congress of Philosophy, pp. 497–498. Athens (2013)
17. Nakayama, Y.: The ontological basis for the extended mind thesis. In: Moyal-Sharrock, D., Minz, V.A., Coliva, A. (eds.) Mind, Language and Action: Papers of the 36nd international Wittgenstein Symposium, vol. XXI, pp. 282–284. The Austrian LuDwig wittgenstein society, Kirchberg am Wechsel (2013)
18. Nakayama, Y.: The extended mind and the extended agent. Procedia Soc. Behav. Sci. **9**, 503–510 (2013)
19. Nakayama, Y.: Dynamic normative logic and information update. In: Yamada, T. (ed.) SOCREAL 2013: 3rd International Workshop on Philosophy and Ethics of Social Reality. Abstracts, pp. 23–27. Hokkaido University, Sapporo (2013)
20. Searle, J.R.: Speech Acts: An Essay in the Philosophy of Language. Cambridge University Press, Cambridge (1969)
21. Searle, J.R.: Expression and Meaning: Studies in the Theory of Speech Acts. Cambridge University Press, Cambridge (1979)
22. Searle, J.R.: The Construction of Social Reality. The Free Press, New York (1995)
23. Seligman, J., Liu, F., Girard, P.: Facebook and the epistemic logic of friendship. In: TARK 2013, Fourteenth conference on Theoretical Aspects of Rationality and Knowledge, pp. 229–238 (2013)
24. Sider, T.: Four-Dimensionalism: An Ontology of Persistence and Time. Oxford University Press, Oxford (2001)

25. Simons, P.M.: Parts: A Study in Ontology. Clarendon, Oxford (1987)
26. Vanderveken, D.: Meaning and Speech Acts, vol. 1, 2. Cambridge University Press, Cambridge (1991)
27. Varzi, A.C.: Mereology. In: Stanford Encyclopedia of Philosophy (2009)
28. Wittgenstein, L.: Philosophical Investigations (1953) (In German)
29. Yamada, T.: Logical dynamics of some speech acts that affect obligations and preferences. Synthese 165(2), 295–315 (2008)

Constructive Generalized Quantifiers Revisited

Ribeka Tanaka[1]([⊠]), Yuki Nakano[1], and Daisuke Bekki[1,2,3]

[1] Graduate School of Humanities and Sciences, Faculty of Science,
Ochanomizu University, 2-1-1 Ohtsuka, Bunkyo-ku, Tokyo 112–8610, Japan
{tanaka.ribeka,nakano.yuki,bekki}@is.ocha.ac.jp
[2] National Institute of Informatics, Tokyo, Japan
[3] CREST, Japan Science and Technology Agency, Tokyo, Japan

Abstract. This paper proposes a proof-theoretic definition for generalized quantifiers (GQs). Sundholm first proposed a proof-theoretic definition of GQs in the framework of constructive type theory. However, that definition is associated with three problems: the proportion problem, absence of strong interpretation and lack of definitional uniformity. This paper presents an alternative definition for "most" based on polymorphic dependent type theory and shows strong potential to serve as an alternative to the traditional model-theoretic approach.

1 Introduction

Since Sundholm [19], the type-theoretic approach in natural language semantics has progressed considerably (Ranta [17], Dávila-Pérez [5]). To examine whether the type-theoretic approach can serve as an alternative for the traditional model-theoretic approach, one of the important issues that must be addressed is its empirical coverage. From this point of view, the type-theoretic and constructive formalization for generalized quantifiers (GQs) in Sundholm [18] is of interest. By making use of the advantages of type theory, Sundholm presented a definition of GQs which avoids the proportion problem while retaining internal dynamics, thus providing a solution to a common problem associated with *donkey sentences* (Geach [8]). However, three problems remain in his analysis. In this paper, we resolve these problems and present a revised definition while maintaining the proof-theoretic properties of the original definition.

2 Constructive GQs in Sundholm [18]

It is known that naive formalization for GQs results in the proportion problem (Heim [9], Kadmon [12], Kanazawa [13]). We quote from Kanazawa [13]:[1]

> Of the two types of donkey sentences, I confine myself to donkey sentences with determiners and relative clauses, which I will simply call 'donkey sentences' in what follows. (1)–(3) are standard examples involving farmers

---

[1] Linguistics and Philosophy 17 (1994), pp. 109–111.

© Springer International Publishing Switzerland 2014
Y. Nakano et al. (Eds.): JSAI-isAI 2013, LNAI 8417, pp. 115–124, 2014.
DOI: 10.1007/978-3-319-10061-6_8

and donkeys, which share the general form (4) (Q stands for a quantificational determiner):

(1) Every farmer who owns a donkey beats it.
(2) No farmer who owns a donkey beats it.
(3) Most farmers who own a donkey beat it.
(4) Q farmer who owns a donkey beats it.

For convenience, let us use (4) as the general form of donkey sentences
with relative clauses in the following preliminary discussion. Of course,
the choice of specific lexical items farmer, own, donkey, and beat is not
important.[...]In the literature, one can recognize at least four proposals
as to the interpretation of sentences of the form (4):[...]

(6) Pair quantification reading:
$Q\{\langle x, y\rangle | farmer(x) \wedge donkey(y) \approx own(x, y)\}\{\langle x, y\rangle | beat(x, y)\}$.

[...]The pair quantification reading (6) is the interpretation assigned to
(4) by the classical DRT/file change semantics (Kamp 1981, Heim 1982).
Here, the quantifier Q counts farmer-donkey pairs: (6) is true if and only
if Q holds of the set consisting of farmer-donkey pairs which stand in the
owning relation and the set consisting of pairs which stand in the beating
relation. This seems to give the correct truth conditions for sentences like
(1) and (2), but, as is well-known, it does not work with (3). Consider
the following situation:

(9) $[farmer] = \{f_1, f_2\}$
$[donkey] = \{d_1, d_2, d_3\}$
$[own] = \{(f_1, d_1), (f_1, d_2), (f_2, d_3)\}$
$[beat] = \{(f_1, d_1), (f_1, d_2)\}$

In this model, there are three farmer-donkey pairs that stand in the owning relation, and two of them, the majority, stand in the beating relation.
However, (3) is intuitively false in such a situation. In the literature, this
is known as the 'proportion problem' for the classical DRT/file change
semantics account of quantification and indefinites.

Sundholm [18] defined the GQ "most" in terms of constructive type theory
(Martin-Löf [16]) and proposed a way to avoid the proportion problem in that
setting.

Most men who own a donkey beat it.

This sentence is analyzed as "most donkey owners beat their donkeys" and the
restrictor is of the form "set of Bs such that C". Thus, in the following definition
by Sundholm [18], restrictor A is assumed to be a set of the form $\Sigma(B, C)$.
$Most(A, \phi)$, which is a semantic representation for "Most A are ϕ", is defined as
follows:

$A : \texttt{set} \quad \phi : A \to \texttt{Prop} \quad a : \text{A-finite}(A)$

$$\text{Most}(A, \phi) \overset{\text{def}}{\equiv} (\Sigma k : N)(k \geq [\pi_1(a)/2] + 1 \ \& \ (\Sigma f : M(k) \to A)$$
$$(f \text{ is an injection} \ \& \ (\Pi y : M(k))\phi(\pi_1(fy)))) : \texttt{Prop}$$

Here, we accommodate parts of this notation into our definition. Being conformable to a notation in constructive type theory, Σ stands for an existential quantifier and a conjunction, while Π stands for a universal quantifier and an entailment.

Constant a is a proof object for A-finite(A), that A's first projection inhabits at most a finite number of elements, which is defined as follows. Operator eq means that the elements for the second and third arguments are equal.

$B : \texttt{type} \quad C : \texttt{type} \quad (\Sigma x : B)C : \texttt{set}$

$$\text{A-finite}((\Sigma x : B)C) \overset{\text{def}}{\equiv}$$
$$(\Sigma k' : N)(\Sigma f : M(k') \to (\Sigma x : B)C)(f \text{ is an A-bijection}) : \texttt{type}$$

$B : \texttt{type} \quad C : \texttt{type} \quad D : \texttt{type} \quad f : D \to (\Sigma x : B)C$

$$f \text{ is an A-injection} \overset{\text{def}}{\equiv}$$
$$(\Pi y : D)(\Pi z : D)(eq(B, \pi_1(fy), \pi_1(fz)) \to eq(D, y, z)) : \texttt{type}$$

Since a is a proof of A-finite(A), its first projection $\pi_1(a)$ is the number of elements. Note that A-bijections are bijections which use A-injections in place of general injections[2]. In the codomain of A-injections, only the first projection of elements matters, namely, two elements of A whose first projections match are identified. Using this A-injection in A-finite(A) means that $\pi_1(a)$ is the number of pairs whose first projections differ, rather than simply the cardinality of A. In the above case of the donkey sentence, $\pi_1(a)$ is the number of farmers rather than the number of farmer-donkey pairs. The term $[m]$ is the largest natural number smaller than m. Thus, $[\pi_1(a)/2] + 1$ represents the size of the least possible majority in A and k is a number which can be interpreted as "most".

$M(k)$ is a set with cardinality k. Injection is defined as follows, and f maps k elements in $M(k)$ to elements of A which satisfy certain conditions. This intuitively means counting the number of elements in A and checking whether there are at least k elements. According to the last condition for f in the representation, if f maps one element in $M(k)$ to element $(p, q) \in A$, then the first projection of the pair, p, must satisfy $\phi(p)$. In the above case of the donkey sentence, a farmer-donkey pair (p, q) in A must satisfy the relation $\phi = $"$p$ beats

---

[2] The fact that, in Sundholm [18], A-injections are defined as in the Appendix. Since the type for f is obviously unsuitable for Sundholm's purposes, we understand that he intended to define it as we show in this section. What is more, Sundholm mentioned that injections should be simply replaced with A-injections, and did not account for surjections. Therefore, we also define what we call A-surjections, whose domain is of the form $(\Sigma x : B)C$, and address that point. All of these definitions are available in the Appendix.

donkey q", and thus f counts the number of donkey-owning farmers who beat their donkey.

$$f : D \to A$$

$$f \text{ is an injection} \overset{\text{def}}{\equiv} (\Pi x : D)(\Pi y : D)(eq(A, fx, fy) \to eq(D, x, y)) : \texttt{Prop}$$

Sundholm also showed that other GQs, such as "finitely many" and "at least as many ... as ...", can be defined in a similar way. However, three problems remain with this analysis.

Problem 1. The above definition cannot avoid the proportion problem. Pairs in A are identified with their first projection in A-finite(A), but f is defined simply as an injection in the definition of "most", which simply counts farmer-donkey pairs and leads to wrong prediction. In case there is a farmer who owns several donkeys and beats all of them, f may map $M(k)$'s elements to all possible pairs, which seems to lead to a situation which is the same as the proportion problem.

Problem 2. The strong interpretation is not taken into account. Although it is arguable whether the interpretation of a sentence with a quantificational determiner has a weak reading and a strong reading, we follow the analysis in Chierchia [4], Kanazawa [13] and Krifka [15]. According to them, "most farmers who own a donkey beat it" has two readings. In the weak reading, it means that "most farmers who own a donkey beat at least one donkey they own", while the strong reading leads to the analysis that "most farmers who own a donkey beat all donkeys they own". The definition by Sundholm [18] accounts for the weak reading. When f maps elements in $M(k)$ to those in A, such farmer-donkey pairs surely satisfy a beating relation, and this ensures that those farmers beat at least one donkey they own. To account for the strong reading, we additionally have to check whether the farmers have other donkeys and whether they beat all of them.

Problem 3. The definition is not uniform. As Sundholm himself pointed out, the definition is not uniform. The above definition of "most" only applies to the case where A is of the form $(\Sigma x : B)C$. When A is of the form $Farmer(x)$ or contains another relative clause, we require a separate definition for each case. This redundancy is not empirically motivated, as criticized in Fox and Lappin [7].

3   Proposal

3.1   A Revised Definition

We define the quantifier "most" as follows. The definition is based on polymorphic dependent type theory (PDTT) (Jacobs [11]), which is well-known as an instance of a λ-cube (Barendregt [2]). Definitions for $M(k)$ and [ ] are provided in PDTT in the same way as in Sundholm [18].

$A : (\Pi x : \mathtt{Entity})(\Pi \delta : \mathtt{type})(\Pi c : \delta)\mathtt{type}$
$B : (\Pi x : \mathtt{Entity})(\Pi \delta : \mathtt{type})(\Pi c : \delta)\mathtt{type}$

$Most\ A\ B \overset{\mathrm{def}}{\equiv}$

$(\lambda \delta : \mathtt{type})(\lambda c : \delta)(\Sigma k : N)(k \geq [\frac{\pi_1(sel_{A-\mathrm{finite}((\Sigma x : \mathtt{Entity})Ax\delta c)}(c))}{2}] + 1$
$\wedge (\Sigma f : M(k) \to (\Sigma x : \mathtt{Entity})Ax\delta c)$
$(f$ is an A-injection
$\wedge(\Pi y : M(k))(B(\pi_1(fy))(\delta \wedge (\Sigma x : \mathtt{Entity})Ax\delta c)(c, fy))))$
$: (\Pi \delta : \mathtt{type})(\Pi c : \delta)\mathtt{type}$

The first argument for type of A, x, is an entity defined in the same manner as above. The third argument, c, is a variable for context and represents previous sentences. The type of c is passed together as the second argument, δ. $(\Sigma x : \mathtt{Entity})Ax\delta c$ is a representation for the restrictor, which is "a set of entities such that $Ax\delta c$". Although the condition for the natural number k seems somewhat complex, it is in fact defined as in Sundholm [18], and the proof object for A-finite(A) provides the cardinality of the restrictor, as we explain below.

As mentioned above, f maps $M(k)$ to $(\Sigma x : \mathtt{Entity})Ax\delta c$, which intuitively means counting the number of elements in $(\Sigma x : \mathtt{Entity})Ax\delta c$. Here, the last condition for f is defined in the same manner as in Sundholm [18] except for the difference in notation.

3.2 Proportion Problem

Next, we show how the three remaining problems are solved in the above definition. In fact, the solution to the first problem is rather simple. We use A-injection in place of injection in the condition for f. Thus, pairs whose first projections are the same are identified when f counts the elements in the set $(\Sigma x : \mathtt{Entity})Ax\delta c$, which entirely avoids the proportion problem.

3.3 Strong Reading

The representation for strong reading is given by adding simple conditions for f as in the underlined part.

$A : (\Pi x : \mathtt{Entity})(\Pi \delta : \mathtt{type})(\Pi c : \delta)\mathtt{type}$
$B : (\Pi x : \mathtt{Entity})(\Pi \delta : \mathtt{type})(\Pi c : \delta)\mathtt{type}$

$Most\ A\ B \overset{\mathrm{def}}{\equiv}$

$(\lambda \delta : \mathtt{type})(\lambda c : \delta)(\Sigma k : N)(k \geq [\frac{\pi_1(sel_{A-\mathrm{finite}((\Sigma x : \mathtt{Entity})Ax\delta c)}(c))}{2}] + 1$
$\wedge (\Sigma f : M(k) \to (\Sigma x : \mathtt{Entity})Ax\delta c)$
$(f$ is an A-injection
$\wedge(\Pi y : M(k))(B(\pi_1(fy))(\delta \wedge (\Sigma x : \mathtt{Entity})Ax\delta c)(c, fy)$
$\wedge(\Pi z : (\Sigma x : \mathtt{Entity})Ax\delta c)$
$(eq(\mathtt{Entity}, \pi_1 z, \pi_1(f(y))) \to B(\pi_1 z)(\delta \wedge (\Sigma x : \mathtt{Entity})Ax\delta c)(c, z)))))$

$: (\Pi \delta : \mathtt{type})(\Pi c : \delta)\mathtt{type}$

Consider the sentence "Most farmers who own a donkey beat it". Its strong interpretation is that most farmers who own at least one donkey beat every donkey they own. Since it is already known that farmer $\pi_1(fy)$ beats one of the donkeys he owns, the added condition checks whether he beats all other donkeys he owns. For all z in $(\Sigma x : \mathtt{Entity})Ax\delta c$, if its first projection, farmer $\pi_1 z$, is identical to farmer $\pi_1(fy)$ then farmer $\pi_1 z$ should beat the donkey he owns, too, to make for a successful strong reading. Thus, the underlined part enables f to map $M(k)$ to the elements in $(\Sigma x : \mathtt{Entity})Ax\delta c$, which are suitable for strong reading.

3.4   Uniformity

Regarding the third problem, a restrictor A has a type $(\Pi x : \mathtt{Entity})(\Pi\delta : \mathtt{type})(\Pi c : \delta)\mathtt{type}$ in our treatment. Given the context and its type, the restrictor is of the form $(\lambda x : \mathtt{Entity})\phi$. For the noun phrase "farmers", for example, its semantic representation is $(\lambda x : \mathtt{Entity})Farmer(x)$, while $(\lambda x : \mathtt{Entity})(Farmer(x) \wedge (\Sigma y : \mathtt{Entity})(Donkey(y) \wedge own(x,y)))$ for "farmers who own a donkey". Since A is of the form $(\lambda x : \mathtt{Entity})\phi$ and $(\Sigma x : \mathtt{Entity})Ax\delta c$ is calculated in the definition, noun phrases are always treated as a form $(\Sigma x : \mathtt{Entity})\phi$. This always allows us to use A-injection, and thus we have definitional uniformity.

   Now, we find that the abovementioned three problems remaining in Sundholm [18] are eliminated in our formalization.

3.5   Existential Presupposition

In addition, we use the selection function (Bekki [3]) in the condition for k. The selection function consists of projections. The index of the selection function represents conditions which should be satisfied by the proof term extracted from a previous context c by the selection function, and this represents presupposition. The selection function here has A-finite$((\Sigma x : \mathtt{Entity})Ax\delta c)$ as its index, and this presupposes that the set of $(\Sigma x : \mathtt{Entity})Ax\delta c$, the restrictor, is a finite set. It is known that strong quantifiers such as "most" exhibit existential presupposition (Heim [10]), and the sentence "most farmers who own a donkey beat it" presupposes that there is at least one farmer who owns a donkey. This presupposition can be predicted by replacing the following with the condition for k and specifying that the cardinality is not less than 1. The number provided by $|(\Sigma x : \mathtt{Entity})Ax\delta c| \geq 1$ is 1 less than the cardinality of the set rather than the actual cardinality, and thus we need to add "1" to the numerator.

$$k \geq \lceil \frac{\pi_1(sel_{|(\Sigma x:\mathtt{Entity})Ax\delta c| \geq 1}(c)) + 1}{2} \rceil + 1$$

The representation for the presuppositions enables us to explain why the following discourse is not preferred. We skip the details in this paper.

   There are no unicorns. Most unicorns have special powers.

3.6   Internal Dynamics

Internal dynamics is also considered in our definition. Variable c, the third argument of A, is a variable for context and is passed through the sentence. With c and the selection function mentioned above, the pronouns in donkey sentences can be interpreted properly by choosing the appropriate antecedent from previous context. See Bekki [3] for details about context passing.

4   Previous Approaches

Fox and Lappin [7], Kievit [14] and Fernando [6] also considered GQs in type-theoretic frameworks.

Fox and Lappin [7] define "most" in the framework of property theory with Curry typing. The proportion problem does not arise in their definition.

Most farmers who own a donkey beat it.

$$|\{x \in B.^{\mathsf{T}}\mathsf{man}'(x) \wedge (|\{y \in B.^{\mathsf{T}}\mathsf{own}'(x,y) \wedge {}^{\mathsf{T}}\mathsf{donkey}'(y)\}|_B >_{Num} 0)$$
$$\wedge \forall z(z \in A \rightarrow \neg^{\mathsf{T}}\mathsf{beat}')(\mathsf{x},\mathsf{z}))\}|_B$$
$$<_{Num}$$
$$|\{x \in B.^{\mathsf{T}}\mathsf{man}'(x) \wedge (|\{y \in B.^{\mathsf{T}}\mathsf{own}'(x,y) \wedge {}^{\mathsf{T}}\mathsf{donkey}'(y)\}|_B >_{Num} 0)$$
$$\wedge \forall z(z \in A \rightarrow {}^{\mathsf{T}}\mathsf{beat}')(\mathsf{x},\mathsf{z}))\}|_B$$
$$where\ A = \{y \in B.^{\mathsf{T}}\mathsf{own}'(x,y) \wedge {}^{\mathsf{T}}\mathsf{donkey}'(y)\}$$

The antecedent for "it" is provided by A, whose contents are copied from the previous representation. However, this formalization is associated with certain problems. For example, it cannot provide a proper antecedent for the pronoun "it" in the following sentence.

Every man who owns a cat which has [a nice tail]$_1$ loves it$_1$.

$$|\{x \in B.^{\mathsf{T}}\mathsf{man}'(x) \wedge (|y \in B.^{\mathsf{T}}\mathsf{cat}'(y) \wedge (|z \in B.^{\mathsf{T}}\mathsf{tail}'(z) \wedge {}^{\mathsf{T}}\mathsf{has}'(y,z)|_B >_{Num} 0)$$
$$\wedge^{\mathsf{T}}\mathsf{own}'(x,y)|_B >_{Num} 0) \wedge \forall u(u \in A \rightarrow {}^{\mathsf{T}}\mathsf{love}'(x,u))\}|_B$$

What we can copy from the previous representation for the place of A is as shown below, where variable y is unbound.

$$A = \{z \in B.^{\mathsf{T}}\mathsf{tail}'(z) \wedge {}^{\mathsf{T}}\mathsf{has}'(y,z)\}$$

What we see here is unboundness of variables, which can also be seen in traditional donkey sentences. Many approaches have been developed in the past to account for this type of anaphora. Sundholm [18] and Ranta [17] also proposed a proper representation of donkey anaphora by making use of the advantages of type theory. Therefore, it might be necessary to resolve the proportion problem while providing correct analysis for anaphora at the same time.

We show the representation in our analysis for reference.

$$\Pi u : ((\Sigma x : \texttt{Entity})(Man(x) \wedge (\Sigma y : \texttt{Entity})(Cat(y)$$
$$\wedge(\Sigma z : \texttt{Entity})(Tail(z) \wedge has(y,z)) \wedge own(x,y)))(loves\ it)$$

It is not difficult to extract the entity "a nice tail" by combining projections.

A similar discussion is provided in Kievit [14], who showed that his formalization can derive "More than two men run" from "Most men run. There are five men". However, his definition does not support internal dynamics. Although he provided successful treatment of cardinality, we cannot consider that he made any substantial progress.

Fernando [6] defined *left-projection* for types, which allows us to consider only the first projection of pairs. To implement this, a form of sub-typing is used in his analysis. It is known, however, that introducing sub-typing generally makes systems unnecessarily complex and eliminates determinacy. Therefore, it is better to adopt other systems without sub-typing if they can achieve similar performance. In addition, the meaning of GQs is defined only in a model-theoretic way. Since such model-theoretic analysis for GQs has been conducted in previous approaches, the next step would be to show how the corresponding complete proof-theoretic definitions can be constructed in type theory.

The following table shows a comparison of different approaches, where "*" stands for "not mentioned" and "X" stands for "fail".

	Sundholm (1989)	Fox and Lappin (2005)	Fernando (2001)	Kievit (1995)
Cardinality	ok	ok	ok	ok
Proportion problem	X	ok	ok	*
Weak/strong reading	*	ok	ok	*
Existential presupposition	*	*	*	*
Internal dynamics	ok	X	ok	X

5   Conclusion

Sundholm [18] attempted formalization of GQs which avoids the proportion problem while retaining internal dynamics. In this paper, we resolved the problems remaining in Sundholm [18] and presented proof-theoretic definitions for GQs. We made use of advantages of proof theory and showed that other linguistic properties of GQs can be proven as well in this setting. Our formalization has a broader empirical coverage of natural language than that in Sundholm [18] and shows strong potential to serve as an alternative to the model-theoretic approach.

Acknowledgments. We would like to express our deepest gratitude to the anonymous reviewers of LENLS10, whose comments and suggestions were of tremendous value. We would also like to thank Katsuhiko Yabushita, Matthijs Westera and Yuyu So, who provided invaluable comments and kind encouragement. Daisuke Bekki is partially supported by JST, CREST.

Appendix

Definition 1 (Construction of the Natural Numbers). *Sundholm [18] presented the construction of natural numbers with the M_r sequence proposed by Aczel [1]. First put*

$$f(a) = R(a, eq(N,0,0), (x,y)(y + eq(N,0,0))) : U.$$
$$(a : N, U \text{ is the universe of small types, } + \text{ is a disjoint sum.})$$

and

$$f(0) = eq(N,0,0) : U$$
$$f(s(a)) = f(a) + eq(N,0,0) : U \ (a : N).$$

Finally, put

$$M(a) = R(a, \bot, (x,y)f(x)) : U \ (a : N)$$

then

$$M(0) = \bot : U$$
$$M(1) = eq(N,0,0) : U$$
$$M(s(s(a))) = M(s(a)) + eq(N,0,0) : U \ (a : N).$$

Definition 2 (Primitive Recursive Functions).

$$\begin{array}{ll} sg(0) & = 1 \\ sg(s(a)) & = 0 \end{array} \qquad \left\{ \begin{array}{l} rem(0/2) = 0 \\ rem(s(a)/2) = sg(rem(a/2)) \end{array} \right.$$

$$\begin{array}{l} [0/2] = 0 \\ [s(a)/2] = [a/2] + rem(a/2) \end{array}$$

Definition 3 ($|\_|$).

$$\frac{A : \mathsf{type} \quad B : \mathsf{type}}{|(\Sigma x : A)B| \stackrel{\mathrm{def}}{\equiv} \pi_1(A\text{--finite}((\Sigma x : A)B)) : N}$$

Definition 4 ($\geq$).

$$\frac{m : N \quad n : N}{m \geq n \stackrel{\mathrm{def}}{\equiv} (\Sigma k : N)eq(N, m, n+k) : \mathsf{type}}$$

Definition 5 (Surjection).

$$\frac{f : A \to B}{f \text{ is a surjection} \stackrel{\mathrm{def}}{\equiv} (\Pi y : A)(\Sigma x : A)eq(B, fx, y) : \mathsf{Prop}}$$

Definition 6 (Bijection).

$$\frac{f : A \to B}{f \text{ is a bijection} \stackrel{\mathrm{def}}{\equiv} f \text{ is an injection \& } f \text{ is a surjection} : \mathsf{Prop}}$$

Definition 7 (Finite).

$$\frac{A : set}{Finite(A) \overset{\text{def}}{\equiv} (\Sigma k : N)(\Sigma f : M(k) \to A)(f \ is \ a \ bijection) : \texttt{Prop}}$$

References

1. Aczel, P.: Frege structures and the notions of proposition truth and set. In: Barwise, J., Keisler, H.J., Kunen, K. (eds.) The Kleene Symposium, pp. 31–59. North-Holland Publishing Company, Amsterdam (1980)
2. Barendregt, H., et al.: Introduction to generalized type systems. J. Funct. Program. **1**(2), 125–154 (1991)
3. Bekki, D.: Dependent type semantics: an introduction. In: Christoff, Z., Galeazzi, P., Gierasimczuk, N., Marcoci, A., Smet, S. (eds.) Logic and Interactive RAtionality (LIRa) Yearbook 2012, vol. I, pp. 277–300. University of Amsterdam, The Netherlands (2014)
4. Chierchia, G.: Anaphora and dynamic binding. Linguist. Philos. **15**(2), 111–183 (1992)
5. Dávila-Pérez, R.: Semantics and Parsing in Intuitionistic Categorial Grammar. Ph.d. thesis, University of Essex (1995)
6. Fernando, T.: Conservative generalized quantifiers and presupposition. In: Proceedings of the SALT, vol. 11, pp. 172–191 (2001)
7. Fox, C., Lappin, S.: Foundations of Intensional Semantics. Wiley-Blackwell, Malden (2005)
8. Geach, P.T.: Reference and Generality: An Examination of Some Medieval and Modern Theories. Cornell University Press Ithaca, Ithaca (1962)
9. Heim, I.: E-Type pronouns and donkey anaphora. Linguist. Philos. **13**(2), 137–177 (1990)
10. Heim, I., Kratzer, A.: Semantics in Generative Grammar, vol. 13. Blackwell, Oxford (1998)
11. Jacobs, B.: Studies in Logic and the Foundations of Mathematics, vol. 141. Elsevier, Amsterdam (1999)
12. Kadmon, N.: Uniqueness. Linguist. Philos. **13**(3), 273–324 (1990)
13. Kanazawa, M.: Weak vs strong readings of donkey sentences and monotonicity inference in a dynamic setting. Linguist. Philos. **17**(2), 109–158 (1994)
14. Kievit, L.: Sets and quantification in a constructive type theoretical semantics for natural language (manuscript) (1995)
15. Krifka, M.: Pragmatic strengthening in plural predications and donkey sentences. In: Proceedings of the SALT, vol. 6, pp. 136–153 (1996)
16. Martin-Lof, P., Sambin, G.: Intuitionistic Type Theory. Bibliopolis, Naples (1984)
17. Ranta, A.: Type-Theoretical Grammar. Oxford University Press, Oxford (1995)
18. Sundholm, G.: Constructive generalized quantifiers. Synthese **79**(1), 1–12 (1989)
19. Sundholm, G.: Proof theory and meaning. In: Gabbay, D., Guenthner, F. (eds.) Handbook of Philosophical Logic, pp. 165–198. Springer, New York (1986)

Argumentative Insights from an Opinion Classification Task on a French Corpus

Marc Vincent[1]([✉]) and Grégoire Winterstein[2,3]

[1] Université Paris 5, UMR S775, Paris, France
marc.r.vincent@gmail.com
http://www.dsi.unifi.it/~vincent/
[2] Laboratoire Parole et Langage, Aix Marseille Université, Marseille, France
gregoire.winterstein@linguist.univ-paris-diderot.fr
http://gregoire.winterstein.free.fr
[3] Nanyang Technological University, Singapore, Singapore

Abstract. This work deals with sentiment analysis on a corpus of French product reviews. We first introduce the corpus and how it was built. Then we present the results of two classification tasks that aimed at automatically detecting positive, negative and neutral reviews by using various machine learning techniques. We focus on methods that make use of feature selection techniques. This is done in order to facilitate the interpretation of the models produced so as to get some insights on the relative importance of linguistic items for marking sentiment and opinion. We develop this topic by looking at the output of the selection processes on various classes of lexical items and providing an explanation of the selection in argumentative terms.

Sentiment analysis and opinion mining cover a wide range of techniques and tasks that are oriented towards the classification and extraction of the opinions and sentiments that can be found in a text (see e.g. Pang and Lee (2008) for an extensive review). Interestingly, these tasks are sufficiently different from those of information extraction that they deserve specific approaches. For example, sentiments are seldom expressed overtly in a text, and a keyword based approach for sentiment detection is not very effective, even though it yields some results for information extraction, cf. Cambria and Hussain (2012). One classical task in opinion mining is that of opinion classification. Given a text, the aim of the task is to assign it a label from a pre-determined set (e.g. *positive, negative* or *neutral/balanced*). Successful attempts usually involve the use of machine learning techniques, see e.g. Pang et al. (2002) and Pang and Lee (2008).

In this work we pursue two objectives. First, we deal with the task of opinion classification on a corpus of French texts extracted from the web (Sect. 1). We begin by presenting the results of a binary classification task which aims at setting apart positive and negative reviews. In a second experiment, the classification is ternary with the introduction a middle class of neutral (or balanced)

This research was supported in part by the Erasmus Mundus Action 2 program MULTI of the European Union, grant agreement number 2010-5094-7.

Y. Nakano et al. (Eds.): JSAI-isAI 2013, LNAI 8417, pp. 125–140, 2014.
DOI: 10.1007/978-3-319-10061-6_9

reviews. To enhance the performances of our classifiers we use dimension reduction techniques and show that they have a positive impact.

In the second part of the paper, we try to interpret the output of these selection algorithms from a linguistic point of view (Sect. 2). To carry this out, we look at the elements that "survive" the selection process, and show that there are some similarities between the elements selected in various lexical classes such as coordinating conjunctions, prepositions and adverbs.

1   Opinion Classification

1.1   Corpus

The corpus used for the opinion classification task is based on the automatic extraction of 14 000 product reviews taken from three websites that allow their users to post their opinion online. Along with the textual content of the reviews, the score attributed by the users to the product was also extracted. All three websites use a 5 point scale to measure the product quality (1 being the lowest grade and 5 the highest). The origin and number of reviews per grade is given in Table 1.

Table 1. Contents of the corpus (total number of reviews: 14 000)

Product type	Source	N. reviews (per grade)
Hotels	`tripadvisor.fr`	1000
Movies	`allocine.fr`	1000
Books	`amazon.fr`	800

Besides the contents of the reviews and the grade (or score) attributed by the author, we also extracted other information for future use (only from the Amazon and TripAdvisor reviews):

- A one sentence summary of the review (as written by the author of the review).
- A measure of usefulness of the review, indicated by the number of users who judged the review useful.

The TripAdvisor part of the corpus also offers some scores on specific attributes of the hotels reviewed such as the cleanliness of the rooms, the service etc. The complete corpus is available upon request to the authors.

For each grade, the diversity of products and authors was maximized, i.e. one given class of notation contains as many different products and authors as possible. This is to ensure the generality of the models produced by the learning algorithms.

1.2   Classification Method

We tried two different opinion classification tasks on this corpora:

1. A binary classification task to differentiate positive and negative reviews. For this task only reviews with a score of 1 (*negative*) and 5 (*positive*) were considered.
2. A ternary classification task with three possible levels: *positive* (scores 4 and 5), *negative* (scores 1 and 2) and *neutral* (score 3).

The set of features used in each task was determined in identical ways:[1]

- Each review was first lemmatized using the state of the art POS tagger and lemmatizer `MElt` by Denis and Sagot (2012).
- Only the lemmas that were successfully recognized by the tagger were used to produce a bag of words representation of the reviews.
- In order to minimize the domain sensitivity of the models produced, all items tagged as proper nouns were removed from the feature set.
- The lemmas that appeared less than 10 times were also ignored because they were deemed too specific.

1.3   Results

The binary classification task was carried out by using three different techniques:

1. *Support Vector Machines* (using SVM^{light}, Joachims (1999)).
2. Logistic regression with *elastic net* regularization (cf. Zou and Hastie (2005)).
3. *SVM* on a reduced feature set obtained with the output of the elastic net regularization.

For both tasks the performances were estimated by 10-fold cross validation. The parameters for each approach (i.e the gaussian kernel size and the c coefficient for SVM, and (α, λ) for the elastic net regularization) were optimized inside each fold by a subsequent 5-fold cross validation.

The results of each approach are given in Table 2. As can be seen, the regularization with elastic net not only greatly reduces the number of initial features (by more than half) but also helps to improve the performance of the classifiers.

Given the results of the binary task, we focused on logistic regression with elastic net regularization for the ternary classification task.

The approach we used is to learn a multinomial logistic regression model with an *elastic net* penalty, meaning that three binary classifiers were produced concurrently, each classifying one class (its associated positive class) against the other two and so that the output class probabilities sum to one. The final classifier is a combination of these three, it predicts the class which is associated

---

[1] Some approaches use the presence of negation as a feature. This was experimented with, but it did not improve the results and it added a great number of features which slowed down the learning. Therefore it was abandoned.

Table 2. Binary classification task: results

	N. features	Precision	Recall	F-value
SVM	2829	88.18 %	89.54 %	88.84
Logistic reg. + *elastic net* sel.	1219	**88.78 %**	**91.61 %**	**90.16**
SVM + *elastic net* sel.	1219	88.22 %	90.32 %	89.25

to the binary classifier that outputs the highest probability. By analogy to the *one-vs-rest* multiclass approach (cf. Bishop (2006)), we will refer to the performances obtained on each binary classification subtasks as *one-vs-rest* classifier perfomances.

Models were fitted using the `glmnet` package by Friedman et al. (2010) available for the R environment (R Development Core Team (2011)). The results are given in Table 3.

Table 3. Ternary classification task: results. Precision, recall and F1-score are macro-averaged over classes. The micro averaged F1-score is also given (F1 μ). All measures are averaged over the 10 final test folds.

	N. features	Precision	Recall	F1	F1 μ
1,2 vs. 3 vs. 4,5	2082.5	63.56	61.35	60.11	69.94
1,2 vs. *rest*	1147.3	71.52	80.57	75.77	-
3 vs. *rest*	645.2	46.80	18.39	26.35	-
4,5 vs. *rest*	1026.6	72.37	85.09	78.20	-

1.4   Discussion

The results of Table 2 show the great benefit in using feature selection techniques both for reasons of dimension reduction and for the improvement of the final performance. The results prove superior to the baselines usually reported for English (e.g. by Pang et al. (2002), who report a $F1$ score of about 83 on a similar task).

The results of the ternary task appear poorer. This is essentially due to the poor performance of the *3 vs. rest* classifier who sports a very low recall. To explain these poor performances we had a closer look at the reviews scored 3 by the users and manually re-labeled them. This manual reclassification was done on a subset of 1 667 reviews (mainly for reasons of time) and it led to the reclassification of about 24 % of the reviews. This means that on average, one review out of four that was labeled "neutral" because of its grade of 3 was manually reassigned either to the "positive" or "negative" class due to its content. The Table 4 gives the number of reviews with score 3 whose labels were manually checked, for each of part of the corpus.

Table 4. Manual classification of a subset of reviews with a score of 3

Manual tag	Allocine	Amazon	TripAdvisor	Total
Neutral	458	418	394	1270
Negative	15	36	61	112
Positive	97	135	53	285

It is worth noting that depending on the origins of the reviews, the relabeling is different. Reviews from Allocine and Amazon were mainly done from the neutral towards the positive, with much fewer towards the negative. In contrast, reviews from TripAdvisor are roughly equally divided between the positive and negative.

Decomposing the classification performances of our classifiers with respect to the source of the test data provides some further insights into the meaning of these relabeling statistics. As showed in Table 5, the three corpora are not equal in that matter, with TripAdvisor reviews being clearly better classified than reviews of the other two sources. Although the differences in lexicons used to characterize cultural products and travel accommodations may be responsible, this also correlates intriguingly well with the previous observation that while "mistaken" labels of Amazon and Allocine are biased toward the positive class, those same labels are fairly well balanced between positive and negative class for TripAdvisor. This provides further motivation to investigate the effect of potential mislabeling.

Table 5. Performances by corpus measured by the F1 score (macro averaged for the ternary classification problem). AMZ: Amazon reviews, TA: TripAdvisor reviews, AC: Allocine reviews

F1	1,2 vs. *rest*	3 vs. *rest*	45 vs. *rest*	12 vs. 3 vs. 45
AMZ	73.39	19.27	76.13	56.26
TA	83.62	35.50	84.41	67.84
AC	69.81	21.43	73.90	55.05

To assess the usefulness of our recoding, we ran a quick comparison between the results of the ternary classification task using a maximum entropy algorithm (using the megam software, Daumé III (2004)). The choice was mainly due to the fact that this algorithm is fast, even if less efficient than the other techniques we used. The results comparing the overall performance before and after the manual relabeling are given in Table 6.

Although the improvement is small, it seems that manually reclassifying the reviews with a score of 3 has a positive effect on the classification task. We therefore ran a classification similar as the previous one on the relabeled corpus. The results are however not up to the expectations.

Table 6. Ternary classification task: original vs. partially relabelled corpus (Max. Ent. algorithm)

	Precision	Recall	F1
Original corpus	64.03	48.45	55.16
Partially manually relabeled corpus	64.52	49.74	56.00

Table 7. Ternary classification task after reclassification of reviews with a score 3: results. Precision, recall and F1-score are macro-averaged over classes. The micro averaged F1-score is also given (F1 μ) All measures are averaged overed the 10 final test folds.

	N. features	Precision	Recall	F1	F1 μ
1,2 vs. 3 vs. 4,5	2136	64.67	61.05	59.94	72.19
1,2 vs. *rest*	1171.1	73.26	81.35	77.07	-
3 vs. *rest*	632.5	46.51	15.23	22.84	-
4,5 vs. *rest*	1083	74.23	86.56	79.90	-

As can be seen from Table 7 by looking at macro scores, the prediction performance of the three classes altogether is almost identical to the one before the relabeling. In detail, looking at binary classifiers it is more difficult to predict the partially re-labelled third class while the prediction of the two other classes improves, although by a smaller margin. Looking at the micro-F1 scores gives the other side of the story, i.e. since these latter classes are more populated, the modest improvement in their prediction performances is sufficient to increase the number of instances that are correctly classified. Still, our original goal to improve the classification of the middle class is not satisfied and we lose in recall compared to what we had before reencoding (while keeping the same precision).

However, it must be remembered that all these observations have to be relativized by the fact that only 1 667 of the 2 800 reviews scored 3 have been manually checked. This means that about 300 reviews are still incorrectly labeled as "neutral". The completion of the manual check should therefore further help to improve the classifiers. Another factor to take into account is that the classes are fairly imbalanced while the models produced aim at producing an overall best fit, therefore favorishing these classes. It is possible that giving a reasonably bigger weight to the middle class examples would help improve the classification of its instances.

On a final note, it should however be pointed out that even with the poor performance of one of the classifiers, the global classifier achieves results that can be compared with the usual baselines for French on this particular task (cf. the results of the DEFT'07 challenge by Grouin et al. (2007) who report a $F1$ of 60.3 on a corpus that is comparable, although less general in the range of topics covered). In order to confirm this observation and to provide a fair comparison, we obtained the DEFT'07 challenge dataset, then trained and tested our classifier with the same restrictions than participants in the challenge (i.e. using the same test set, previously unseen during training and validation). Our classifier performance on this unique test set, with a $F1$ of 58.3, was less than reported by Grouin et al. (2007) but this difference was reasonnably small and the quality of the models produced can be seen as roughly equivalent. This is all the more true given that our classifier's performance displayed a standard deviation of 3.5 (measured by cross validation using all the DEFT dataset). Moreover, the main advantage of this classifier is the use of the elastic net penalty that dramatically reduces the number of features used (460 features for the DEFT dataset after selection, vs. 17246 before) and allows us to give an interpretation of the models produced. By comparison, the solution proposed by Grouin et al. (2007), while performing better, is a mixture of experts based on six classifiers, such as SVMs, producing black box models much less amenable to interpretation.

2   Interpreting the Models

The reviews that form the corpus used here express more than just opinions: they are also *argumentative* because they (usually) provide rationales to back up the opinions expressed by their authors. Therefore, the study of these reviews can be of some interest for the study of the way people use argumentative connectives and schemes to convey a positive or a negative opinion. One of the upshots of the elastic net regularization is a reduction of the feature space that retains only those features that are relevant for the classification task. Therefore, by studying the features that come out of this feature selection phase, one can try to get an idea of the argumentative strategies employed by the authors, or at least use them as a way to profile classes of expressive items (in the same vein as has been done by Constant et al. (2009) on the topic of expressive items, although with a different methodology).

We also compared the output of the selection derived from the elastic net with another selection method based on *bootstrapping*. This section first introduces the technique of bootstrapping and then presents the output of the selection processes by distinguishing between elements belonging to open categories and those belonging to closed categories.

2.1   Bootstrapping

Following the general method of the non parametric bootstrap (cf. Efron (1979)), 200 bootstrap samples were generated by sampling with replacement from the original dataset. Sampling was done for each stratum separately to ensure that bootstrap samples had the same number of examples from each stratum as in the original dataset. Logistic regression models were learnt from each of these samples, using an elastic net penalization with α and λ parameters chosen using a 3 folds cross-validation. Distribution of each regression coefficient β_i with respect to the bootstrap samples is used to qualify the robustness of the corresponding feature f_i.

The results of the bootstrap offer a way to test the robustness of a feature: if the feature gets consistently selected over the samples, this means that its contribution is general. Therefore, to determine the general relevance and impact of a feature, we first begin by looking at the percentages of bootstrap samples where its coefficient is non null. If this percentage is high enough, we look at two values:

1. the value of the coefficient coming from the elastic net regularization
2. the average of the coefficients coming from each of the bootstrap samples.

It is expected that these two values are rather similar, but for reasons of completeness we report both of them in the following tables.

2.2   Closed Categories

In this section we focus on two specific closed categories: coordinating conjunctions on one hand and prepositions on the other. The elements in those classes are few in number and usually very frequent. We are thus mainly interested in knowing which of these elements are the most relevant for the classification task.

Coordinating Conjunctions. Coordinating conjunctions are obvious discourse connectives, and as such it is interesting to check which of those prove to be the most relevant for sentiment analysis.

We begin by looking at the binary classification task. Table 8 shows for each conjunction: the proportion of bootstrap samples where its coefficient was not null, the average of its bootstrap coefficient, its elastic net coefficient and the number of occurrences of the conjunction in the corpus.

Table 8. Coordinating conjunctions: coefficients selection (Binary task).

Conjunction	Proportion	Bootstrap avg.	Elastic Net	N. occ.
et	0.97	0.157	0.139	4284
ou	0.27	0.019	0.0	864
donc	0.15	−0.049	0.0	11
sinon	0.30	−0.052	0.0	44
voire	0.23	−0.069	0.0	33
soit	0.4	−0.095	−0.028	69
car	0.73	−0.103	−0.076	549
puis	0.57	−0.129	−0.075	120
mais	1	−0.335	−0.245	1889
ni	0.99	−0.511	−0.464	169
or	0.83	−0.528	−0.693	21

Only four conjunctions seem to have a significant contribution here, i.e. get selected in more than 75 % of the bootstrap samples and have non-null coefficients. On the positive side there is the conjunction *et* (≈*and*), while on the negative are *mais* (≈*but*), the correlative *ni* (≈*neither/nor*) and the adversative *or* (≈*yet/as it turns out*).[2]

The presence of the negative *ni* is expected to be correlated with negative reviews as it has an intrinsically negative meaning.

Mais and *or* can be grouped together: they both are adversative, i.e. they introduce a sentence that is opposed in one way or another to the left argument of the connective. From the argumentative point of view, it is considered that these items connect opposed arguments (cf. Anscombre and Ducrot (1977); Winterstein (2012)).

On the other hand *et* has been described as a connective that conjoins two arguments that argue for the same goal and are (at least) partly independent arguments for this goal (cf. Jayez and Winterstein (2013)).

Therefore, it seems that negative reviews tend to involve opposed arguments more often than positive reviews (as marked by the significance of adversative connectives for these reviews). On the other hand, positive reviews involve sequences of arguments that target the same goal, but are independent (as marked by *et*).

One way to interpret this is to consider that positive and negative reviews involve different argumentative strategies. Arguing positively requires more effort to convince. A successful positive argumentation will have more chance of being persuasive if it gives several independent arguments in favor of its conclusion. On the other hand, in order to argue negatively, a single negative argument appears to be enough, even if it is put in perspective with a positive one.

This interpretation is further confirmed if one looks at the results of a Naive Bayes approach to the classification task. While such an approach does not give results as good as those reported in Table 2, it can easily be used to detect bigrams that are correlated to positive or negative reviews. Among the ten bigrams whose significance for the classification is the highest one can find the bigram *point positif* ("positive point"). Contrary to what could be expected, this bigram is a strong indicator of a *negative* review. This is in line with our previous observation on the use of *but*: mentioning a positive point usually entails also mentioning a negative one. In case of conflicting arguments, it is expected that the negative one will win.

We now turn to the ternary task. Tables 9 and 10 present the same information as Table 8 but for the ternary task. Table 9 presents the results for the classifier of the positive class (scores 4 and 5) against the rest, whilst Table 10 is for the classifier of the negative class (scores 1 and 2) against the rest. Features for which the elastic net coefficient is null have been omitted from the tables. The middle classifier is ignored because of its poor performances (cf. the discussion on Table 7 above).

For the positive classifier, we see that the positive role of the conjunction *et* remains: our hypothesis about the preference to use additive argumentative

---

[2] We ignore the borderline case of *car* (≈*because/since*).

Table 9. Coordinating conjunctions: coefficients selection (Ternary task, positive classifier).

Conjunction	Proportion	Bootstrap avg.	Elastic Net	N. occ.
et	1	0.108	0.121	10693
car	1	−0.155	−0.075	1539
puis	0.975	−0.207	−0.049	329
mais	1	−0.367	−0.203	6003
ni	0.98	−0.228	−0.003	407

Table 10. Coordinating conjunctions: coefficients selection (Ternary task, negative classifier).

Conjunction	Proportion	Bootstrap avg.	Elastic Net	N. occ.
or	0.99	0.647	0.664	39
voire	0.95	0.344	0.135	89
puis	0.98	0.224	0.135	329
ni	0.1	0.247	0.094	407
comme	0.92	−0.73	−0.328	12

strategies in positive reviews appears confirmed. However the case of *mais* has to be somehow refined. The adversative is still an indicator of a non-positive review as seen in Table 9, but it is no longer an indicator of a negative one. Therefore we can still consider that positive reviews tend to eschew balancing positive and negative arguments, but we can no longer assume that this is the hallmark of negative reviews. This appears quite sensible: middle reviews should form the prototypical case of balanced arguments and thus are good candidates for involving the use of adversative markers. However negative reviews still use adversative strategies: the adversative connective *or* is the strongest indicator for the negative class amongst all conjunctions.

In the end, the study of the output of the feature selection processes on the case of conjunctions outlines the fact that positive, balanced and negative reviews do not use the same argumentative schemes. Further investigation of the reviews, for example at the sentence level and by using a polarity lexicon such as Senticnet (cf. Cambria and Hussain (2012)), should help to strengthen these claims.

Prepositions. We look here at prepositions in the same perspective as the conjunctions. To keep the presentation short, we only present the results of the binary task in Table 11 where we only mention those for which both selection methods produced non-null coefficients.

The main point we wish to underline here is that the selection is consistent with that of the coordinating conjunctions. On one hand the additive preposition

Table 11. Prepositions: selection coefficients (Binary task)

Preposition	Proportion	Bootstrap avg.	Elastic Net	N. occ.
avec	0.97	0.165	0.139	1804
chez	0.58	0.131	0.115	167
selon	0.42	0.101	0.005	74
en	0.88	0.087	0.051	2537
pour	0.67	0.051	0.025	2736
sous	0.39	−0.053	−0.050	210
de	0.76	−0.083	−0.075	4879
jusque	0.5	−0.086	−0.031	193
envers	0.3	−0.089	−0.098	30
sans	1	−0.340	−0.264	1035
malgré	0.94	−0.344	−0.281	208
sauf	0.96	−0.511	−0.416	106

avec (≈*with*) is an indicator of positive reviews, like the additive conjunction *et*. This remains true in the ternary classification task for the positive classifier.

Regarding the prepositions that have a negative impact, the case of the adversative *malgré* (≈*in spite of*) appears similar to the adversative conjunctions of the previous sections. It is a marker of negative review in the binary task, and in the ternary task it is a mark of a non-positive review, but it does not specifically mark negative reviews.

Finally, the case of *sauf* (≈*except*) and *sans* (≈*without*) also prove to be interesting. These two prepositions have the same profile as the adversative elements. So far these elements have not been described in these terms, but the results presented here suggest that these elements might also be appropriately be described in argumentative terms as carrying an adversative value. Roughly both these prepositions are exceptive, i.e. they indicate that an element is not included in some predication. If the excepted element is important, then it is expected that the use of these prepositions carries an argumentative reversal effect similar to what the exclusive adverb *only* conveys in some contexts.

2.3   Open Categories

We briefly focus here on the case of elements belonging to open categories, i.e. on lemmas that are either verbs, nouns, adjectives or adverbs.

Class Distribution. First, we look at how the relative importance of each of the four open categories is affected by the selection process. For this we look at the number of items in each class before and after the selection process. The numbers in Table 12 correspond to the binary task. The number of items after

Table 12. Open categories: number of items before and after selection.

Category	Before selection	Proportion	After selection	Proportion
Adjective	543	20.03 %	148	35.41 %
Adverb	158	5.83 %	28	6.7 %
Noun	1398	51.57 %	164	39.23 %
Verb	612	22.57 %	78	18.66 %
Total	2711	100 %	418	100 %

selection correspond to the number of items which have non null coefficients in more than 75 % of the bootstrap samples and for which the elastic net coefficient is not null.

The differences in the distribution of the categories before and after the selection are quite significant ($\chi^2 = 57.71, p$-value $\ll 1.0^{-10}$). They show that adverbs and adjectives are more represented after the selection process, whereas nouns and verbs see a strong decrease in their frequencies. This strongly supports the opinion that adjectives and adverbs are the most likely elements to convey sentiment in a text, as has been claimed previously (e.g. by Turney (2002) or Benamara et al. (2007)).

Adverbs. For reasons of space and relevance, we only develop the case of adverbs here. Table 13 gives the list of the 28 adverbs that were selected.

Negation. A first feature to be noted is that these results are consistent with those found about English by Potts (2011) concerning the "negativity" of negation. Potts underlines that negation is more than just a logical switch for truth-values, but also seems to carry an intrinsic negative tone. He shows how this is confirmed by the fact that the distribution of negative markers is not homogeneous across notations: elements like *not* appear more often in negative reviews than in positive ones. According to Potts, this is explained by the fact that a negative sentence is usually less informative than a positive one, and is thus more likely to be used as a rebuttal rather than as an informative statement.

We find the same situation for French in the reviews of our corpus: the markers of negation *ne* and *pas* both appear as strong indicators of negative reviews, selected in all the bootstrap samples, with relatively strong coefficients.

From the methodological point of view our approach slightly differs from the one of Potts. In both cases, we adopt a reader-oriented perspective: given a lexical item, we evaluate which kind of opinion is the most likely (i.e. positive or negative), so the general goal is the same for Potts and us.

However, Potts typically starts by selecting some elements which he assumes have a specific profile and uses the data as a way to confirm his hypotheses (e.g. as was done with negation). In a related work and using similar data, Constant et al. (2009) use the profiles of known elements as a way to discover

Table 13. Selected adverbs (Binary task)

Adverb	Proportion	Bootstrap avg.	Elastic Net	N. Occ.
magnifiquement	0.99	1.44	1.05	13
agréablement	0.92	0.71	0.70	21
bientôt	0.97	0.78	0.69	34
absolument	1.00	0.62	0.42	253
très	1.00	0.51	0.37	2247
vivement	0.96	0.43	0.37	118
bien	1.00	0.37	0.29	1543
toujours	0.95	0.25	0.21	406
aussi	0.99	0.25	0.21	588
peu	0.85	−0.15	−0.09	710
même	0.93	−0.18	−0.16	749
là	0.87	−0.22	−0.17	358
mieux	0.95	−0.31	−0.24	319
totalement	0.87	−0.32	−0.26	130
bref	0.96	−0.30	−0.27	212
alors	0.98	−0.31	−0.28	381
sûrement	0.87	−0.46	−0.34	53
ne	1.00	−0.40	−0.35	2915
franchement	0.89	−0.33	−0.36	129
vite	0.97	−0.48	−0.40	128
non	1.00	−0.54	−0.40	421
pourtant	0.99	−0.53	−0.41	182
pas	1.00	−0.66	−0.49	2791
plutôt	0.99	−0.59	−0.50	211
trop	1.00	−0.82	−0.62	465
strictement	0.90	−1.03	−0.73	16
heureusement	1.00	−1.08	−0.88	102
mal	1.00	−1.21	−0.93	308

other elements which share the same profile, aiming at automatically discovering classes of expressive elements.

Our approach is different in that we do not make any preliminary assumption on the profile of lexical items. The learning and selection processes automatically provide us with classes of elements which behave similarly regarding the task at hand. One drawback is that the classes have to be coarser than the ones one can obtain by Potts's approach. This comes as a consequence of the fact that predicting notations beyond the positive/negative dual case is difficult (cf. the discussion on the ternary classification task). This means that elements that are not characteristic of either the positive or negative notation class will be harder to detect since this implies dealing with three or more classes of notation.

Nevertheless, the fact that our approach and that of Potts give similar results for negation gives credence to both as ways to get some pragmatic insights by relying on large corpora and meta-textual data.

Other elements. Apart from the case of negation other features of Table 13 appear striking:

- Positivity intensifying adverbs such as *magnifiquement* (≈*beautifully*) or *agré-ablement* (≈*pleasantly*) are strong positive indicators. This is expected since those elements are non-controversially positive.
- *Heureusement* (≈*fortunately*) might appear as a counter-example because of its apparent positive undertone. However, the use of this adverb usually marks a counter expectation akin to an adversative reading. This is again consistent with our observations on conjunctions and prepositions. The presence of *pourtant* (≈*yet*) as a negative indicator is also coherent.
- Purely intensifying adverbs have mixed profiles:
  - *très* (≈*very*) and *absolument* (≈*absolutely*) are positive indicators.
  - *totalement* (≈*totally*) is negative.
  Initially, one could have thought that those intensifying adverbs have no polarity bias since their essential meaning is to indicate a high degree of the property they modify, without further constraints on the kind of property it can act on (in the same way that one could initially expect negation to have no specific orientation on its own). However, it seems that speakers have preferences for using some adverbs for intensifying positive properties and others for negative properties. We leave the question of why this happens to future work.
- Finally, the adverb *aussi* (≈*too*) is often described as being additive (e.g. by Winterstein and Zeevat (2012)) and is shown here to be a positive marker. This is consistent with the previous hypothesis about positive reviews involving multiple parallel arguments since *aussi* indicates that the speaker is using two sentences that are related and argumentatively co-oriented.

3   Conclusion

The work reported here underlined the importance and usefulness of feature selection techniques when tackling a problem like opinion classification. Not only do these techniques improve the performances of the classifiers, they also offer some insight on the way the classifiers work and on which elements have the same profile regarding the task at hand.

In future work, we intend to try to further enhance the ternary classifier. First, we will complete the manual check of the reviews scored 3 to get a final evaluation of the usefulness of this relabeling. Another improvement direction is in the detection of irony which appears rather common, especially in negative reviews. Finally, an analysis of the reviews at the sentence level, by using a polarity lexicon, is also a potential solution to improve the performances. However, such a resource is not readily available yet for French and needs to be constructed beforehand.

Another direction of research is testing the general character of the argumentative strategies we have characterized. First, we intend to determine whether the same conclusions can be reached on other languages, notably English for which resources of the type we need already abound. Another fruitful comparison is to compare our results with insights gathered on reviews of a different kind. For example, reviews of scientific papers might exhibit different profiles. There we would expect negative reviews to be more thorough and involve parallel, independent negative arguments. This is because, at least intuitively, a negative review should be strongly motivated and usually cannot be reduced to a single negative point.

Finally, regarding the interpretation of the models, a further investigation of the ternary models should be carried out once they have been improved, especially regarding the middle classifier.

References

Anscombre, J.-C., Ducrot, O.: Deux mais en français. J. Lingua **43**, 23–40 (1977)

Benamara, F., Cesarano, C., Picariello, A., Reforgiato, D., Subrahmanian, V.S.: Sentiment analysis: adjectives and adverbs are better than adjectives alone. In: Proceedings of the International Conference on Weblogs and Social Media (ICWSM) (2007)

Bishop, C.M.: Pattern Recognition and Machine Learning. Springer, Berlin (2006)

Cambria, E., Hussain, A.: Sentic Computing: Techniques, Tools, and Applications. Springer, Heidelberg (2012)

Constant, N., Davis, C., Potts, C., Schwarz, F.: The pragmatics of expressive content evidence from large corpora. Sprache und Datenverarbeitung **33**(1–2), 5–21 (2009)

Daumé III, H.: Notes on CG and LM-BFGS Optimization of Logistic Regression (2004). Paper available at http://pub.hal3.name#daume04cg-bfgs, implementation available at http://hal3.name/mcgam/

Denis, P., Sagot, B.: Coupling an annotated corpus and a lexicon for state-of-the-art POS tagging. Lang. Resour. Eval. **46**, 721–746 (2012)

Efron, B.: Another look at the jackknife. Ann. Stat. **7**, 1–26 (1979)

Friedman, J., Hastie, T., Tibshirani, R.: Regularization paths for generalized linear models via coordinate descent. J. Stat. Softw. **33**(1), 1–22 (2010). http://www.jstatsoft.org/v33/i01/

Grouin, C., et al.: Présentation de l'édition 2007 du Défi fouille de textes (DEFT07). In: Actes de l'atelier de clôture du 3ème Défi Fouille de Textes (DEFT07), pp. 1–8 (2007)

Jayez, J., Winterstein, G.: Additivity and Probability. Lingua **132**, 85–102 (2013)

Joachims, T.: Making large-scale SVM learning practical. In: Schölkopf, B., Burges, C.J.C.B., Smola, A.J. (eds.) Advances in Kernel Methods - Support Vector Learning, pp. 41–56. MIT Press, Cambridge (1999)

Pang, B., Lee, L.: Opinion mining and sentiment analysis. Found. Trends Inf. Retrieval **2**(1–2), 1–135 (2008)

Pang, B., Lee, L., Vaithyanathan, S.: Thumbs up? Sentiment classification using machine learning techniques. In: Proceedings of the Conference on Empirical Methods in Natural Language Processing (EMNLP). Association for Computational Linguistics, pp. 79–86 (2002)

Potts, C.: On the negativity of negation. In: Li, N., Lutz, D. (eds.) Semantics and Linguistic Theory (SALT) 20, pp. 636–659. eLanguage (2011)

R Development Core Team: R: A Language and Environment for Statistical Computing. organization R Foundation for Statistical Computing, Vienna, Austria (2011)

Turney, P.: Thumbs up or thumbs down? Semantic orientation applied to unsupervised classification of reviews. In: Proceedings of the ACL (2002)

Winterstein, G.: What but-sentences argue for: a modern argumentative analysis of but. Lingua **122**(15), 1864–1885 (2012)

Winterstein, G., Zeevat, H.: Empirical constraints on accounts of too. Lingua **122**(15), 1787–1800 (2012)

Zou, H., Hastie, T.: Regularization and variable selection via the Elastic Net. J. R. Stat. Soc. Ser. B **67**(2), 301–320 (2005)

Exhaustivity Through the Maxim of Relation

Matthijs Westera[(✉)]

Institute for Logic, Language and Computation,
University of Amsterdam, Amsterdam, The Netherlands
matthijs.westera@gmail.com

Abstract. I show that the exhaustive interpretation of answers can be explained as a conversational implicature through the Maxim of Relation, dealing with the problematic epistemic step (Sauerland 2004). I assume a fairly standard Maxim of Relation, that captures the same intuition as Roberts' (1996) contextual entailment. I show that if a richer notion of meaning is adopted, in particular that of attentive semantics (Roelofsen 2011), this Maxim of Relation automatically becomes strong enough to enable exhaustivity implicatures. The results suggest that pragmatic reasoning is sensitive not only to the information an utterance provides, but also to the possibilities it draws attention to. Foremost, it shows that exhaustivity implicatures can be genuine conversational implicatures.

(Cannot access the full version? Download it from the author's website.)

Keywords: Exhaustivity · Competence assumption · Maxim of relation · Attentive content

1 Introduction

Responding to a question with one of its possible answers may convey that the answer is *exhaustive*, i.e., that the other possible answers are false.

(1) Of blue, red and green, which colours does John like?
    John likes blue. ⤳ *He doesn't like red*

The exhaustive interpretation of answers is often considered a prime example of Gricean *conversational implicature*, but so far no theory exists that wholly explains it as such. A conversational implicature of an utterance is information that must be supposed in order to maintain the assumption that the speaker is cooperative [8]. The typical 'Gricean' approach to the exhaustive interpretation of the response in (1) goes as follows:[1]

---

[1] For brevity, I will typically use the word 'know' (and, likewise, 'knowledge') as if saying 'taking oneself to know', i.e., without requiring the usual *factivity* associated with knowledge as being 'true belief'.

© Springer International Publishing Switzerland 2014
Y. Nakano et al. (Eds.): JSAI-isAI 2013, LNAI 8417, pp. 141–153, 2014.
DOI: 10.1007/978-3-319-10061-6_10

1. The speaker didn't say John likes red.
2. She should have said so, had she been able to.
3. She must lack the knowledge that John likes red.
   . . .
4. She knows that John *doesn't* like red.

The exhaustivity in 4. is obtained from the *Quantity implicature* in 3. through a strengthening known as the *epistemic step* [15]. Chierchia et al. [3] argue that the epistemic step does not follow from the assumption of cooperativity, i.e., that exhaustivity is not a case of conversational implicature (instead, they defend a 'grammatical' approach to exhaustivity, which I briefly discuss in Sect. 4). I argue, instead, that the epistemic step *does* follow from the assumption of cooperativity, i.e., that exhaustivity *is* a genuine case of Gricean conversational implicature.

My starting point is the idea that a pragmatic theory can be only as good as the stuff one feeds into it, i.e., meanings. Perhaps our pragmatic theories are fine as they are; rather, it is the underlying, classical semantics that is too coarse for an account of exhaustivity implicatures. Classical semantics models only the *informative content* of utterances, but the following example suggests that this is insufficient foothold for a theory of exhaustivity:

(2) Which colours does John like?
     John likes {at least blue / blue or red and blue}. $\not\leadsto$ *John doesn't like red.*

The response in (1) is just as informative as the responses in (2), but only the former is interpreted exhaustively.[2] Intuitively, the difference between (2) and (1) lies not in the informative content, but in the possibilities that the responses *draw attention to*, in particular, whether the response draws attention to the possibility that John also likes red. The responses in (2), but not (1), draw attention to this possibility, and perhaps pragmatic reasoning is sensitive to this. This intuition also underlies the account by Alonso-Ovalle [1], who proposes that only in (1), and not in (2), that John likes red can be *innocently excluded*.

The idea that pragmatic reasoning is sensitive to attentive content has been entertained before by [5], in their account of 'might'. It entails that, if we study pragmatic phenomena, we should be using a semantics that models not only the informative content of utterances, but also their attentive content. I show that if we thus enrich the underlying semantics, the Maxim of Relation as it occurs in the literature [13] automatically inherits this sensitivity. That is, the response in (1) will come out as being *not entirely related* to the question, because it leaves the possibility that John also likes red, unattended. This increased sensitivity of the Maxim of Relation will result in a Relation implicature that will enable us to take the epistemic step.

---

[2] Groenendijk and Stokhof [10] already argued that exhaustivity cannot be derived through the Maxim of Quantity alone. After all, the Quantity implicature says 'this is as informative as I can safely be', hence it can never be used to *strengthen* what is said, as is required for the epistemic step. The contrast in (2) supports this view from a different angle: it's not only quantity of information that matters.

Section 2 introduces the main building blocks - an *attentive* semantics and a set of fairly standard conversational maxims. Section 3 shows how exhaustivity implicatures are accounted for. Section 4 discusses the results in a broader context.

2 Ingredients

2.1 Attentive Semantics

As the enriched semantic backbone for pragmatics, I adopt Roelofsen's [14] *attentive semantics*, designed to model informative and attentive content. Attentive semantics is closely related to *basic inquisitive semantics* [9], which is aimed at modeling informative content and *inquisitive content*, and *unrestricted inquisitive semantics* [4,5], which has been taken to model all three components at once.

In attentive semantics, the meaning of a sentence, called a *proposition*, is a *set of sets of worlds*, i.e., a set of classical propositions. The proposition $[\phi]$ expressed by a sentence ϕ is conceived of as the set of possibilities that the sentence *draws attention to*. The union of these possibilities corresponds to the sentence's *informative content*, i.e., the information provided by the sentence. Hence, attentive semantics models two semantic components - attention and information - in a single semantic object, which is called a proposition.

Let $\mathbf{W}$ be the set of all possible worlds, assigning truth values to the atomic formulae of a language of choice. Following [14], I define:

- A *possibility* α is a set of worlds, $\alpha \in \wp\mathbf{W}$.
- A *proposition* A is a set of non-empty possibilities, $A \in \wp\wp\mathbf{W}$.
- $[\phi]$ denotes the proposition expressed by ϕ.
- *Informative content*: $|\varphi| := \bigcup[\varphi]$
- A *restricted* to a set of worlds s: $A_s := \{\alpha \cap s \mid \alpha \in A, \alpha \cap s \neq \emptyset\}$

The relevant fragment of propositional logic (without implication) is given in Backus-Naur Form:

Definition 1 (Syntax). *For p a propositional letter, ϕ formulae: $\phi :: \bot \mid p \mid \neg\phi \mid (\phi \wedge \phi) \mid (\phi \vee \phi)$*

For all formulae ϕ, the proposition it expresses, $[\phi]$, is defined recursively as:

Definition 2 (Attentive semantics). *For p a proposition letter, ϕ, ψ formulae:*

1. $[p] = \{\{w \in \mathbf{W} \mid w(p) = \text{TRUE}\}\}$
2. $[\neg\varphi] = \{\overline{\bigcup[\varphi]} \mid \overline{\bigcup[\varphi]} \neq \emptyset\}$
3. $[\varphi \vee \psi] = ([\varphi] \cup [\psi])_{|\varphi|\cup|\psi|}$ $(= [\varphi] \cup [\psi])$
4. $[\varphi \wedge \psi] = ([\varphi] \cup [\psi])_{|\varphi|\cap|\psi|}$

I will briefly translate this definition into natural language: atomic formulae draw attention only to one possibility, namely the possibility that the formula is true; the negation of a formula draws attention only to the single possibility where its argument is false; a disjunction draws attention to what both disjuncts draw attention to, and provides the information that at least one of them is true; and a conjunction draws attention to each possibility that a conjunct draws attention to, when restricted to the information of the other conjunct.

With this richer-than-usual semantics, entailment becomes sparser than usual:

(3) For all propositions A, B, A *entails* B, $A \models B$, iff:
    a. $\bigcup A \subseteq \bigcup B$; and
    b. $\forall \beta \in B$, if $\beta \cap \bigcup A \neq \emptyset$, $\beta \cap \bigcup A \in A$

Item a. requires, just like classical entailment, that A is *at least as informative* as B. Item b. requires that A is, in addition, *at least as attentive* as B. That means that every possibility that B draws attention to, must be a possibility that A draws attention to, insofar as this is compatible with the information provided by A. Intuitively, A entails B if you can get from B to A by *removing* worlds and *adding* possibilities.

Despite the richer semantics, attentive semantics treats informative content fully classically, in the following sense:

Fact 1 (Classical treatment of informative content). *For all formulae ϕ, $|\phi|$ $(= \bigcup[\phi])$ is its classical meaning.*

This shows that, entirely in the spirit of Grice [8], I try to account for implicatures without giving up the classical treatment of information.

2.2   Translating the Examples into Logic

To avoid having to assume anything about the semantics of 'which'-questions, I will replace the 'which'-question in (1) by the existential sentence in (4), for which the same exhaustivity implicature obtains:

(4) Among blue, red and green, there are colours that John likes.
    Yes, he likes blue. $\rightsquigarrow$ *He doesn't like red*

Of course, the fact that this existential sentence behaves completely like the 'which'-question does suggest that 'which'-questions, likewise, have existential force, i.e., draw attention to the set of their non-exhaustive, mention-some answers. I will briefly return to this in Sect. 4.

Without loss of generality, we can furthermore assume that there exist only two colours, blue and red. Since existential quantification over a finite domain can be translated using a disjunction, this enables the following minimal translations into propositional logic:

(5) a. There are colours that John likes. $p \lor q \lor (p \land q)$
    b. John likes blue. p
    c. John likes at least blue / blue or red and blue. $p \lor (p \land q)$

Translations (5b) and (5c) are straightforward, where the latter is in line with Coppock and Brochhagen's account of 'at least' in inquisitive semantics [6].

Attentive semantics assigns to the formulae in (5) the propositions depicted in Fig. 1. In the figure, circles represent possible worlds, where, e.g., $p\bar{q}$ marks the world where p is true but q is false, i.e., where John only likes blue. Shaded regions represent possibilities, i.e., sets of worlds. It should be clear that, in these simple examples, the possibilities that a sentence draws attention to correspond exactly to the sentence's disjuncts. This is true more generally for formulae in *disjunctive normal form*.

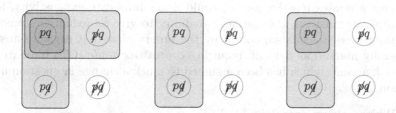

Fig. 1. The attentive propositions expressed by, from left to right, $p \vee q \vee (p \wedge q)$, p and $p \vee (p \wedge q)$. Circles represent worlds, shaded areas represent possibilities.

With a richer semantics, entailment is more sparse than usual. Whereas classically (5b) and (5c) would both entail (5a), now only (5c) does. This is easily seen: one cannot get from the left diagram in Fig. 1 to the middle one by removing worlds or adding possibilities, due to the $|p \wedge q|$ possibility (which would have to be removed). This reflects the fact that (5b), though more informative than (5a), is strictly *less attentive*. This is not the case for (5c).

2.3   The Maxims

Following Grice [8], what it means to be cooperative in a typical context is spelled out in terms of a set of *conversational maxims* that require the speaker to be truthful (Quality), as informative as possible and required (Quantity), relevant (Relation), and clear (Manner). Following a long tradition (at least since [10]; see also [13]), the maxims of Quantity and Relation are defined in terms of a *question under discussion Q* (which in the examples under consideration is always given as the initiative).

Definition 3 (The relevant maxims). *For all propositions A, Q, a speaker with information state s (a set of worlds), responding A to Q, is cooperative iff:*

- *Quality:* $s \subseteq \bigcup A$;
- *Quantity:* for all $Q' \subseteq Q$, if $\bigcup A \nsubseteq \bigcup Q'$ then $s \nsubseteq \bigcup Q'$;
- *Relation:* $A_s \models Q$; and
- *Manner:* the speaker thinks she is being clear, concise, etc.

I will briefly explain the maxims of Quantity and Relation.

The Maxim of Quantity. The Maxim of Quantity requires that the speaker gives enough information to establish all (sets of) possibilities that she knows (or takes herself to know) contain the actual world. If a speaker is fully informed, this means that the Maxim of Quantity requires that she gives a *positively* exhaustive answer (i.e., an answer that decides for every possibility of the question that is the case, that it is the case), and not necessarily a positively *and* negatively exhaustive answer (i.e., an answer that decides for every possibility whether it is the case). This choice (i.e., that the Maxim of Quantity asks only for a positively exhaustive answer) is based on a remark by Groenendijk and Stokhof [10] that, if it would ask for a positively *and* negatively exhaustive answer, then giving a non-exhaustive answer could never implicate exhaustivity, for it would implicate, instead, the speaker's inability to give an exhaustive answer.[3] In addition, this choice is supported by the observation that negative answers are generally marked as deviant, requiring *contrastive topic* (CT) intonation, a rising pitch accent, which has been assumed to mark a change in question under discussion [2]:

(6) A: Which colours does John like?
     B: [Not]$_{CT}$ blue...

The present Maxim of Quantity explains why this is deviant: the fact that John does not like blue does not support any possibility, or set of possibilities, of the question under discussion.

The Maxim of Relation. The Maxim of Relation requires that an utterance, together with some piece of information known by the speaker, entails the question under discussion. This maxim is meant to account for examples like the following:

(7) A: Did John come to the party?
     B: It was raining.

Here, the response is only relevant if it is known that (and how) John's attendance depends the rain, for instance that if it was raining, he wasn't at the party. The Maxim of Relation requires that such a dependency is known, because only together with such a dependency does the response entail the question (as I will formally show later).

My Maxim of Relation is slightly different from two very similar notions in the literature:

---

[3] The difference between positively and negatively exhaustive answers fades if the question under discussion is a partition. Given Groenendijk and Stokhof's argument, I take this as strong evidence that, in general, questions (at least those in response to which a non-exhaustive answer would implicate exhaustivity) are not partitions; i.e., they have existential force, i.e., draw attention to their mention-some answers, rather than universal force.

- Groenendijk and Stokhof's [10] notion of 'pragmatic answer', which requires that the utterance, relative to the *hearer's* knowledge state, entails an answer to the question; and
- Roberts's [13] 'contextual entailment', which requires that the utterance, relative to the *common ground*, entails an answer to the question.

First, note that my Maxim of Relation follows logically from Roberts's notion, for if the required dependency is *common* knowledge, then it is also the *speaker's* knowledge. Second, my maxim can be said to 'typically' follow from Groenendijk and Stokhof's, in the following sense: if the speaker has to know that the hearer knows a dependency between answer and question, then under normal conditions of thrust, the speaker will share this knowledge. In other words, my maxim is *not that different*.

What my Maxim of Relation lacks, however, is a requirement that the hearer should be able to figure out how the speaker intends her response to be related to the question. I will leave such a requirement implicit (though one might think of it as part of the Maxim of Manner, which is left informal anyway), with the additional remarks that (i) compared to [10] and [13], I think it is cleaner to separate this kind of 'transparency' requirement from the Maxim of Relation, and (ii) the requirements imposed by [10,13] are definitely too strong: as I will show, exhaustivity implicatures are a prime example of a case where the listener need not *already know* how the Maxim of Relation is complied with, but, rather, can *figure it out* on the spot by combining it with the Quality and Quantity implicatures.

3 Results

3.1 Examples

Example (7): The Rainy Party. Before generalizing, let us look at the examples discussed so far. Example (7), which I used solely to illustrate how the Maxim of Relation works, is accounted for as follows. Let the question (whether John was at the party) translate as $p \vee \neg p$, and the response (that it was raining) as r. Attentive semantics assigns to these formulae the following meanings:

(8) $[p \vee \neg p] = \{|p|, \overline{|p|}\}$
    $[r] = \{|r|\}$

The response does not entail the question. For the response to be related to the question, therefore, the speaker must have some knowledge, relative to which it does entail the question. This requirement can be met in two ways:

- The speaker thinks that if it rained, John was there ($s \subseteq \overline{|r|} \cup |p|$; the response restricted to this information yields $\{|r| \cap |p|\}$, which entails $\{|p|, \overline{|p|}\}$); or
- The speaker thinks that if it rained, John wasn't there ($s \subseteq |r| \cup |p|$; the response restricted to this information yields $\{|r| \cap \overline{|p|}\}$, which entails $\{|p|, \overline{|p|}\}$).

Furthermore, given that the speaker thinks that it was raining ($s \subseteq |r|$), it follows that she must know whether John was at the party ($s \subseteq |p|$ or $s \subseteq |\overline{p}|$).[4]

Example (5c): No Exhaustivity. Example (5c) is predicted to yield the following implicatures:

(9) 1. The speaker knows that $p(s \subseteq |p|)$ (Quality)
    2. She lacks the knowledge that $q(s \not\subseteq |q|)$ (Quantity)

Because the response entails the question, it automatically complies with the Maxim of Relation as it is, i.e., no Relation implicature occurs, and as a consequence the epistemic step cannot be taken, and no exhaustivity is implicated.

Example (5b): Exhaustivity! Example (5b) yields the same Quality and Quantity implicatures as (5c) (to which it is, after all, informatively equivalent). Unlike in (5c), however, here the response doesn't entail the question, hence it does not automatically comply with the Maxim of Relation. For this to be the case, the response must entail the question relative to the speaker's information state. This can be complied with in two ways:

- The speaker thinks that if John likes blue, he likes red ($s \subseteq \overline{|p|} \cup |q|$; the response restricted to this information yields $\{|p| \cap |q|\}$, which entails the question); or
- The speaker thinks that if John likes blue, he doesn't like red ($s \subseteq \overline{|p|} \cup \overline{|q|}$; the response restricted to this information yields $\{|p| \cap \overline{|q|}\}$, which entails the question).

Combining this disjunctive implicature with the Quality and Quantity implicatures yields the following:

1. The speaker knows that p ($s \subseteq |p|$)   (Quality)
2. She lacks the knowledge that q ($s \not\subseteq |q|$)   (Quantity)
3. $s \subseteq \overline{|p|} \cup |q|$ or $s \subseteq \overline{|p|} \cup \overline{|q|}$ (Relation)

---

4. She knows $\neg q$ ($s \subseteq \overline{|q|}$)   = exhaustivity!

That is, exhaustivity is derived as a conversational implicature.

---

[4] This shows that, in this case, the Relation implicature in fact clashes with the Quantity implicature: for if the speaker really knows whether John was at the party, then why didn't she just say so? As a consequence, the example will likely trigger additional implicatures to explain this clash, e.g., that the speaker wants to test the hearer's knowledge. (Understandably, therefore, the example under discussion is more natural with a final rising pitch contour, which has been argued to convey *uncertain* relevance/relatedness [18].) Regardless of this, the example suffices for merely illustrating how the Maxim of Relation works.

At a more intuitive level, the following is going on in this example. For any possibility of the question that the speaker, in her response, leaves unattended (in this case $|q|$), the Maxim of Relation, with its increased sensitivity due to attentive semantics, requires that the speaker knows whether it or its negation follows from the information that she did give (i.e., the information that $|p|$). Or in other words, *the Maxim of Relation requires that the speaker is* competent *about every possibility she leaves unattended.* This implicated competence is what enables the epistemic step.[5]

3.2   General Results

While the implicatures due to the maxims of Quantity and Quality are quite easy to spell out, the Maxim of Relation is rather more complex - and yet that is where the key to exhaustivity lies. In order to better understand it, the implicatures due to the Maxim of Relation can be generally characterized as follows:

(10) **Relation implicature**

A cooperative speaker with information state s, responding A to Q, complies with the Maxim of Relation iff:
a. $s \subseteq \overline{\bigcup A} \cup \bigcup Q$; and
b. for all $\gamma \in Q$: $s \subseteq \overline{\bigcup A} \cup \overline{\gamma}$, or there is an $\alpha \in A$ s.t.
$s \subseteq (\gamma \cap \overline{\bigcup A \cap \overline{\alpha}}) \cup (\gamma \cap \bigcup A \cap \alpha)$

Item (a) requires that the speaker knows the material implication from A to Q, which ensures that the response together with the speaker's information, is at least as *informative* as the question. Item (b) requires that the response, together with the speaker's information, is at least as *attentive* as the question. This amounts to the following: for each possibility that the question draws attention to, the speaker knows either that it is incompatible with the response, or that, when restricted to the response and the speaker's information, it coincides with a possibility α that the response draws attention to.

For responses that select exactly one of the possibilities of the question, which are interesting from the viewpoint of exhaustivity, this amounts to the following more readable result:

(11) **Relation implicature for singleton answer**

A cooperative speaker with information state s, responding $\{\alpha\}$ to Q for some $\alpha \in Q$, complies with the Maxim of Relation iff:
for all $\gamma \in Q$, $s \subseteq \overline{\alpha} \cup \overline{\gamma}$; or $s \subseteq \overline{\alpha} \cup \gamma$.

---

[5] The speaker can be competent in two ways, and in the case of exhaustivity, the Quality and Quantity implicatures together settle *how* (something which did not happen in the rainy party example). This shows that, as mentioned above, it would be too strict to require that the hearer *already knows* how a response is related to the question; in the case of exhaustivity, the Quality and Quantity implicatures enable her to *figure it out*.

That is, for every possibility that the question draws attention to, the speaker should know how it depends on the information provided - which is exactly the kind of Relation implicature that enables the epistemic step.[6]

4   Discussion

Using standard conversational maxims, but assuming that they are sensitive to attentive content, I have shown that exhaustivity implicatures are genuine conversational implicatures. The current section discusses further links to the literature.

4.1   Against the Competence Assumption

In the literature, the epistemic step is often taken by invoking the contextual assumption that the speaker is *competent*, e.g. [15–17], but basically every 'Gricean' approach since John Stuart Mill [12].

1. She lacks the knowledge that q (Quantity)
2. She knows either q or $\neg q$ (Competence)

---

3. She knows $\neg q$

Aside from being very unsatisfying (it demotes exhaustivity to a case of underspecification), these approaches predict that removing the competence assumption from the context would cancel the exhaustivity implicature, which is not in fact borne out:

(12) A: I'm probably asking the wrong person - you're probably not competent about this at all - but of red, green, blue, and yellow, which colours does John like?
B: He likes red and blue. ⤳ *He doesn't like green, yellow.*

This shows that the competence is not a contextual assumption, but, rather, something conveyed by the speaker herself. Indeed, in my approach, the Quality and Relation implicatures together entail that the speaker is competent about every possibility she leaves unattended. To my awareness, my approach is the first Gricean approach of exhaustivity that can account for this.

4.2   Other Suitable Semantics

Attentive semantics is not the only suitable backbone for a pragmatic account of exhaustivity, and thinking of these richer meanings in terms of attentive content

---

[6] In the future I hope to give such general descriptions of the Relation implicature also for cases where the response is any (potentially non-singleton) subset of the question, as these are all and only the cases that may yield exhaustivity implicatures. However, so far the results for such question-response pairs have not turned out any more readable than the general characterisation in (10), and I will omit them.

is not the only way. For instance, in previous work [19] I used Ciardelli's [4] *unrestricted inquisitive semantics* (I chose attentive semantics this time because it is conceptually better motivated). A detailed comparison, also with, e.g., *alternative semantics* [1], will have to wait for another occasion. Minimally, however, a suitable semantic backbone for a pragmatic theory of exhaustivity must lack the *absorption laws*:

Definition 4 (Absorption laws). $\phi \wedge (\phi \vee \psi) \equiv \phi \equiv \phi \vee (\phi \wedge \psi)$

This is seen very easily: (5b) and (5c) can be semantically distinguished only if the absorption laws do not hold, and this is precisely the distinction to which the maxims should be sensitive for an account of exhaustivity.

4.3 Alternatives and Scales

All existing Gricean approaches in the literature treat exhaustivity as an answer to the question: 'why did the speaker not utter a *more informative* alternative?'. The problem is that mere ignorance is sufficient excuse for not giving a more informative answer, hence no exhaustivity implicature is predicted. One way to think of my approach, however, is that exhaustivity is treated as an answer to the question: 'why did the speaker not utter a *more attentive* alternative?'. Because one can draw attention to possibilities without committing to them (such as $|p \wedge q|$ in (5c)), mere ignorance is insufficient reason for not uttering the more attentive alternative, and therefore something stronger is implicated: exhaustivity.

One can (but need not) think of the present approach as relying on *scales* of alternatives, that are ordered by entailment just as in the original work of Horn [11], with the difference that entailment is now sensitive to attentive content. However, the notion of a scale is unnecessary: what counts as an alternative in a particular context is fully determined by what counts as cooperative in that context. This is defined by the maxims, which do not themselves refer to this kind of alternative. It also entails that speakers need not reason in terms of alternatives; they can just apply the maxims directly (as I did for the examples above). This perspective might be relevant for experimental work on the processing of implicatures.

4.4 The Grammatical Approach to Exhaustivity

In the introduction I mentioned the 'grammatical approach' to exhaustivity. This approach aims to attribute exhaustivity to the presence of optional, silent 'exhaustivity operators' in the grammar. The motivation for this kind of approach, which is certainly dispreferred compared to a Gricean approach for reasons of parsimony, comes mainly from two claims (as formulated by Chierchia et al. [3]). First, that the Gricean approach cannot solve the epistemic step. Second, that there are cases of exhaustivity that seem to 'arise from' an embedded position (such as the 'not all' interpretation of an existential embedded under a universal quantifier) that a Gricean theory cannot account for. I will not spend too much time on this debate, but I will briefly show why I think that both claims, in fact, bite their owner (see also [7]).

Regarding the first claim, I have of course shown at length that a Gricean approach *can*, in fact, account for the epistemic step. As for the grammatical approach, it is completely unclear to me how it could avoid the epistemic step, as its followers seem to believe it could. For how can a speaker convey, or a listener know, when and where an optional, silent exhaustivity operator should be inserted? The answer, I think, can only be: through some kind of Gricean inference schema - but this makes the grammatical approach just as susceptible to the problem of the epistemic step as existing Gricean approaches.

The second claim is of course susceptible to the same criticism: how does a speaker convey, or a hearer infer, that a silent, optional exhaustivity operator should be inserted in an *embedded* position? Again, the only answer can be: through some kind of Gricean inference schema (and the same objection holds to approaches that assume exhaustivity operators are inserted by *default* – how should the set of default usages be characterised?).[7] (Ironically, this shows, quite unintended by those who have been gathering evidence for embedded implicatures, that a Gricean theory *must* be able to account for 'embedded implicatures'.) It has never been shown that a Gricean theory cannot in principle account for implicatures that target embedded positions. The general intuition seems to be that because Gricean reasoning is post-compositional (it operates on speech acts), it cannot have access to sub-sentential structure, which would be required for 'embedded implicatures'. But clearly this doesn't hold. For instance, in attentive semantics, the possibilities in a proposition reflect the disjuncts in a sentence, and the maxims have access to those, even if the disjuncts are embedded under, say, a conjunction. More generally, why would a post-compositional pragmatic reasoner ever wilfully *ignore* the constituents of a sentence, or its derivational history, or the words that were used?

A wide range of cases of 'embedded implicature' is reported on in the literature, but also met with scepticism, and space does not permit a more detailed discussion. I am confident that my theory already accounts for some cases of embedded implicature, but also that it cannot *yet* account for all cases. At the same time, I am confident that the grammatical approach does not explain *any* of them – even though it may well *describe* them.

Acknowledgments. I am grateful to Jeroen Groenendijk, Floris Roelofsen, Donka Farkas, Kai von Fintel, Chris Cummins, the audiences of SPE6 (St. Petersburg), S-Circle (UC Santa Cruz), ESSLLI Student Session 2013 (Düsseldorf), LIRA (Amsterdam), TbiLLC 2013 (Gudauri) and many anonymous reviewers for very helpful comments. Financial support from the Netherlands Organisation for Scientific Research is gratefully acknowledged.

---

[7] This possible objection was brought to my attention by Donka Farkas.

References

1. Alonso-Ovalle, L.: Disjunction in alternative semantics. Unpublished doctoral dissertation, University of Massachusetts, Amherst (2006)
2. Büring, D.: On d-trees, beans, and accents. Linguist. Philos. **26**, 511–545 (2003)
3. Chierchia, G., Fox, D., Spector, B.: The grammatical view of scalar implicatures and the relationship between semantics and pragmatics. In: Maienborn, C., Portner, P., von Heusinger, K. (eds.) Semantics: An International Handbook of Natural Language Meaning, vol. 2, pp. 2297–2332. Mouton de Gruyter, Boston (2012)
4. Ciardelli, I.: Inquisitive semantics and intermediate logics. Master Thesis, University of Amsterdam (2009)
5. Ciardelli, I., Groenendijk, J., Roelofsen, F.: Attention! Might in inquisitive semantics. In: Ito, S., Cormany, E., (eds.) Proceedings of Semantics and Linguistic Theory, SALT XIX (2009)
6. Coppock, E., Brochhagen, T.: Raising and resolving issues with scalar modifiers. Semant. Pragmat. **6**, 1–57 (2013)
7. Geurts, B.: Quantity Implicatures. Cambridge University Press, Cambridge (2011)
8. Grice, H.: Logic and conversation. In: Cole, P., Morgan, J. (eds.) Syntax and Semantics, vol. 3, pp. 41–58. Academic Press, New York (1975)
9. Groenendijk, J., Roelofsen, F.: Inquisitive semantics and pragmatics. In: Larrazabal, J.M., Zubeldia, L., (eds.) Meaning, Content, and Argument: Proceedings of the ILCLI International Workshop on Semantics, Pragmatics, and Rhetoric (2009)
10. Groenendijk, J., Stokhof, M.: Studies on the semantics of questions and the pragmatics of answers. Unpublished doctoral dissertation, University of Amsterdam (1984)
11. Horn, L.: Towards a new taxonomy of pragmatic inference: Q-based and R-based implicatures. In: Schiffrin, D. (ed.) Meaning, Form, and Use in Context, pp. 11–42. Georgetown University Press, Washington (1984)
12. Mill, J.S.: An examination of Sir William Hamiltons philosophy (1867)
13. Roberts, C.: Information structure in discourse. In: Yoon, J., Kathol, A. (eds.) OSU Working Papers in Linguistics, vol. 49, pp. 91–136. Ohio State University, Columbus (1996)
14. Roelofsen, F.: Information and attention (Manuscript, ILLC University of Amsterdam (2011)
15. Sauerland, U.: Scalar implicatures in complex sentences. Linguist. Philos. **27**(3), 367–391 (2004)
16. Schulz, K., van Rooij, R.: Pragmatic meaning and non-monotonic reasoning: the case of exhaustive interpretation. Linguist. Philos. **29**, 205–250 (2006)
17. Spector, B.: Scalar implicatures: exhaustivity and Gricean reasoning. In: Aloni, M., Butler, A., Dekker, P. (eds.) Questions in Dynamic Semantics, pp. 225–250. Elsevier, Oxford (2007)
18. Ward, G., Hirschberg, J.: Implicating uncertainty: the pragmatics of fall-rise intonation. Language **61**(4), 747–776 (1985)
19. Westera, M.: Meanings as proposals: a new semantic foundation for Gricean pragmatics. In: Proceedings of SemDial 2012 (2012)

First-Order Conditional Logic
and Neighborhood-Sheaf Semantics
for Analysis of Conditional Sentences

Hanako Yamamoto[✉] and Daisuke Bekki

Ochanomizu University, Bunkyo, Japan
{yamamoto.hanako,bekki}@is.ocha.ac.jp

1   Introduction

In this study, we define the neighborhood-sheaf semantics (NSS) of VC^+, a first-order conditional logic system [10, Chaps. 5 and 19]. NSS was proposed in [5]. Additionally, we prove that the traditional Kripke semantics of VC^+ can be constructed in terms of NSS.

Neighborhood semantics is more general than Kripke semantics because it allows a family of sets of possible worlds "near" a certain world to have more than two members. On the other hand, Kripke semantics of conditional logic uses special Kripke frames in which accessibility relations of a world vary depending on formula. This special frame is properly represented by NSS.

Sections 2 and 3 provide an overview of NSS and first-order conditional logic. In Sect. 4 we define NSS of VC^+ and prove the equivalence between NSS and Kripke semantics of VC^+.

2   First-Order Conditional Logic

Conditional logic [7,8,12] is a kind of modal logic. It can represent implicit premises, which is impossible with classical logic. This is necessary for the analysis of conditional sentences because of the examples given below.

Classical logic has a property known as *antecedent strengthening*:

$$A \rightarrow B \vdash A \wedge C \rightarrow B$$

According to this property, the following inference is predicated to be valid [10]:

(1) If it does not rain tomorrow we will go to the cricket. Hence, if it does not rain tomorrow and I am killed in a car accident tonight then we will go to the cricket.

However, this is unnatural because a dead person cannot go to the cricket. Formally, the unnaturalness arises from adding an antecedent which includes states that contradict the consequence.

Conditional sentences omit states implicitly contained in the antecedent. Classical logic cannot express implicit states, and thus fails to exclude unnatural

© Springer International Publishing Switzerland 2014
Y. Nakano et al. (Eds.): JSAI-isAI 2013, LNAI 8417, pp. 154–164, 2014.
DOI: 10.1007/978-3-319-10061-6_11

infcrences such as in the example above. Conditional logic represents clauses that include implicit premises known as *ceteris paribus clauses*.

C was proposed as a conditional logic system [7,8,12], and different systems have been designed by extending C.

2.1 Modal Logic and Possible-World Semantics

This section reviews the definitions of three systems of propositional modal logic and their semantics.

Modal logic represents necessity and possibility. It introduces a propositional modal language L with two unary operators:

- $\Box\varphi$: necessarily φ
- $\Diamond\varphi$: possibly φ ($\Diamond\varphi \equiv \neg\Box\neg\varphi$)

Generally, possible-world semantics is used as semantics of modal logic [6]. The interpretation of the modal operator $[\![\Box]\!]$ varies in accordance with the semantics.

Definition2.1

Possible-world semantics of propositional logic is defined as follows:

- $(X, [\![\cdot]\!])$: model
  X is a set of possible worlds. $[\![\cdot]\!]$ is an interpretation. For an arbitrary formula φ, $[\![\varphi]\!] \subseteq X$. The truth value of a certain formula is a subset of X consisting of the possible worlds in which the formula is true. If the interpretation is trivial, the model is simply written as X.
- Interpret the n-ary operator $\otimes$ as follows:
  $[\![\otimes(\varphi_1, ...\varphi_n)]\!] = [\![\otimes]\!]([\![\varphi_1]\!], ...[\![\varphi_n]\!])$
- Interpret propositional connectives as operators on sets:
  $[\![\wedge]\!] = \cap$, $[\![\vee]\!] = \cup$, $[\![\neg]\!] = X\backslash-$

Validity is defined as follows:

- A formula φ is valid in $(X, [\![\cdot]\!])$ iff $[\![\varphi]\!] = X$
- An inference $\varphi \models \psi$ is valid iff $[\![\varphi]\!] \subseteq [\![\psi]\!]$ in any model

---

$[\![\Box]\!]$ in each semantics is defined as follows:

- In Kripke semantics, equip X with a binary relation $R_X \subseteq X \times X$ referred to as *accessibility relation*. Define $[\![\Box]\!] : \mathcal{P}X \to \mathcal{P}X$ as follows:

$$w \in [\![\Box]\!](A) \text{ iff } u \in A \text{ for all } u \subset X \text{ such that } R_X wu$$

(X, R_X) is known as a Kripke frame.
- In topological semantics, define $[\![\Box]\!] = \mathbf{int} : \mathcal{P}X \to \mathcal{P}X$ (an interior operation).

$$w \in \mathbf{int}(A) \text{ iff there exists } U \in \mathcal{O}X \text{ such that } w \in U \subseteq A$$

– In neighborhood semantics, define $[\![\Box]\!] = \mathbf{int}$ as follows by using the neighborhood function $\mathcal{N}_X : X \to \mathcal{PP}X$:

$$w \in \mathbf{int}(A) \text{ iff } A \in \mathcal{N}_X(w)$$

$(X, \mathcal{N}_X)$ is known as a neighborhood frame.

The hierarchy among systems of modal logic arises depending on whether a system employs each of the following five rules:

$$\frac{\varphi \vdash \psi}{\Box\varphi \vdash \Box\psi} \qquad\qquad \mathbf{M}$$

$$\frac{\vdash \varphi}{\vdash \Box\varphi} \qquad\qquad \mathbf{N}$$

$$\Box\varphi \wedge \Box\psi \vdash \Box(\varphi \wedge \psi) \qquad\qquad \mathbf{C}$$

$$\Box\varphi \vdash \varphi \qquad\qquad \mathbf{T}$$

$$\Box\varphi \vdash \Box\Box\varphi \qquad\qquad \mathbf{4}$$

Since each of the five rules can be rephrased as a property of a neighborhood function, each of the various modal logic systems has a corresponding neighborhood frame. Accessibility relations and interior operations can be constructed by neighborhood functions.

– An accessibility relation in Kripke semantics is represented by a Kripke neighborhood frame, which has the following property [1]:

There exists a map $\mathbf{R} : X \to \mathcal{P}X$ such that $(A \in \mathcal{N}(w)$ iff $\mathbf{R}(w) \subseteq A)$

This equivalence implies

$$w \in \mathbf{int}(A) \text{ iff } \mathbf{R}(w) \subseteq A \text{ iff } u \in A \text{ for all } u \in X \text{ such that } R_X wu$$

– An interior operation in topological semantics can be replaced with a neighborhood function which satisfies five rules given above.

$$w \in \mathbf{int}(A) \text{ iff } A \in \mathcal{N}(w)$$

Neighborhood frames add certain properties that can represent accessibility relations or interior operations. As mentioned above, neighborhood semantics generalizes Kripke and topological semantics.

2.2 First-Order Conditional Logic Systems

Priest [10][1] introduces several conditional logic systems. First, consider a set of formulas $\mathcal{L}$ in C, which is a language with a binary connective $>$. Its semantics is defined by a Kripke frame formed as $(X, \{R_{X,A} : A \in \mathcal{F}\})$. Unlike an ordinary Kripke frame, the accessibility relation changes in accordance with the formula. Define $N^w_{X,A} = \{x \in X : R_{X,A}wx\}$. Then the truth condition for $A > B$ is as follows:

$$w \in [\![A > B]\!] \text{ iff } N^w_{X,A} \subseteq [\![B]\!]$$

The system C^+ is constructed by adding the following conditions to C [3]:

1. $N^w_{X,A} \subseteq [\![A]\!]$
2. If $w \in [\![A]\!]$, then $w \in N^w_{X,A}$.

CC and CC^+ are first-order versions of C and C^+, respectively, each of which has a constant domain that does not vary with respect to possible worlds. VC and VC^+ are systems constructed by extending CC and CC^+ by allowing the domain to vary according to possible worlds. Such systems can consider situations where existing objects vary with the possible worlds. For the variable domain, introduce an existential predicate $\mathfrak{E}$. $\mathfrak{E}a$ can be considered as 'a exists'. Interpretation of quantifiers is allowed only within $[\![\mathfrak{E}]\!]$. An advantage of such systems is that rules of quantifiers are formalized for nonexistent objects as well.

3 Neighborhood-Sheaf Semantics (NSS)

Neighborhood-sheaf semantics (NSS) is possible-world semantics of first-order modal logic and generalizes Kripke-sheaf and topological-sheaf semantics.

There are propositional versions of these three types of semantics, and similar generalizations hold. This relation is extended to the first-order level in [5]. Figure 1 shows nine semantics and their extended relations. In [5], NSS (2 in the figure) is proposed as an extension of 5 and a generalization of 1 and 3. The third row corresponds to semantics of propositional logic, and the other rows correspond to semantics of quantified logic. In this regard, [1] extended 7 to 5 via 8. [6] extended 7 to 4, and [4] extended 4 to 1. Furthermore, [11] and [9] extended 9 to 8, and [2] extended 9 to 5 and 9 to 3 via 6.

In Fig. 1(a)–(d) next to the arrows indicate how systems are extended or generalized:

(a) Generalization of topological semantics by considering interior operations that are more general than topological ones.

---

[1] In this section, conditional logic is described along the lines of NSS for convenience in order to represent the semantics of conditional logic using NSS after this section.

(b) Generalization of accessibility relations in Kripke semantics with the notion accessibility in neighborhood semantics.

(c) Interpretation of first-order vocabulary with a domain D of possible individuals; in particular, interpreting the "transworld identity" of individuals with the identity of elements of D.

(d) Interpretation of first-order vocabulary and transworld identity on the basis the structure of a sheaf over the set of possible worlds.

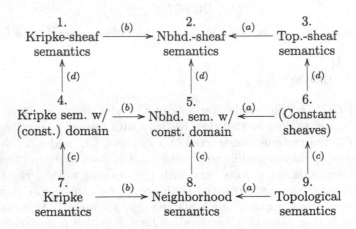

Fig. 1. Semantics of modal logic

3.1 Bundle Interpretation

The semantics of first-order modal logic corresponding to Kripke, topological and neighborhood semantics are Kripke-sheaf, topological sheaf and neighborhood-sheaf semantics, respectively. Generalizations hold in the three semantics based on generalizations at the propositional level [5]. To define the three semantics, we start by introducing semantics for operators in first-order logic. This is referred to as bundle interpretation, which is common to all three semantics systems, as described in the top row of Fig. 1. The interpretation of the modal operator $[\![\Box]\!]$ varies depending on the semantics. This section introduces the concept of bundle interpretation.

In bundle interpretation, a domain of first-order logic is interpreted on the basis of the structure of a sheaf over the set of possible worlds. Bundle interpretation uses a slice category $\mathbf{Sets}/X$, where X is a set of possible worlds. Consider a surjection $\pi : D \to X \in ob(\mathbf{Sets}/X)^2$. Each $w \in X$ has its inverse image $D_w = \pi^{-1}[\{w\}]$ (known as *fibre* over w). A domain D is the *bundle* of all fibres taken over X, written as $D = \Sigma_{w \in X} D_w$. This means that D is the disjoint

---

[2] For a category $\mathcal{C}$, $ob(\mathcal{C})$ means a set of all objects of $\mathcal{C}$ and $mor(\mathcal{C})$ means a set of all morphisms of $\mathcal{C}$.

union of all D_w. Fixing a world amounts to semantics of ordinary first-order logic. Products and operators are similarly defined as follows:

- n-fold product of $D : D^n = \Sigma_{w \in X} D_w^n$
  (n-fold *fibred* product of D over X)
- An operator $f \in mor(\mathbf{Sets}/X)$
  (an arrow from $\pi_D : D \to X$ to $\pi_E : E \to X$) is a map represented as
  $f = \Sigma_{w \in X} f_w$ where $f_w : D_w \to E_w$
- A structure $\mathfrak{M}$ is defined as $\mathfrak{M} = \Sigma_{w \in X} \mathfrak{M}_w$
  ($\mathfrak{M}_w$ is a structure of first-order logic on w)
- An interpretation is defined as $[\![\cdot]\!] = \Sigma_{w \in X} [\![\cdot]\!]_w$
  ($[\![\cdot]\!]_w$ is an interpretation on w)

Bundle interpretation is determined by a model $(\mathfrak{M}, [\![\cdot]\!])$.

$$\mathfrak{M} \begin{cases} \text{A surjection} \pi : D \to X \\ \quad F^{\mathfrak{M}} \subseteq D^n (F \text{ is an } n\text{-ary predicate}) \\ \text{Arrows of } \mathbf{Sets}/X \\ \quad f^{\mathfrak{M}} : D^n \to D (f \text{ is } n\text{-ary function}) \\ \text{A constant } c \\ \quad \text{An arrow of } \mathbf{Sets}/X \; c^{\mathfrak{M}} : D^0 (= X) \to D \end{cases}$$

$$[\![\cdot]\!]^3 \begin{cases} \text{An } n\text{-ary predicate } F : [\![\bar{x}|F\bar{x}]\!] = F^{\mathfrak{M}} \; (\bar{x} \in D^n) \\ n\text{-ary operators } \otimes : \\ \quad [\![\otimes(\varphi_1, \ldots \varphi_n)]\!] = [\![\otimes]\!]([\![\varphi_1]\!], \ldots [\![\varphi_n]\!]) \\ \text{Connectives} : \\ \quad [\![\wedge]\!] = \cap, \; [\![\vee]\!] = \cup, \; [\![\neg]\!] = D^n \backslash - (n \in \mathbb{N}) \\ \text{Quantifiers} : [\![\bar{x}|\exists y \varphi]\!] = p_n[\![\bar{x}, y|\varphi]\!] \\ \quad (\; p_n : D^{n+1} \to D^n :: (a_1, \ldots, a_n, b) \mapsto \bar{a} \;) \end{cases}$$

Validity is the same as described in Definition 2.1.

3.2 Sheaf Semantics

Kripke-sheaf, topological-sheaf and neighborhood -sheaf semantics are constructed as follows. Conditions vary depending on the semantics.

- Replace $\mathbf{Sets}/X$ with a more limited slice category. This limitation adds conditions to components of the structure.
- Define $[\![\Box]\!]$.
  To interpret $\Box$, define *transworld identity*, which is the relation between identified objects which live in different worlds.

Kripke-sheaf semantics uses $\mathbf{Krsh}/(X, R_X)$, a slice category of Kripke sheaves over a Kripke frame (X, R_X). All objects and arrows in the structure are Kripke sheaves.

---

[3] $\bar{a}$ means $(a_1, \ldots, a_n)$.

Definition 3.1

Given Kripke frames (X, R_X) and (D, R_D), a map $\pi : D \to X$ is a p-morphism under the following conditions.

- π is monotone, i.e., if $R_D ab$, then $R_X f(a) f(b)$
- If $R_X \pi(a) w$, there exists $b \in D$ such that $R_D ab$ and $w = \pi(b)$

A p-morphism $\pi : D \to X$ is a *Kripke sheaf* over (X, R_X) if such $b \in D$ as in the definition of p-morphism above is unique, i.e., if every pair $u, w \in X$ such that $R_X uw$ has a map $C_{uw} : D_u \to D_w$ which satisfies the following equivalence:

$$\text{For any pair } a \in D_u \text{ and } b \in D_w, \; R_D ab \text{ iff } C_{uw}(a) = b$$

The definition of $[\![\Box]\!]$ is the same as for Kripke semantics. Accessibility relations on D^n, R_{D^n} are defined by using R_D:

$$R_{D^n}(\bar{a}, \bar{b}) \text{ iff for all } i \leq n, \; R_D a_i b_i$$

Topological-sheaf semantics uses $\mathbf{LH}/X$, which is a slice category of sheaves over a space X. All objects and arrows in the structure are sheaves.

Definition 3.2

Given topological spaces D and X, a map $\pi : D \to X$ is a homeomorphism if it is a continuous bijection with a continuous inverse.
A continuous map $\pi : D \to X$ is a *local homeomorphism* if every $a \in D$ has some $U \in \mathcal{O}D$ such that $a \in U$, $\pi[U] \in \mathcal{O}X$, and $\pi|_U : U \to \pi[U]$ is a homeomorphism.
Such a pair (D, π) is known as a *sheaf* over X.

Definition of $[\![\Box]\!] : \mathcal{P}D^n \to \mathcal{P}D^n :: [\![\bar{x}|\varphi]\!] \mapsto [\![\bar{x}|\Box\varphi]\!]$ is $\mathbf{int}_{D^n}$.

NSS uses $\mathbf{LI}/X$ (a slice category of neighborhood sheaves over MC neighborhood frame $(X, \mathcal{N}_X)$, which satisfies modal logic rules $\mathbf{M}$ and $\mathbf{C}$), and all objects and arrows in the structure are neighborhood sheaves.

Defenition 3.3

Given MC neighborhood frames $(X, \mathcal{N}_X)$, $(D, \mathcal{N}_D)$, a map $\pi : D \to X$ is a *local isomorphism* under the following conditions.

- π is an open map, i.e., for all $B \subseteq X$, $\pi^{-1}[\mathbf{int}_X(B)] = \mathbf{int}_D(\pi^{-1}[B])$
- For every $a \in D$ such that $\mathcal{N}_D(a) \neq \emptyset$, there is $U \in \mathcal{N}_D(a)$ such that $\pi|_U : U \to \pi[U]$ is bijective.

Such a pair (D, π) is known as a *neighborhood sheaf over* $(X, \mathcal{N}_X)$.

$[\![\Box]\!]$ on D^n is defined in the same way as for neighborhood semantics, $[\![\Box]\!] = \mathbf{int}_{D^n}$. $\mathcal{N}_{D^n}$ is defined by using $\mathcal{N}_D$ as follows:

Given $\bar{a} \in D^n$, define $\mathcal{B} : D^n \to \mathcal{PPD}^n$ such that

$$\mathcal{B}(\bar{a}) = \{\cap_{i \leq n}(q_n^i)^{-1}[U_i] | \forall i \leq n, U_i \in \mathcal{N}_D(a_i)\}$$
$$(q_n^i : D^n \to D :: (a_1, \ldots, a_n) \mapsto a_i)$$

And then set.

For a set A, $A \in \mathcal{N}_{D^n}(\bar{a})$ iff there is $B \in \mathcal{B}(\bar{a})$ such that $B \subseteq A$.

MC neighborhood frames subsume Kripke frames and topological spaces, and neighborhood sheaves subsume Kripke sheaves and topological sheaves. Similar generalization holds for transworld identities. From the above, we can conclude that NSS generalizes Kripke-sheaf and topological-sheaf semantics. As in the case of propositional logic, NSS can freely decide what modal logic rules it contains, and thus corresponds to the semantics of various systems of first-order modal logic.

4   Neighborhood-Sheaf Semantics of First-Order Conditional Logic

This section defines NSS of VC^+ by assigning formula indices to neighborhoods of possible worlds. From Sect. 2, the high generality of NSS is based on the following features:

- The family of neighborhoods for each world is allowed to have more than two members
  (in Kripke semantics, every family is a singleton.)
- Neighborhood frames requires few rules
  (topological spaces do not require all the five rules.)

As mentioned in Sect. 2, the semantics of conditional logic uses a special Kripke frame in which there is an accessibility relation for each formula. NSS and Kripke semantics of conditional logic shows a similarity that leads us to the idea of describing semantics of first-order conditional logic by using NSS. Furthermore, NSS has the advantage that it can immediately formalize the existential predicate and the variable domain. These notions are obviously important for formalization of natural language semantics because otherwise at a world we cannot formalize sentences including non-existent objects. But in Kripke semantics in [10], these notions are defined additionally by introducing a set of objects and defining the relation between the set and truth value. In this respect, NSS is more proper semantics than Kripke semantics for natural language.

The construction of NSS of first-order conditional logic proves the higher generality of NSS, namely that it subsumes Kripke semantics of conditional logic as well as ordinary Kripke semantics. In the next section, we define NSS of VC^+ and prove that Kripke semantics of VC^+ can be constructed from proposed NSS.

4.1 NSS of VC^+

The basic idea behind the definition of NSS of VC^+ is to assign formula indices to neighborhoods such that the accessibility relation in each world varies depending on the formula.

Let us restate the syntax of first-order conditional logic:

$$\mathcal{F} ::= p \mid P(\tau_1, \ldots \tau_n)(P : n\text{-ary predicate}) \mid \neg\mathcal{F} \mid \mathcal{F} \wedge \mathcal{F} \mid \mathcal{F} \vee \mathcal{F} \mid \mathcal{F} \rightarrow \mathcal{F}$$
$$\mid \Box\mathcal{F} \mid \Diamond\mathcal{F} \mid \mathcal{F} > \mathcal{F} \mid \forall x\mathcal{F} \mid \exists x\mathcal{F} \mid \mathfrak{E}\tau$$
$$\tau ::= c \mid x$$

Several additional conditions (denoted with (*)) must be added to the rules in Definition 3.1 to obtain NSS.

- For every $\bar{a} \in D^n$ $(n \in \mathbb{N})$, $\mathcal{N}_{D^n}(\bar{a})$ has at most countable members. This condition stems from the fact that formulas are at most countable.
- Formulas assigned as indices to each point in D^n $(D^0 = X)$ have n free variables. This is because the interpretation of each formula with n free variable is a subset of D^n in NSS, whereas [10] defines the accessibility relation only in X. Thus NSS differs from Kripke semantics in this respect.
- The interpretations of formulas assigning a constant to a free variable are defined as follows. Consider a formula A whose only free variables are $(\bar{x}, y)$ $(\bar{x} \in D^n)$. Let $A(\bar{x}, y)[c/y]$ be a formula such that a constant c is assigned to a free variable y in A. Then

$$[\![\bar{x}|A(\bar{x}, y)[c/y]]\!] = \Sigma_{w \in X}\{\bar{x} \in D^n_w | (\bar{x}, c^{\mathfrak{M}}(w)) \in [\![\bar{x}|A\bar{x}]\!]\}$$

Definition 4.1

Consider that $\pi : D \rightarrow X$ is a neighborhood sheaf satisfying (*) and following conditions:

- Take $N^{\bar{a}}_{D^n, A}$ be the neighborhood of a point $\bar{a} \in D^n$ with formula A as an index, A has n free variables. For every $\bar{a} \in D^n(n \geq 1)$ and every formula A, the following relation holds between $N^{\bar{a}}_{D^n, A} \in \mathcal{N}_{D^n}(\bar{a})$ and $N^{\pi^n(\bar{a})}_{X, A\bar{b}} \in \mathcal{N}_X(\pi(a))$:

$$N^{\bar{a}}_{D^n, A} = \left(\bigcup_{\bar{b} \in [\![\bar{x}|A\bar{x}]\!]} M_{\bar{b}}\right)$$

$(M_{\bar{b}} \in \mathcal{N}_{D^n}(\bar{a})$ satisfing $\pi^n(M_{\bar{b}}) = N^{\pi^n(\bar{a})}_{X, A\bar{b}} \in \mathcal{N}_X(\pi^n(\bar{a})))$
- $D^0 = X$ satisfies the conditions of C^+.

The interpretation of $>$ is defined as follows:

$$\bar{a} \in [\![\bar{x}|(A > B)\bar{x}]\!] \text{ iff } N^{\bar{a}}_{D^n, A} \subseteq [\![\bar{x}|B\bar{x}]\!]$$

The following lemma holds immediately:

Lemma 4.1

In NSS, if X satisfies the conditions of C^+ **1** and **2**, then D^n $(n \geq 1)$ also satisfies these conditions.

Next, we construct Kripke-sheaf semantics of VC^+ from NSS in a similar way as mentioned in Sect. 3.1.

Consider a family of maps as follows:

$$\left\{ \begin{array}{l} \boldsymbol{R}_A : D^n \to \mathcal{P}D^n| \\ A \text{ has } n \text{ free variables } (n \in \mathbb{N}), \text{ for } \bar{a} \in D^n, \ \boldsymbol{R}_A(\bar{a}) = N_{D^n,A}^a \end{array} \right\}_{A \in \mathcal{F}}$$

Define $\bar{a} \in [\![\bar{x}|(A > B)\bar{x}]\!]$ iff $\boldsymbol{R}_A(\bar{a}) \subseteq [\![\bar{x}|B\bar{x}]\!]$. Then, the following equivalence holds:

$$\bar{a} \in [\![\bar{x}|(A > B)\bar{x}]\!] \text{ iff for all } \bar{b} \in D^n \text{ such that } R_{D^n,A}\bar{a}\bar{b}, \ \bar{b} \in [\![\bar{x}|B\bar{x}]\!]$$

It is constructed in a similar way as 3.1 such that in every $D^n (n \in \mathbb{N})$, and thus semantic equivalence between Kripke semantics and NSS holds. This completes the proof of equivalence between NSS and Kripke-sheaf semantics.

Theorem 4.1

Kripke-sheaf semantics can be constructed from NSS for any VC^+ system.

---

5   Conclusion and Future Work

In this study, we defined NSS for first-order conditional logic. Kripke-sheaf semantics for first-order conditional logic was constructed from the NSS. The two systems of semantic systems were proven to be equivalent. A further investigation would be needed in order to give NSS for stronger conditional logics than VC^+, to see if NSS is also applicable to such extended systems.

References

1. Arló-Costa, H., Pacuit, E.: First-order classical modal logic. Stud. Log. **84**(2), 171–210 (2006)
2. Awodey, S., Kishida, K.: Topology and modality: the topological interpretation of first-order modal logic. Rev. Symbolic Log. **1**, 146–166 (2008)
3. Chellas, B.: Modal Logic : An Introduction. Cambridge University Press, Cambridge (1980)
4. Goldblatt, R.: Topoi: The Categorial Analysis of Logic. Dover Publications, Mineola (2006)
5. Kishida, K.: Neighborhood-sheaf semantics for first-order modal logic. Electron. Notes Theor. Comput. Sci. **278**, 129–143 (2011)

6. Kripke, S.: Semantical considerations on modal logic. Acta Philosophica Fennica **16**, 83–94 (1963)
7. Lewis, D.K.: Counterfactuals. Blackwell, Oxford (1973)
8. Lewis, D.K.: Counterfactuals and comparative possibility. Journal of Philosophical Logic 2 (1973).
9. Montague, R.: Universal grammar. Theoria **36**(3), 373–398 (1970)
10. Priest, G.: An Introduction to Non-classical Logic: From If to Is. Cambridge Introductions to Philosophy, 2nd edn. Cambridge University Press, Cambridge (2008)
11. Scott, D.: Advice on modal logic. In: Lambert, K. (ed.) Philosophical Problems in Logic: Some Recent Developments, pp. 143–173. Reidel, Dordrecht (1970)
12. Stalnaker, R.: A Theory of Conditionals, in Studies in Logical Theory. American Philosophical Quarterly Monograph Series, vol. 2. Basil Blackwell, Oxford (1968)

JURISIN

Requirements of Legal Knowledge Management Systems to Aid Normative Reasoning in Specialist Domains

Alessio Antonini[1](✉), Guido Boella[1], Joris Hulstijn[2], and Llio Humphreys[3]

[1] Department of Informatics, Università Degli Studi di Torino, Turin, Italy
{antonini,boella}@di.unito.it
[2] Faculty of Technology, Policy, and Management,
Delft University of Technology, Delft, The Netherlands
J.Hulstijn@tudelft.nl
[3] ICR, University of Luxembourg, Luxembourg, Luxembourg
llio.humphreys@unilu.lu

Abstract. This paper discusses the challenges of legal norms in specialist domains - the interplay between industry/professional standards and legal norms, the information gap between legal and specialist domains and the need for interpretation at all stages of compliance - design, operation and justification. We propose extensions to the Eunomos legal knowledge management tool to help address the information gap, with particular attention to aligning norms with operational procedures, and the use of domain-specific specialist ontologies from multiple domains to help users understand and reason with norms on specialist topics. The paper focuses mainly on medical law and clinical guidelines.

Keywords: Legal ontology · Knowledge representation · Norm compliance · Ontology alignment · Clinical guidelines

1 Introduction

The objects of legal norms are non-legal, real-life entities. In practice, complying with norms involves interpreting real life from a legal perspective. It follows that legal practitioners need to understand real life entities to evaluate legal compliance. However, specialist domains such as healthcare, finance and architecture have their own lexicon, ontologies and standards of behaviour. In practice, standards of behaviour have different normative status (absolute, defeasible, advice), and this difference can be attributed to the relative unpredictability of events in a particular domain as well as the level of authority and expertise expected of particular roles. In general, industrial or professional standards and legal norms can exist separately and independently, until some event triggers the need to analyse the specialist standards from a legal point of view.

This research is part of the research project ITxLaw supported by the Compagnia di San Paolo.

© Springer International Publishing Switzerland 2014
Y. Nakano et al. (Eds.): JSAI-isAI 2013, LNAI 8417, pp. 167–182, 2014.
DOI: 10.1007/978-3-319-10061-6_12

The motivation for our research is to find a pragmatic way to bridge the gap between legal and specialist knowledge by enriching legal knowledge management systems with domain-specific ontologies and rules.

In this paper we address the following research questions:

1. What is the relationship between specialist and legal domains?
2. How does legal reasoning in specialist domains differ from legal reasoning in general?
3. How can legal knowledge management systems provide authoritative and comprehensive information to assist legal reasoning in complex specialist areas of law?

The paper is structured as follows. In Sect. 2 we describe different situations where standards from legal and specialist domains interact. In Sect. 3 we describe a legal knowledge management system that is a suitable starting point for cross-domain analysis, norm representation and retrieval, mapping to business processes and professional standards, and viewing ontologies from different domains. In Sect. 4 we compare our proposed solution with other related work. Finally, in Sect. 5 we summarise the contribution of the paper.

2 Specialist Domains in Law

There are many specialist legal rules governing aspects such as quality of products, procedures or organizational roles. They are often complemented by industry-wide guidelines. Failure to comply with such guidelines may or may not have legal consequences. Other possible consequences are social disapproval or professional expulsion. Such guidelines may influence legal reasoning, but they must not explicitly contravene norms or legal principles. Guidelines may be written for a range of different reasons - to introduce new improved standards, to standardise and justify cost-cutting procedures or to make practical recommendations on how to fulfill legal obligations. The latter is becoming increasingly important due to a recent trend for goal-based legal norms, where the details of compliance are largely left to industry and monitoring bodies to resolve [1–3].

2.1 Independent Specialist and Legal Norms

Industry standards and legal norms can exist separately and independently, until some event triggers the need for industry standards to become legally relevant.

Example 1. A cookie factory may prescribe a rule stating that *a cookie is only acceptable if its shape is almost perfectly round with a 0.1 % margin of error.* This rule concerns a feature that is usually irrelevant for the law e.g. the relative roundness of a cookie is unlikely to be the subject of consumer complaint. However, if an employee is dismissed because (s)he cannot produce cookies to the required standard of roundness, the reasonableness of the roundness rule could be analysed in a trial for unfair dismissal.

2.2   Goal-Based Norms for Specialist Domains

Regulatory compliance in the financial sector has been increasing in volume and complexity in recent years. It is a highly international sector, and the fact that norms can come from different sources (European directives, national legislation, regulators), and are frequently updated, makes it very difficult for organisations to keep track of their legal obligations. Moreover, organisations have to interpret and adapt norms to their own business processes, which are multiple and differ from those of other organisations. Companies and auditors, who carry out audits to verify that a company is adhering to regulations, have to navigate compliance problems in an environment of uncertainty [4]. Companies and auditors have to decide between different interpretations of norms and legal concepts used in the domain and how these apply to specific business processes. Moreover, both parties have to decide which aspects of legislation to focus on in adoption or enforcement. The details of such decisions matter. These details need to be tracked regularly, as both the business processes or the legislation are subject to frequent change. From a compliance point of view, interpretation of how norms apply to specific processes takes place at two separate stages: first by the auditee in designing and adapting business processes, and secondly by the auditor when assessing the compliance of the auditee. The auditee usually attempts to interpret the law in a way that anticipates how the auditor is going to interpret it. The auditees task is made difficult by the fact that auditors can interpret norms more or less strictly, depending on public opinion and the policies of the day. Occasionally, auditees may decide it is not worth conforming to certain norms when the cost of implementation is great compared to the fine or loss of reputation they would face in case of non-compliance.

Compliance involves assessing the organisation's business processes to see whether they comply with norms - which is often difficult to do with certainty. Norms are purposefully general so that they can cover a range of different scenarios, including unanticipated future developments [5,6]. By contrast, business processes and the computer systems that support them are specific to a situation.

Example 2. The Markets in Financial Instruments Directive (MIFID) is an EU directive that aims to increase competition and provide consumer protection in investment services. The directive covers most tradeable financial products, such as brokerage, advice, dealing, portfolio management, underwriting etc. MIFID requires financial institutions to archive and provide frequent reports about transactions. The following article has been subject to much debate to determine how much information and advice organisations are required to ensure the most favourable terms:

> (33) It is necessary to impose an effective best execution obligation to ensure that investment firms execute client orders on terms that are most favourable to the client. This obligation should apply to the firm which owes contractual or agency obligations to the client (Directive 2004/39/EC).

2.3 Soft Legal Norms in Complex, Unpredictable Specialist Domains

The interplay between legal and medical standards is complex. In most jurisdictions, physicians have to adhere to a set of rules on professional behaviour
defined by an Order of Physicians, and the Order of Physicians can expel those
who fail to follow these rules. Victims of medical malpractice or maladministra-
tion also have the right to compensation, and this is a matter to be determined
by a court of law. The required standard of care, however, refers to professional
standards (as is the case for accountants, notaries and lawyers). In recent years,
best practice is increasingly codified by domain experts in the form of clinical
guidelines.

Clinical guidelines (CG) are carefully designed procedures that have been
subject to scrutiny by domain experts. The key reason for their popularity is the
difficulty for medical practitioners to keep abreast with latest developments in
medical research, as medical knowledge bases continue to expand and the scruti-
nability of scientific research suffers. Clinical guidelines also seek to address the
issue of variable patient care, and the presumption that at least some of this variation is due to inferior service [7, p. 527], by standardising procedures. Other
benefits attributed to clinical guidelines are that they "offer explicit recommendations for clinicians who are uncertain about how to proceed, overturn the
beliefs of doctors accustomed to outdated practices, improve the consistency
of care, and provide authoritative recommendations that reassure practitioners about the appropriateness of their treatment policies." Nevertheless, clinical
guidelines also pose risks. Woolf et al. [7, p. 529] claim that "[g]uideline development groups often lack the time, resources, and skills to gather and scrutinise
every last piece of evidence. Even when the data are certain, recommendations
for or against interventions will involve subjective value judgements when the
benefits are weighed against the harms. The value judgement made by a guideline development group may be the wrong choice for individual patients."

The development, usage and prescriptive status of clinical guidelines varies
according to medical topic and jurisdiction. The United Kingdom develops
guidelines in consensus conferences. In the United States, some evidence-based
guidelines are produced by government panels and medical societies but many
healthcare organisations purchase commercially produced guidelines that empha-
sise shortened lengths of stay and other resource savings. In common law countries such as Canada, adherence to clinical guidelines is not necessarily regarded
as an indication of reasonable standard of care, since the guidelines may be over-
idealistic, novel, or less than the expected standard [8]; nevertheless, there have
been some cases where clinical guidelines have been used successfully both by
plaintiff and defendant lawyers [9] to support arguments for and against convic-
tion for malpractice.

Italy is an interesting example of how recommendations from clinical
guidelines can become legal norms. While there are *national* legislation about
medical practice, most norms come from *regional* law, and they often incorporate international CGs while taking into account the local healthcare system.

Each hospital trust has internal regulations and procedures designed to be compliant with these regional norms. The norms are written from a management rather than medical point of view - they are more about performance, outcomes, and efficient use of resources than the precise details of what physicians should do.

In practice, norms from clinical guidelines are treated in different ways by different actors. Hospital managers and clinicians have different permissions of interpretation. Managers have to show that their operations strictly align with the letter of the norm(s) as can be seen in the following example from a clinical guideline on minimal staff requirements for intensive care departments [10]:

Example 3. Minimal requirement for intensive care department - Functional criteria - Multidisciplinary approach. Besides the 24 h coverage of the medical staff of the ICU (Intensive Care Unit), the following physicians should be on call and available:

Specialization	1	2	3
Anaesthesiologist	E	E	E
General surgeon	E	E	E
Neurosurgeon	E	E	O
Cardiovascular surgeon	E	D	O
Thoracic surgeon	E	D	O
...	.	.	.

E = essential, D = desirable and O = optional, and 1, 2 and 3 level of intensity of care (1 is the highest).

The CG specifies staff requirements for three levels of intensity for intensive care units. Those requirements specify the competence of physicians and the strength of the requirement, defined as essential, desirable or optional. In a typical hospital, it is always possible to find most of the specialisations above from other units, but it can be difficult to find a neurosurgeon. If the hospital claims that it has an ICU 1 or 2, it implies that the hospital has one or more physicians with a specialization in neurosurgery who have a contract as a neurosurgeon which contains a clause of availability. We can read this guideline as a way to say that an intensive care unit 1 or 2 cannot be opened in hospitals without all the units marked as essential.

In contrast, physicians are permitted to reason more freely. They can choose to implement the guidelines or adhere to the overall goal in another reasonable way. Indeed, they are obliged not to follow the guidelines if there are good medical (but not efficiency) reasons for doing so. Bottrighi et al. [11] point out that in practice, CGs are integrated with basic medical knowledge (BMK) to cope with real cases and uncertain conditions. The following example from [11] shows a conflict arising in recommendations from clinical guidelines and basic medical knowledge (BMK):

Example 4. CG: Patient with acute myocardial infarction presenting with acute pulmonary edema; before performing coronary angiography it is mandatory to treat the acute heart failure.

BMK: The execution of any CG may be suspended, if a problem threatening the patient's life suddenly arise. Such a problem has to be treated first.

If the physician does not adhere to the guidelines with negative consequences, the physician may seek to justify his/her actions by claiming that the relevant clinical guidelines do not convey a 'reasonable standard of care'. Such a claim that would only be accepted in law if also accepted by the medical community. However, there may be different opinions in the medical community, leading to a "battle of experts" [12]. The legal domain would become the ultimate arbiter among different medical standards in this scenario.

   Traditionally, legal norms are often classified [13] as prescriptive norms (obligations), constitutive norms and specialist norms (procedures). Many guidelines are more in the form of advice than obligations, and for open norms in particular, there is much professional judgement required to implement them in the best way depending on the context. The current tendency in law is to interpret CGs as terms of a contract between health care service providers and patients. This interpretation only partially captures the meaning of CGs. CG describe ideal conditions, which may be fine for hospital management, but can cause problems in the application of medical standards. In real life, a patient can have more than one illness, a physician can concurrently follow more than one CG, there can be unexpected reactions, and so on. Moreover, physicians are supposed to keep up to date with the latest research so that they employ the best practice, even if these practises are not yet implemented as CGs. It is evident that medical procedures in clinical guidelines are not created to be used like legal norms, and following CGs is not sufficient to avoid negligent behaviour. CGs are intended to reduce the waste of resources generated by bad organization and, more importantly, to reduce risks for patients [14] by providing recommendations based on ideal conditions.

3   The Eunomos Legal Knowledge Management System in Specialist Domains

The above scenarios are challenging for legal research. Legal professionals have to evaluate not only the law but also local and specialist rules. The latter involves evaluating whether the rules are reasonable and legally relevant, whether the rules have been applied, and - where local rules conflict with each other or with legal norms or legal principles - determining which rules should have priority. Moreover, evaluating whether the rules are reasonable and legally relevant means looking at different aspects of the rule (preconditions, rule, goals) from a legal perspective (scope of the norm, relation to legal norms, and legal principles), as schematized in Fig. 1. Are the preconditions such that the rule can reasonably be followed e.g. does the factory have suitable equipment, does the hospital have the right medicine? Do the specialist rules and their goals correspond in any way with legal rules or legal principles, or are they beyond the scope of the law? Do the specialist rules contravene legal norms or principles? What are the

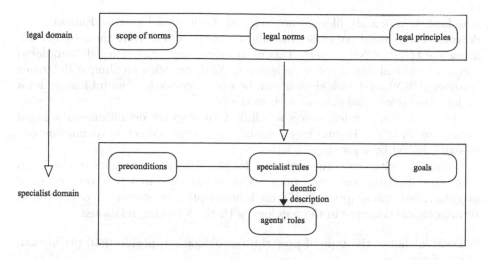

Fig. 1. Evaluating specialist rules from a legal perspective

rights, duties and responsibilities attributed to different roles, and who has legal accountability for any transgressions?

This is all compounded by the fact that finding and interpreting any area of the law is difficult, as it requires understanding of legislative text taking into account jurisprudence and legal doctrine. Moreover, the increasing level of regulation in specialist domains makes it difficult for professionals and organisations to keep up to date with the law and ensure compliance with requirements for ethical and legal conduct.

To address this problem, the Eunomos Legal Management System [15] was developed - the result of a collaboration between the University of Turin's Law and Computer Science Departments in the context of the ICT4LAW1 project. Eunomos is a web-based system enabling legal researchers and practitioners to manage knowledge about laws and legal concepts in different sectors and different jurisdictions. The system currently focusses on providing authoritative explanations of legal norms and concepts. The question this paper seeks to address is: how can legal knowledge management systems provide authoritative and comprehensive information to assist legal reasoning in complex specialist areas of law? Below we outline the main elements of the Eunomos system as well as extensions required to handle specialist domains.

3.1 Norm Retrieval

The Eunomos system makes it easier to find and understand norms on specific issues. The system makes clever use of legislative XML and ontologies to enable expert contributors to analyse legislation and organise norms and legal definitions. Legislation in the Eunomos database are downloaded from government websites and stored in accordance with the Norme in Rete (NIR) legislative XML standard. Where XML files are not available from official portals, Eunomos can download text-based laws from the official portals and convert them into XML, using the ITTIG CNR parser[1]. This means that each piece of legislation, document, article and paragraph is assigned an XML tag with an Unique Reference Number (URN), and each element can be easily accessed - a useful feature for a highly cross-referential domain such as the law.

Terms within the legislation are linked to relevant definitions in a Legal Taxonomy Syllabus [16] ontology. Legislation-specific and generic definitions can co-exist, linked by a group by relation.

While constitutive norms are used for definitions of concepts in the Legal Taxonomy Syllabus ontology, prescriptive norms are used to create complex objects called "prescriptions" [17]. Each prescription is necessarily connected to terminological concepts in the ontology with the following relations:

- Deontic clause: the type of prescription: obligation, prohibition, permission, exception.
- Active role: the addressee of the norm (e.g., director, employee).
- Passive role: the beneficiary of the norm (e.g., customer).
- Crime: the type of crime resulting from violation of the prescription, often defined in other legislation such as the Penal Code.
- Sanction: the sanction resulting from violation (e.g., a fine of 1 *quote*, where *quote* is defined in other legislation).

The arrangement of prescriptions and their constituent elements makes it possible to conduct semantic, and not just keyword, searches on norms. The Eunomos system also uses classification and similarity tools to help users find related and similar norms [18].

3.2   Compliance Monitoring

Applying norms to specific and variable contexts is a complex process involving many actors. In [19], we provided a methodological compliance decision-making workflow involving several key actors - compliance managers, lawyers, business process management designers, operational managers and senior managers - and showed how modelling alternative interpretations of prescriptions in Eunomos systems can facilitate discussions among key actors on what kind of changes should be done to business processes. Moreover, in dynamic environments, it can be useful to map norms to business processes, so that they can be re-evaluated when either legal norms or business processes change.

3.3   Cross-Domain Information Management

The normative search and compliance monitoring features described above are well-suited for resource and general management issues in hospital trusts, where

---

[1] http://www.ittig.cnr.it

procedures are planned to adhere to compliance requirements in advance. Clinical decisions, on the other hand, involve complex analysis on individual patients in specific situations. Planning is important here too, but the norms are essentially soft constrains [20].

Clinical guidelines management systems such as GLARE [21] help physicians work effectively by formalising clinical guidelines in a workflow fashion. The GLARE system is suitable for use by physicians after brief training. The system uses knowledge from specialist databases - the Clinical Guidelines DataBase (CG-DB) where the guidelines are modelled, the Pharmacological DB, storing a structured list of drugs and their cost; the Resources DB, listing the resources available in a given hospital; the ICD DB, containing an international coding system of diseases; and the Clinical DB, providing a standard restricted terminology to be used when building a new guideline; and a Patient DB which contains data about individual patients. Different kinds of control relations establish which actions should be executed next and in what order - sequence, any-order, alternative and repetition - with temporal reasoning to ensure actions respect temporal constraints specified in the guidelines. The GLARE system enables physicians to make and record decisions taken for individual patients, and tie this with patient records. The tool checks the well-formedness of guidelines when input into the system, and highlights any inconsistencies arising from using more than one guideline. It shows alternative paths available to physicians and can suggest the most appropriate course of action based on decision theory classical algorithms for calculating utility and evaluating optimal policy. Most interestingly, the tool provides a mechanism for handling the discrepancy between scenarios envisaged in the design of clinical guidelines (ideal patients with only one disease, ideal physicians and ideal resource availability) and the real world where physicians use their Basic Medical Knowledge (BMK) to make alternative decisions. In GLARE, the BMK is modelled in Prolog as logical rules. They can be used formally to "defeat" prescriptions in clinical guidelines where preconditions for actions prescribed are not satisfied or some other abnormality occurs.

Eunomos can complement this system, most effectively to assist with *post facto* normative reasoning. Medical lawyers can use the valuable evidence accumulated in GLARE to analyse decisions taken at the time based on guidelines or basic medical knowledge. Eunomos can be used to map generic norms from medical law as modelled in the prescriptions ontology to norms in clinical guidelines. Regional guidelines and source international clinical guidelines can be processed as legislative XML, and clinical terms linked to specialist medical ontologies. The availability of knowledge from both legal and medical domains enables lawyers to construct arguments based on substantive evidence to justify actions taken.

3.4 Cross-Domain Ontological Alignment

Medical norms use established concepts from the domains of law and medicine. There is an intersection of concepts used in both domains, due to 'borrowing' from either direction, but usually from medicine to law. There are several different branches of medicine, and they may use terms in the same or different ways.

Ontology alignment means showing the relationship between terminology in different domains [22]. Legal Taxonomy Syllabus, which is part of the Eunomos legal management system, is a suitable framework for managing and visualising multiple ontologies [16].

Legal Taxonomy Syllabus was originally designed to define legal terms in different jurisdictions. It contains a suite of different ontologies for national jurisdictions as well as an ontology for European law. Concepts within European law are linked to terms in different languages. Recognising that the relationship between terms and concepts are rarely one-to-one, concepts may be linked to multiple terms, and vice versa. Moreover, concepts from national jurisdictions can be mapped to European concepts. We assert that this multi-level ontology framework is highly suitable for mapping ontologies from different domains, such as a legal ontology to a medical ontology. Just as in comparative law, the mapping between ontologies can help an expert from one jurisdiction understand concepts from another jurisdiction, Syllabus could help lawyers understand concepts from the medical domain.

To work effectively, the Eunomos system needs to integrate data about several sub-domains: medical practice (MP), patient conditions (PC) and the administration of healthcare systems (HCS). Note that:

- the medical domain is highly structured, and MP, PC and HCS have their own ontologies;
- MP norms are included in Basic Medical Knowledge and Clinical Guidelines, their purpose is to reduce risks;
- rules for PC are expected and defeasible, causal relations supported by practitioner experience and research;
- CGs are applied to HCS as the minimum level of resources required to provide health care services;
- CGs are applied to MP as suggestions of safe behaviour and taken into account in medical decisions;
- CGs are applied in PC as explanations that connect medical decisions, medical evidence of the patient conditions and treatments.

To give an intuition of the entities in a CG, we consider GLIF3: a representation format for shareable computer-interpretable clinical practice guidelines [23].

In GLIF3, guidelines are represented as flowcharts of temporally ordered nodes called *Guideline_Step*. This class has the following subclasses:

- The *Decision_Step* class represents decision points in the guideline. Decisions can be automatic, like information delivering, or approved by an external agent, such as a physician, or another healthcare employee. A non-automatic decision may involve significant risks, ambiguous criteria and other clinical considerations. A decision step may have multiple options with their own selection criteria.
- The *Action_Step* class is used for modelling recommended actions or tasks. The action specification hierarchy is divided into two major types of actions: automatic actions such as invoking a subguideline, performing inferencing

(e.g., stage of a tumor from a set of observations), or computing values for data (e.g., age of patient); clinical actions that are carried out by a care-provider, such as changing a medication for a patient.

- *Patient_State_Step*, the clinical state that characterizes a patient: the clinical state of a patient and an entry point into the flowchart.
- Multiple concurrent paths through the guideline.
- Flow control entities like trigger events or exceptions.

The flowchart structure defines the relations among the medical concepts (actions, decisions, patient state). It describes the agents' involvement: responsibility (the decision) and evaluation (decision criteria) of different options. This view is relevant for legal reasoning, for instance to understand what is negligence and what is a wrong but legitimate decision. Where can norms and CG entities relations be founded? For instance:

- from norms for medical procedures, norms references can be extended with the medical entities in the CGs related with the ruled procedures;
- from judgements about health care trials, including procedures, patient conditions and other medical concepts;
- from domain norms that refers to legal concepts like roles in health care organizations, timing of procedures or resources.

We believe that the ontology that provides the widest coverage of the sub-domains above is SNOMED[2], a clinical healthcare multilingual terminology ontology used in over 50 countries to provide the core general terminology for electronic health records (EHR). SNOMED CT descriptions link appropriate human-readable terms to concepts. A concept can have several associated descriptions, each representing a synonym that describes the same clinical idea. Concepts are related via traditional "is-a" relations as well as semantic relations such as 'causative agent', linking e.g. the concept "viral pneumonia "to the concept "virus". SNOMED CT is a terminology that can cross-map to other international standards and classifications. Specific language editions are available which augment the international edition and can contain language translations, as well as additional national terms. Despite its strengths, it has been noted that "neither the labels nor the description logic definitions in SNOMED CT are easy for users to understand" [24]. It would therefore be useful to access additional ontologies such as the Unified Medical Language System (UMLS)[3], which integrates a number of standard medical terminologies and contains straightforward definitions. For example, the definition for Diagnostic Procedure is 'A method, procedure or technique used to determine the nature or identity of a disease or disorder. This excludes procedures which are primarily carried out on specimens in a laboratory' [25, p. 83]. Fortunately, the UMLS Metathesaurus now contains the complete International Release of SNOMED CT in English and Spanish[4].

---

[2] http://www.ihtsdo.org/snomed-ct/snomed-ct0/
[3] http://www.nlm.nih.gov/research/umls/
[4] http://www.nlm.nih.gov/research/umls/Snomed/snomed_faq.html

Medical lawyers will also need to be able to refer to legal conceptualisations of legal and medical terms from legislation. To extract terms from a new specialist area of law we could integrate an unsupervised technique called TMine [26], which is able to automatically bootstrap a domain ontology from a set of plain texts making use of statistical techniques such as Latent Semantic Analysis. This can help both with exploration of the data and creation of initial categorizations to be verified by experts. The technique could be improved by considering entire noun phrases as well as single terms.

There are a number of software tools that aid ontology matching such as SAMBO, Falcon, DSim, RIMOM, ASMOV, Anchor-Flood and AgreementMaker which were analysed in [27]. Most of these systems require input ontologies to be formulated in OWL or RDFS. Entities in different ontologies are compared for similarity based on terminological similarity (using edit distance or Wordnet), structural similarity (based on is-a hierarchies or sibling similarities) and background knowledge from corpora. The [27] analysis of the results of recent ontology matching evaluations concluded that the best performing of these systems was AgreementMaker[5]. We envisage that this tool would be useful for semi-automatically mapping the GLARE, UMLS and Medical Law ontology in Legal Taxonomy Syllabus. We expect that much manual work would be required to evaluate the results, and therefore its sophisticated user interface for evaluating alignment quality is an important feature. Eunomos has an OWL conversion tool which would facilitate the import of ontologies into Legal Taxonomy Syllabus and alignment of domain-specific ontologies.

3.5   Evaluation

The assumption behind this work is that the improved knowledge from medical (specialistic) domain can help lawyers to defend their clients, judges to evaluate cases and in general legal scholars to reconstruct legal cases from a medical point of view. As far as we know, there are no benchmarks for legal knowledge management systems integrating knowledge from other domains or other systems that we can reasonably compare with. Evaluation must therefore be on the feasibility or usefulness of such a system in a general sense. We propose to involve expert users at each stage of design and realization steps to evaluate the implementation decisions and the system results. We can include some quantitative measures in our evaluation. For instance, we could look at norms from legislation and clinical guidelines and evaluate how many terms in the norms can be linked to specialist glossaries or ontologies such as SNOMED. This should be followed by qualitative analysis with legal professional end users of just how useful such definitions are to their understanding of the medical point of view.

4   Related Work

The Eunomos system has some similarities with Bianchi et al. [28] in that it is a legal knowledge management system designed to help users view laws and

---

[5] http://agreementmaker.org/

classify terms. While Bianchi et al. [28] takes XML files as input, Eunomos can download text-based laws made available in official portals and convert them into XML, where XML files are not available. The use of ontology in the two systems are also quite different. Bianchi et al. [28] use the Semantic Turkey ontology [29], where definitions can be taken from any source and arranged in any order. The Eunomos product is more careful, designed as it was to address real problems in accessing and managing information by lawyers and their strict demand for accuracy and transparency. Eunomos has user-friendly interfaces to help expert contributors create links to definitions in legislation, judgment and official journals, and to track the evolution of terms in a systematic manner.

Eunomos makes much use of the NormeInRete legislative XML and Unique Reference Numbers to cross-reference different parts of legislation and ontological definitions. Legislative XML is an important technology in legal informatics. A review of various legislative XML standards developed for different jurisdictions is provided in "Legal informatics and management of legislative documents" [30].

Ontology alignment is an important topic in medical ontologies, and a number of systems have been tried and tested in this domain. Please refer above (Sect. 3.4) to a fuller discussion. In the legal domain, SEKT [31] is an example of ontology alignment in the legal domain using Semantically Enabled Knowledge Technologies (SEKT) to semi-automatically learn ontologies and extract metadata. Ontologies derived from general jurisprudence (Ontology of Professional Judicial Knowledge (OPJK) were merged with ontologies derived from Frequency Asked Questions (Question Topic Ontology, QTO), the overall goal being to assist young judges find information based on queries in natural text. The EU Employee Legal Ontology by Despres and Szulman [32] is an example of an ontology developed with a bottom-up approach with terms selected from two European directives about employees, and structured into two different micro-ontologies. The two micro-ontologies were then merged, and connected by subsumption to higher-level concepts defined formally in the LRI-Core Legal Ontology [33], and the DOLCE ontology [34]. The ontologies were constructed using the SYNTEXT term extractor in the TERMINAE system, and the merging was carried out manually within the same system. We do not know of any work aligning legal ontologies with specialist domains.

In our proposal, we don't merge ontologies from the same domain nor collapse two domains into one. Even when there is a connection, legal terms and specialist terms differ in their meaning, usage and context. Eunomos provides a framework for aligning ontologies to describe relations between classes from domains that remain separate. In particular, we propose to align the restricted vocabulary in GLARE with equivalent terms and elaborate definitions in general purpose medical ontologies, as well as to link the terms to legal terms which have similar meanings. This will help mainly legal experts to evaluate medical cases from a legal and medical perspective. Eunomos is the only legal knowledge management system we know of which enables users to explore terms from different independent domains.

5   Conclusions

This paper discussed the interplay between specialist and legal domains using case scenarios to show factors that influence normative reasoning. Our discussion brings out three main issues for integration between law and specialist domains:

a. ontological differences in domain description,
b. different status of norms (absolute, defeasible, advice),
c. different permissions for normative reasoning (goals vs procedures).

It follows that a legal knowledge management system covering specialist domains need to include domain-specific ontologies, model specialist rules and describe their normative status, link rules to goals and legal principles, and describe the level of permissions and accountability assigned to different roles. In this paper we have provided some tentative proposals for extending the Eunomos legal knowledge management systems to facilitate not only monitoring norms [15], mapping norms to business processes [17], but also de facto reasoning on actual cases, but further work is needed on the practicalities of linking rules to permissions and roles as well as providing different views for different roles (hospital managers, physicians, lawyers) to reflect their areas of interests and normative reasoning permissions. Our objectives for future work are:

1. a general methodology for integrating specialist rules into Eunomos,
2. integrating Eunomos with GLARE and possibly other clinical guideline management systems,
3. studying other specialist domains to test the general methodology.

On a theoretical level, this paper has described the issue of cross-domain research only in a general sense. Further research is required to analyse the relationship between different specialist domains and law considering their specific features. Theoretical work is also needed on the different kinds of deontic status - advice as opposed goal-based or procedural obligations.

References

1. Braithwaite, J.: Enforced self-regulation: a new strategy for corporate crime control. Mich. Law Rev. **80**(7), 1466–1507 (1982)
2. Gunningham, N., Rees, J.: Industry self-regulation: an institutional perspective. Law Policy **19**(4), 363–414 (1997)
3. Wildbaum, G., Westermann, J., Maor, G., Karin, N., et al.: A targeted dna vaccine encoding fas ligand defines its dual role in the regulation of experimental autoimmune encephalomyelitis. J. Clin. Invest. **106**(5), 671–679 (2000)
4. Power, M.: Organized Uncertainty: Designing a World of Risk Management. Oxford University Press, Oxford (2007)
5. Dworkin, R.: Taking Rights Seriously. Duckworth, London (1977)
6. Burgemeestre, B., Hulstijn, J., Tan, Y.-H.: Rule-based versus principle-based regulatory compliance. In: Governatori, G. (ed.) Frontiers in Artificial Intelligence and Applications (JURIX 2009), pp. 37–46. IOS Press, Amsterdam (2009)

7. Woolf, S.H., Grol, R., Hutchinson, A., Eccles, M., Grimshaw, J.: Clinical guidelines: potential benefits, limitations, and harms of clinical guidelines. BMJ: Br. Med. J. **318**(7182), 527 (1999)
8. Jutras, D.: Clinical practice guidelines as legal norms. CMAJ: Can. Med. Assoc. J. **148**(6), 905 (1993)
9. Hyams, A.L., Brandenburg, J.A., Lipsitz, S.R., Shapiro, D.W., Brennan, T.A.: Practice guidelines and malpractice litigation: a two-way street. Ann. Intern. Med. **122**(6), 450–455 (1995)
10. Ferdinande, P.: Recommendations on minimal requirements for intensive care departments. Intensive Care Med. **23**(2), 226–232 (1997)
11. Bottrighi, A., Chesani, F., Mello, P., Montali, M., Montani, S., Terenziani, P.: Conformance checking of executed clinical guidelines in presence of basic medical knowledge. In: Daniel, F., Barkaoui, K., Dustdar, S. (eds.) BPM Workshops 2011, Part II. LNBIP, vol. 100, pp. 200–211. Springer, Heidelberg (2012)
12. Brewer, S.: Scientific expert testimony and intellectual due process. Yale Law J. **107**(6), 1535–1681 (1998)
13. Rotolo, A.: Rules typing. In: NorMAS 2013 Proceedings (2013)
14. Programma nazionale per le linee guida (PNLG), Come produrre, diffondere e aggiornare raccomandazioni per la pratica clinica: manuale metodologico. ZADIG editore (2002)
15. Boella, G., Humphreys, L., Martin, M., Rossi, P., van der Torre, L.: Eunomos, a legal document management system based on legislative XML and ontologies. In: Legal Applications of Human Language Technology (AHLTL) at ICAIL'11 (2011)
16. Ajani, G., Boella, G., Lesmo, L., Mazzei, R., Radicioni, D.P., Rossi, P.: Legal taxonomy syllabus: handling multilevel legal ontologies
17. Boella, G., Humphreys, L., Martin, M., Rossi, P., Violato, A., van der Torre, L.: Eunomos, a legal document and knowledge management system for regulatory compliance. In: De Marco, M., Te'eni, D., Albano, V., Za, S. (eds.) Information Systems: Crossroads for Organization, Management, Accounting and Engineering (ITAIS11), pp. 571–578. Springer, Heidelberg (2012)
18. Boella, G., Humphreys, L., Martin, M., Rossi, P., van der Torre, L.: Eunomos, a legal document and knowledge management system to build legal services. In: Palmirani, M., Pagallo, U., Casanovas, P., Sartor, G. (eds.) AICOL-III 2011. LNCS, vol. 7639, pp. 131–146. Springer, Heidelberg (2012)
19. Boella, G., Janssen, M., Hulstijn, J., Humphreys, L., van der Torre, L.: Managing legal interpretation in regulatory compliance. In: Proceedings of the XIV International Conference on Artificial Intelligence and Law (ICAIL2013) (2013)
20. Boella, G., Van Der Torre, L., Verhagen, H.: Introduction tonormative multiagent systems. Comput. Math. Org. Theor. **12**(2–3), 71–79 (2006)
21. Terenziani, P., Montani, S., Bottrighi, A., Molino, G., Torchio, M.: Applying artificial intelligence to clinical guidelines: the glare approach. In: Cappelli, A., Turini, F. (eds.) AI*IA 2003. LNCS, vol. 2829, pp. 536–547. Springer, Heidelberg (2003)
22. Euzenat, J.: Algebras of ontology alignment relations. In: Sheth, A.P., Staab, S., Dean, M., Paolucci, M., Maynard, D., Finin, T., Thirunarayan, K. (eds.) ISWC 2008. LNCS, vol. 5318, pp. 387–402. Springer, Heidelberg (2008)
23. Boxwala, A., Peleg, M., Tu, S., Ogunyemi, O., Zeng, Q., Wang, D., Patel, V., Greenes, R., Shortliffe, E.: GLIF3: a representation format for sharable computer-interpretable clinical practice guidelines. J. Biomed. Inform. **37**(3), 147–161 (2004)

24. Liang, S.F., Stevens, R., Rector, A.: OntoVerbal-M: a multilingual verbaliser for SNOMED CT. In: Proceedings of the 2nd International Workshop on the Multilingual Semantic Web (MSW 2011) in Conjunction with the International Semantic Web Conference (ISWC2011), Bonn, Germany, p. 1324 (2011)
25. Pisanelli, D.M.: Ontologies in Medicine, vol. 102. IOS Press, Amsterdam (2004)
26. Candan, K.S., Di Caro, L., Sapino, M.L.: Creating tag hierarchies for effective navigation in social media. In: Proceedings of the 2008 ACM Workshop on Search in Social Media, pp. 75–82. ACM (2008)
27. Shvaiko, P., Euzenat, J.: Ontology matching: state of the art and future challenges. IEEE Trans. Knowl. Data Eng. **25**, 1 (2012)
28. Bianchi, M., Draoli, M., Gambosi, G., Pazienza, M.T., Scarpato, N., Stellato, A.: ICT tools for the discovery of semantic relations in legal documents. In: Proceedings of the 2nd International Conference on ICT Solutions for Justice (ICT4Justice) (2009)
29. Fiorelli, M., Pazienza, M.T., Petruzza, S., Stellato, A., Turbati, A.: Computer-aided ontology development: an integrated environment. New Challenges for NLP Frameworks 2010 (held jointly with LREC2010) (2010)
30. Biasiotti, M., Francesconi, E., Palmirani, M., Sartor, G., Vitali, F.: Legal informatics and management of legislative documents. Global Center for ICT in Parliament Working Paper, vol. 2 (2008)
31. Ehrig, M.: Semantically enabled knowledge technologies - SEKT. In: Ehrig, M. (ed.) Ontology Alignment: Bridging the Semantic Gap, pp. 175–181. Springer, New York (2007)
32. Desprès, S., Szulman, S.: Merging of legal micro-ontologies from european directives. Artif. Intell. Law **15**(2), 187–200 (2007)
33. Breuker, J.: Constructing a legal core ontology: Lri-core. In: Volz, R., Freitas, F., Stuckenschmidt, H. (eds.) Proceedings WONTO 2004 (2004)
34. Borgo, S., Masolo, C.: Ontological foundations of DOLCE. In: Staab, S., Studer, R. (eds.) Handbook on Ontologies, 2nd edn. Springer, Heidelberg (2009)

ArgPROLEG: A Normative Framework for the JUF Theory

Zohreh Shams[1]([✉]), Marina De Vos[1], and Ken Satoh[2]

[1] Department of Computer Science, University of Bath, Bath, UK
{z.shams,cssmdv}@bath.ac.uk
[2] Principles of Informatics Research Division,
National Institute of Informatics, Tokyo, Japan
ksatoh@nii.ac.jp

Abstract. In this paper we propose ArgPROLEG, a normative framework for legal reasoning based on PROLEG, an implementation of the Japanese "theory of presupposed ultimate facts" (JUF). This theory was mainly developed with the purpose of modelling the process of decision making by judges in the court. Not having complete and accurate information about each case, makes uncertainty an unavoidable part of decision making for judges. In the JUF theory each party that puts forward a claim, due to associated *burden of proof* to each claim, it needs to prove it as well. Not being able to provide such a proof for a claim, enables the judges to discard that claim although they might not be certain about the truth. The framework that we offer benefits from the use of argumentation theory as well as normative framework in multi-agent systems, to bring the reasoning closer to the user. The nature of argumentation in dealing with incomplete information on the one hand and being presentable in the form of dialogues on the other hand, has furthered the emergence and popularity of argumentation in modelling legal disputes. In addition, the use of multiple agents allows more flexibility for the behaviour of the parties involved.

Keywords: Legal reasoning · Normative framework · Argumentation · Agents

1 Introduction

Legal reasoning is a rich application domain for argumentation in which exchanging dialogues and inferencing are combined [17]. On the other hand, legal reasoning is a rich domain for agent modelling in which agent can model individual parties [14]. In the past two decades, the combination of argumentation and agents technology has provided a great modelling tool for legal disputes, in which multiple parties are involved in a dispute and they each try to prove their claims [3,16].

In this work, we offer a normative framework for the JUF theory by means of argumentation and multi-agent systems. This allows an easier presentation of

© Springer International Publishing Switzerland 2014
Y. Nakano et al. (Eds.): JSAI-isAI 2013, LNAI 8417, pp. 183–198, 2014.
DOI: 10.1007/978-3-319-10061-6_13

this theory, compared to the previous implementation in logic called PROLEG [22]. The JUF theory is a decision making tool that has already been successfully used in modelling civil litigation [20]. However, having the users - lawyers and judges - of the system in mind, some of the semantics of logic programming does not seem to be fully accessible to the users. We, therefore, have changed the architecture and algorithm of PROLEG in a way that brings the reasoning process closer to the users. For this purpose, we have used the dialectical proof procedure as a reasoning mechanism for parties involved in an argumentation-based dialogue [25]. The advantage of this mechanism is being close to the human reasoning process as well as being representable in form of dispute trees.

This paper is organised as follows. In Sect. 2 we give an overview of the JUF theory and PROLEG, followed by a brief introduction to argumentation theory and norms. Section 3 provides the main contribution of this paper, which is a normative architecture, called ArgPROLEG. ArgPROLEG is designed for reasoning about JUF theory and in essence, it is an argumentation based implementation of PROLEG. The architecture and algorithm of ArgPROLEG are both included in this section. This section also includes an example of a legal dispute modelled by ArgPROLEG. We then provide a survey of related work in Sect. 4. Finally we conclude and point out some directions for future work in Sect. 5.

2   Background

In this section, we provide a brief introduction to JUF, PROLEG and other key concepts used throughout the paper.

2.1   PROLEG: An Implementation of the Ultimate Fact Theory of Japanese Civil Code

PROLEG [20] is a legal reasoning system based on the Japanese theory of presupposed ultimate facts (JUF). This theory is used for interpreting the Japanese civil code. It was mainly developed to assist judges to make decisions under the incomplete and uncertain information they face in the court. This uncertainty is mainly the result of one party asserting a claim, which is unable to prove due to the lack of evidence. In such a situation, the judge cannot deductively decide whether the claim is true or false since the "deductive" civil code is based on the complete information [22].

The JUF theory helps the judge to handle these cases by attaching a *burden of proof* [17] to each claim. The *burden of proof* is assigned to the party that makes the claim and the judge is not responsible for that. Thus, if a party makes a claim that is unable to prove, the judge can discard the claim without trying to assign a certain true or false value to it. This way the judge can evaluate the correctness of a legal claim under a set of incomplete information.

PROLEG was introduced in an attempt to replace an existing translation of the JUF theory into logic programming [22]. The reason of this shift was the unfamiliarity of the users, namely judges and lawyers, with logic programming and *negation as failure* [5] in particular. According to *negation as failure*, if a claim is unknown or not known to be true, it is considered to be false. By definition, *negation as failure* makes a perfect choice for a mathematical formalisation of the JUF theory in which failing to provide a proof for a claim results in discarding the claim. However, the fact of not being conceptually accessible for the users, led to a new implementation of JUF called PROLEG.

Instead of *negation as failure*, PROLEG uses the Professor Ito's explanation of JUF which is based on the openness of the ultimate facts [20]. In openness theory, facts are divided into two categories; those that result in a conclusion and those that represent an exceptional situation. The latter category are open to challenge meaning they do not have a certain truth value and are therefore undecided form the judge point of view. The *burden of proof* of these facts is on the party claiming them. Judges are therefore able to make decisions based on known facts and exceptions that are explicitly proven by one of the parties.

PROLEG consists of a rulebase and a factbase. The former stores the rules and the exceptions while the later stores the performed actions of both parties as well as the judge's judgement about their truth value. Eqs. 1, 2 and 3 are examples of a rule, an exception and a fact in PROLEG, respectively.

$$deliver\_good(X, Y, Good) <= purchase\_contract(X, Y, Good, Price) \quad (1)$$

$$exception(deliver\_things(X, Y, Good, Price),$$
$$claim\_of\_simultaneous\_performance(Y, X, Price)) \quad (2)$$

$$allege(claim\_of\_simultaneous\_performance, plaintiff) \quad (3)$$

Rule (1) states that party X can expect party Y to deliver a *Good* if there is a purchasing contract between them including the agreed *Price* and *Good*. However there could be an exception to this expectation, which is defined in rule (2). The exception is as follows: if there is a contract between two parties, one may refuse to perform her/his obligation until the other party performs her/his obligation. Moreover, Eq. 3 shows a performed action by the plaintiff party, which is claiming an exception to *deliver_things(X, Y, Good, Price)* by, *claim_of_simultaneous_performance*.

According to the claims and proofs that two parties - Plaintiff and Defendant - assert, PROLEG produces a trace of derivation in the form of an dialogue between them. The plaintiff tries to prove a claim while the defendant tries to find an exception for that claim. If the exception is proven successfully, the plaintiff has to find another exception for the former exception and so on.

2.2 Argumentation

Argumentation theory was initially studied in philosophy and law, and during the past two decades it has been extensively researched in distributed systems. Argumentation Frameworks (AF) have particularly gained a popularity in multi-agent systems as an aid for the agents' reasoning and decision making process.

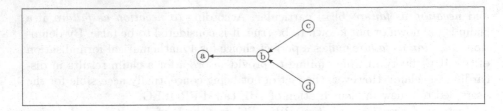

Fig. 1. A graphical representation of *AF*

The first AF was introduced by Dung [7] and it is known as Dung's Argumentation Framework (DAF)[1]. According to DAF, an AF is a pair $AF = \langle Ar, R \rangle$ where $R \subseteq Ar \times Ar$. Ar is a set of arguments and R is a set of attack relations between arguments. We assume a attacks b if $(a, b) \in R$. Figure 1 displays an AF with four arguments and three attack relations between them. Nodes represent the arguments, while edges represent the attack relations among them.

$$AF = \langle \{a, b, c, d\}, \{(b, a), (c, b), (d, b)\} \rangle$$

The evaluation of arguments in an AF depends on the argumentation semantic of choice. The purpose of argumentation semantics is to determine a set of justified and coherent arguments based on the arguments' interactions. If two arguments attack each other then an entity - which could be an agent for example - cannot believe in both of them at the same time. Therefore, the role of argumentation semantics is to examine the acceptability of a set of arguments.

The most basic argumentation semantic is the conflict-free semantics [7] in which none of the arguments attack each other. This is the minimum criteria for a set of arguments to be considered as coherent. The rest of argumentation semantics (e.g. complete extension, preferred extension, stable extension and etc.) are a version of conflict-free semantic that satisfy some form of optimality [6]. As an example, the conflict-free extensions of Fig. 1 are provided below:

$$C\_F : \{\{\} \{a\} \{b\} \{c\} \{d\} \{a, c\} \{a, d\} \{c, d\} \{a, c, d\}\}$$

One of the reasons of developing argumentation theory in multi-agent society is being able to present interactions in the form of dialogues, specially among participants with potentially conflicting viewpoints. Dung [8] states argumentation as a form of reasoning for dispute resolution in which two parties, proponent and opponent, engage in a discussion as a form of proof for their claims. In fact, dialectical proof procedure can be viewed as a reasoning mechanism for parties involved in an argumentation-based dialogue [25]. In such a dialogue, the proponent puts forward an argument with the purpose of proving it. However, the opponent tries to attack this claim. The dispute goes on by the proponent and

---

[1] DAF can also be referred to as an Abstract AF because it abstracts away the internal structure of arguments and instead, it merely focuses on attack relations among arguments.

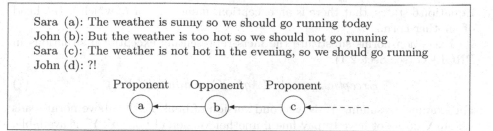

Sara (a): The weather is sunny so we should go running today
John (b): But the weather is too hot so we should not go running
Sara (c): The weather is not hot in the evening, so we should go running
John (d): ?!

Fig. 2. Dialectical Proof Procedure

opponent alternating in attacking each other's previous arguments until one of them runs out of arguments. The winner of the dispute is the party who speaks last. Therefore, the original claim by proponent is proved if the dialectical proof procedure ends with an argument by proponent. Figure 2 shows an example of this nature, in which the proponent claim is accepted.

2.3   Norms

Norms are defined as social rules which control the agent society by regulating agents' behaviour and following them benefits multi-agent systems as a whole [26]. They help multi-agent systems to cope with the heterogeneity, the autonomy and the diversity of interests among agents [27]. Therefore, a normative framework for multi-agent systems, comprises a set of normative agents whose behaviour is governed by norms [27]. If these norms are legal norms, then we have a legal normative framework which is the focus of this work. Free Online Encyclopedia defines legal norms as "mandatory rule of social behaviour established by the state". As this definition suggests, legal norms are normally imposed to the society through an external entity such as the state. We have thus, adopted the same concept and modelled the legal norms external to the agents.

We define each norm as a rule of form (4) consisting of literals L_i.

$$L_0 \leftarrow L_1 \wedge \cdots \wedge L_m \qquad m \geq 0 \tag{4}$$

The left hand side of the arrow L_0 is called head or conclusion of the rule and the right hand side $L_1 \wedge \cdots \wedge L_m$ is called body or premises of the rule. L_0 holds if $L_1, L_2, \cdots$, and L_m are all true. Take for example the norm:

$$payfine(AgX, Y) \leftarrow delay(AgX, Y) \wedge reserved(Y) \tag{5}$$

This norm can be read as: Agent AgX has to pay fine if it delays returning book Y to the library and the book is reserved by someone else.

Since we aim to use norms in a legal reasoning context, as it has been used in the JUF approach presented by PROLOG, we require a second type of norm called an exception norm.

$$Exception(Q, P) \tag{6}$$

Equation 6 states, that there is an exception, namely P for Q which is the head of another norm.

Exception norms substitute the facts representing exceptional situations in PROLEG (see Sect. 2.1).

$$Exception(payfine(AgX, Y), available(Y, Y')) \tag{7}$$

For example, assuming Y' is a second version of book Y, the above norm reads as: AgX does not have to pay fine if another version of book Y, Y' is available.

3   ArgPROLEG: A Normative Framework for Legal Reasoning

The JUF theory was first implemented in logic programming followed by an implementation in prolog called PROLEG. The main advantage of PROLEG over the original system is its accessibility to lawyer and judges. In this section we propose a normative framework to model the JUF theory which is even closer to the natural human reasoning process, since it benefits from multi-agent systems and argumentation theory to represent a legal dispute between two parties, namely plaintiff and defendant.

Arguing is one of the human skills that we learn from early ages. Naturally, the argumentation process between two humans starts with one of them raising an issue which is subject to disagreement of the opposite party. The rest of process is followed by exchanging further arguments with the purpose of reaching an agreement. The agreement could be mutual or could be the result of one party not being able to reject the other party's argument. Similarly, we have tried to reflect the human reasoning process in ArgProleg in a way that even a non expert user can instinctively relate to it. In what follows, We first introduce the overall architecture of ArgPROLEG followed by its algorithm.

3.1   The Framework Architecture

We suggest an architecture (see Fig. 3) in which the two parties in a legal dispute, plaintiff and defendant, are presented by two agents A and B, respectively. The arbitrator plays the role of the judge in the court and the set of norms models the law book. The arbitrator receives the claims and evidences of each parties and judges them by referring to the set of norms.

Please note that the connection between two agents happens through the arbitrator (See Fig. 3) and there is no direct connection between agents. A legal case in the court normally commences with an agent, namely plaintiff raising an issue against another agent, namely defendant. It is then the judge's responsibility to investigate the raised case and ask for defendant's testimony. According to both the original claim by plaintiff and the provided response by defendant, the judge decides whether the dispute is over in favour of one of the parties. If the status of the dispute is still unclear to the judge, it will be more argumentation back and forth between two parties through the arbitrator. Below is a narrative on how the communication works between the various parties:

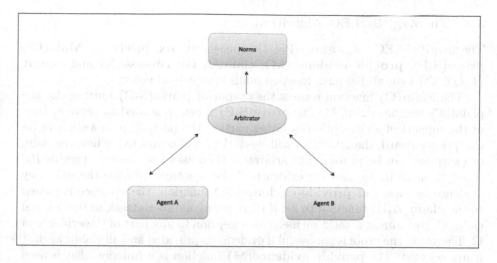

Fig. 3. The framework architecture

- The session starts by agent A submitting a claim to the arbitrator.
- The arbitrator checks the set of norms to find out how agent A should support this claim. In other words, what are the requirements of this claim from the legal viewpoint.
- The arbitrator passes the requirements to agent A.
- If agent A fails in providing the requirements, the claim is rejected.
- If it succeeds then, the arbitrator contacts the set of norms to see if there are any exceptions for this claim. If not, the claim is accepted.
- Otherwise the arbitrator passes the exceptions to agent B to see if it can provide any of them[2].
- If agent B has any of those exceptions, it will then pass it to the arbitrator.
- Subsequently, the arbitrator tries to find out how this exception can be supported from the law viewpoint by referring to the set of norms.
- The arbitrator informs agent B about the required support.
- If agent B cannot provide the necessary support for any of the exceptions, agent A's claim is accepted.
- But if agent B can prove at least one of the exceptions, the arbitrator tries to find out what are the exceptions for that by checking the set of norms.
- If there is any they will be passed to agent A and the same procedure will be repeated.
- This procedure is repeated until either an exception to the original claim cannot be ignored (the plaintiff cannot counteract) or all exceptions to the original claim turn out to be unsupported by the defendant.

---

[2] We assume that a party can use all the exceptions available exhaustively, one-by-one, to make a successful counter attack. Thus, if the party cannot provide the required support for the first exception, it has the opportunity to try the second exception and so on.

3.2   The ArgPROLEG Algorithm

The ArgPROLEG algorithm (Fig. 4) consists of six functions: **Main(C)**, **prove(S,P)**, **provide-evidence(M)**, **claim(A,B)**, **reverse(X)** and **except (F,Q)**. The task(s) that each function fulfils is explained below.

The **Main(C)** function returns the output of **prove(S,P)** function for the plaintiff's original claim, **C**. The **prove(S,P)** function is used to prove a claim or the support of an exception by either parties. If a party **P** puts a claim or an exception forward, the arbitrator will check the set of norms to see how the claim or exception can be proven. The arbitrator then asks the agent to provide the proof of the claim by showing evidence. If the agent can provide the necessary evidence by means of **provide-evidence(M)** function, the evidence is passed to the **claim(A,B)** function to see if there is any indirect attack to the original claim **C**. By indirect attack, we mean an exception to any part of the evidence of **C**. Therefore, the proof is successful if evidence is provided and all claims against it are rejected. The **provide-evidence(M)** function is a function that is used by each single agent collecting all the rules that have **M** as their head. It then recursively, traces back each rule to find all its atoms. The output of this function is a set of sets. Each set provides a possible way to proof the claim. For example in the case provided below, the agent has to provide $\{p1, p2, p3, p4, p5, p6, M\}$ or $\{q1, M\}$.

$$R1 : M \Leftarrow p1, p2 \quad R4 : p2 \Leftarrow p4, p5 \quad R7 : p3 \Leftarrow$$
$$R2 : M \Leftarrow q1 \qquad\quad R5 : p4 \Leftarrow p6 \qquad\; R8 : p5 \Leftarrow$$
$$R3 : p1 \Leftarrow p3 \qquad\;\; R6 : q1 \Leftarrow \qquad\qquad R9 : p6 \Leftarrow$$

Function **claim(A,B)** takes the responsibility of the rest of the dispute after the first claim by plaintiff is proven to be true. This function then gives chances to the defendant and the plaintiff to attack each other's last announcement. If any of the exceptions against an argument remains unattacked by the other party, that means the dispute is over and the winner is the claimer of this argument. The output of this function is **true** if **A** who made the first claim/argument is the last who speaks. Otherwise the output is **false**.

The **reverse(X)** function takes one of the parties, either the plaintiff or the defendant as input and returns the opposite party as output. This function will be called in **claim(A,B)** function, when the parties have to take turn in attacking each other.

The **except(F,Q)** function tries to find the exceptions for a certain claim or exception, **F**. If **F** is provided by one party, the opposite party **Q** needs to show evidence and consequently prove the exceptions for **F**. Thus, the arbitrator checks the set of norms to see whether there is any exceptions for **F**. In case of existence, the exceptions will be passed to **Q**. This party has to firstly show an evidence of such an exception and secondly prove it by calling **prove(S,P)** function. If it fails either of them, then the exception is rejected. The output of

```
Plaintiff-Arbitrator: Main(C)
begin
        return(prove(C, Plaintiff))
end


prove(S,P)
begin
        Arbitrator-P: Provide evidence for S
        V = provide - evidence(s)
        if V = Ø then return(false);
        for every v ∈ V
            begin
                proven = true
                for every v_i ∈ v
                    begin
                        if claim(v_i, P)
                            proven = false
                            break
                    end
                if proven = true
                    return(true)
            end
        return(false)
end

provide-evidence(M)
begin
        Result = {}
        R_u = {M ⇐ D ∈ R}
        if R_u = Ø then return(Ø)
        for every R_i ∈ R_u
            begin
                if D = Ø then add {} to all sets in Result
                else if for all D_i ∈ D
                    begin
                        add provide - evidence(D_i) to all sets in Result
                    end
                    add {D} to all sets in Result
            end
        return(Result)
end

claim(A,B)
begin
        e = except(A, reverse(B))
        if e = Ø then return(true)
        else for all e_i ∈ e
            begin
                result = claim(e_i, reverse(B))
                if result = true then return(false)
            end
        return(true)
end

reverse(X)
begin
        if X = Plaintiff then return(Defendant);
        else return(Plaintiff);
end

except(F,Q)
begin
        Arbitrator-Norms: collect all the exceptions for F: exception(F, E_i) in E
        if E = Ø then return(Ø)
        provenE = Ø
        else for every E_i ∈ E
            begin
                Arbitrator-Q: evidence(E_i)
                if Q can provide the evidence
                    then Arbitrator-Q: prove(E_i, Q)
                        if prove(Q, E_i) = true
                            then provenE = provenE ∪ E_i
                    else return(Ø)
            end
        return(provenE)
end
```

Fig. 4. The ArgPROLEG algorithm

this function is either Ø, which means there is no exception or not any proven one for **F**; or it is set **provenE** which is a set of proven exceptions for **F**.

3.3 Contract Scenario

In this scenario, we aim to model a legal dispute between two parties by means of the architecture and the algorithm we introduced in Sects. 3.1 and 3.2. Imagine a situation in which a lessor wants to cancel his property contract with the lessee. She claims that the lessee has subleased the property to somebody else and therefore, she wants to end the contract. Both the lessor and the lessee agree that there was a contract between them in first place and subsequently the property was handover to the lessee. The lessee also admits her contract of sublease with a third person which was followed by handing over some parts of the property to the sublessee. The lesser believes that the sublease has used the property to make profit, thus she makes the announcement of cancelling the contract. However, the lessee believes that she already informed the lessor and she has approved of the sublease before she made the announcement of cancelling the contract. Moreover, the period of subleasing was so short that does not count as abuse of confidence of the owner. However, the owner considers the case as abuse of confidence since she has received some complaints from the neighbours regarding the noise during the subleasing period. Figure 5 displays the formalisation of this case based on the ArgPROLEG architecture.

Figure 6 illustrates the graphical representation of Contract Scenario based on the ArgProleg algorithm. The plaintiff claims that she wants to cancel the contract. The arbitrator then checks the set of norms to find out the support for this claim. N1 provides this information which will be passed to the plaintiff. Plaintiff is able to provide the required support. Thus the first argument (a) appears. The arbitrator checks the set of norms to see if there is any exceptions for this claim. Exceptions 1 and 2 provide two options for the defendant to make an attack against the plaintiff's claim. The options obtained from the exceptions are $b : get\_approval\_of\_sublease$ and $c : nonabuse\_of\_confidence$. N2 and N3 contains the necessary supports for each of the exceptions, respectively. The attack (b) and (c) to (a) remains as a potential attack unless the defendant can provide the requested support for them. Defendant can only provide this support in case of argument (c). Therefore, the defendant attacks argument (a) by argument (c). Now, based on the algorithm, the arbitrator checks the set of norms to find an exception to this exception. This is going to make a potential case for the plaintiff to perform an attack to the defendant. There is one exception available, namely Exception 3, $abuse\_of\_confidence$. N4 states the requirement for this argument, which is $fact\_of\_abuse\_of\_confidence$. The plaintiff successfully supports this argument which results in an attack from argument (d) to argument (c). The arbitrator looks for another exception to this later exception. Since such an exception is not available the dispute is over.

The last graph in Fig. 6 shows the final argumentation framework for this dispute. Going back to Sect. 2.2, in a dialectical proof procedure, the party who

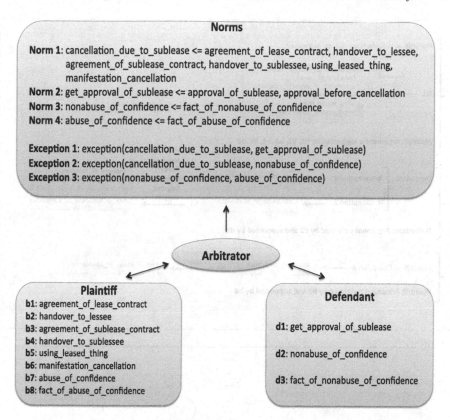

Fig. 5. Contract scenario

makes the last utterance is the winner, which similarly makes the plaintiff the winner of this case.

4   Related Work

The closest work to ours is PROLEG [20] which is an implementation of JUF theory by means of the *burden of proof*. ArgPROLEG has fulfilled two future plans of PROLEG discussed and listed in [20]. These two features are, bringing the knowledge representation closer to the natural human reasoning (see Sect. 3) and also including a diagrammatic representation of reasoning in the JUF theory. Using argumentation in designing ArgPROLEG has served both these purposes.

Apart from PROLEG and ArgPROLEG, another translation of the JUF theory is also available in logic programming [22]. In contrast to PROLEG and ArgPROLEG, this version uses *negation as failure* instead of the *burden of proof*. *Negation as failure* is a non-monotonic form of negation that enables logic programming to formulate problems of non-monotonic reasoning. Kakas [15] has

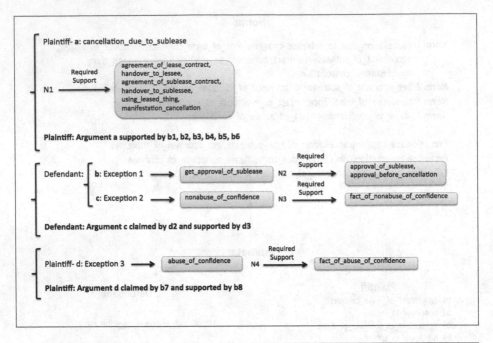

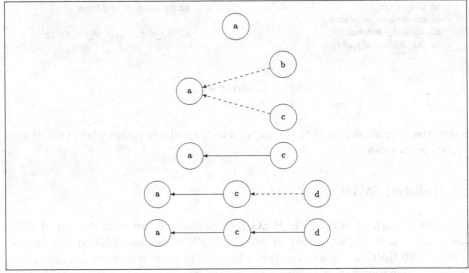

Fig. 6. Contract scenario argumentation framework

already used *negation as failure* for default reasoning. The idea of using negative literals as abductive hypotheses has also been pointed out by Eshghi and Kowalski [9]. However, among *burden of proof* and *negation as failure*, the concept of the former seems to be easier to grasp for lawyers and judges.

In terms of formalisation of the *burden of proof*, other works exist [13, 18, 23, 28]. Gordon et al. offer an argumentation-based system, called Carneades [13], which implements the *burden of proof* as well as the *burden of persuasion*. The main difference of this approach to ours is, that the *burden of proof* for a premise can be assigned to a different party rather than the one who has uttered the claim. The plaintiff has the *burden of production* for the facts of its claim, whereas the defendant has the *burden of production* for exceptions. The same applies to the *burden of persuasion*.

Another example of logic programming being used for expressing and applying legislation is [24]. This work however, focusses on specific legal cases related to British Nationality Act. They describe how complicated regulations such as British Nationality Act can be translated into simple form of logic so that the consequences of each act can easily be determined.

5 Conclusion and Future Work

In this paper we introduced ArgPROLEG, a normative framework for legal reasoning, that uses dialectical proof procedure to support legal parties in resolving their conflicts. ArgPROLEG offers an alternative approach to PROLEG [20]. We believe that ArgPROLEG is closer to human reasoning compared with PROLEG. Additionally, it is able to offer a diagrammatic representation of the plaintiff's and the defendant's reasoning, which enhances the ability of non-expert users to follow the procedure as it unfolds.

For the future, we would like to extend our framework to be able to cope with more than two parties. In real cases, a dispute can involve multiple parties, which all need to be able to bring forward their arguments. If there are more than two agents involved, but we are still able to divide them into two main opposing parties, the argumentation graph keeps its linear shape. However, in each step, there is more than one agent that can put forward an argument. For instance, if it is the defendant party's turn to put forward an argument and the defendant party includes more than one agent, any of them can make the argument. Although from the argumentation graph viewpoint, the dialectical structure does not change, there are some other issues that need to be taken into consideration. The most important issue is the consistency of knowledge bases of different agents belonging to the same party. At the moment we assume each agent's knowledge base is self-consistent, which results in consistency of claims put forward by the agent. However, if a party consists of more than one agent, defending the same viewpoint, their claims have to be consistent too. One possible way of achieving consistency is to merge agents' knowledge bases and resolve possible conflicts to prevent any inconsistent arguments. One example of such an approach is discussed in [10]. Another issue is the method of constructing an argument. Arguments can be put forward by various individual agents belonging to the same party, as well as by a combination of agents. The process of construction of an argument in such cases, results from the reasoning of multiple agents. On the other hand, if there are more than two agents and we cannot simply divide them into two main opposing parties, the argumentation graph would not be linear any more. As a result the argumentation graph can take any shape and the winner of the dispute is not necessarily the party who speaks last. In addition, applying different argumentation semantics, as discussed in Sect. 2.2, would have a different outcome, whereas in linear graphs all the argumentation semantics coincide.

For the implementation, we consider an architecture similar to the Governor approach presented in [1]. Balke et al. use an institution to collect the norms and the normative results of an agent's actions. To make this information accessible via queries, the authors introduce the Governor, an agent that acts as a relay between the norms and their (possible) consequences and the agent's query. In our case, the arbitrator would take the role of the Governor. Apart from simply relaying queries to the institution/norms, the arbitrator will actively retrieve information to pass on to the agents, e.g. the exception to the claim. Reference [1] uses the Jason BDI architecture [4] for setting up the multi-agent system and InstAL based on answer set programming [2, 12] for the institution/norms. The use of a BDI architecture [19] has the added advantage of being able to model agent reasoning in more detail. Currently, our agents' mental model contains only beliefs or a knowledge base. In a BDI architecture, we could express the goals and intentions of the agents more effectively and take them into account when they put forward their claims. In addition, with an institution rather than a set of norms, we would be able to keep track of normative states and allow agents to reason about conflicts that appear after a period of time. Having more expressive agents, gives us the chance to investigate different strategies for agents to deal with norm compliance as well. Agents can check the reward and punishment associated with adhering to a norm, or violating a norm, and decide whether the gain from breaking a norm is worth the sanction. In such cases, the agent has to decide between the importance of individual goals hindered by normative goals compared to individual goals hindered by punishment [27].

For the dialectical proof procedure, we also consider an implementation using answer set programming (ASP). Dung's argumentation framework and semantics have already been implemented in answer set programming [11], giving us a good indication that this approach is worth considering.

Finally, this paper only investigates the use of argumentation for conflict resolution in a legal domain. In particular, it is aimed at mimicking court procedure. However, it has been proven [21], that the PROLEG inference structure can be used for general *rule - exception* patterns (see Sect. 2.3). This argument applies to ArgPROLEG as well since it borrows PROLEG inference structure.

References

1. Balke, T., De Vos, M., Padget, J., Traskas, D.: On-line reasoning for institutionally-situated BDI agents. In: Yolum, P., Tumer, K., Stone, P., Sonenberg, L. (eds.) 10th International Conference on Autonomous Agents and Multiagent Systems (AAMAS 2011), pp. 1109–1110. IF0AAMAS, May 2011

2. Baral, C.: Knowledge Representation, Reasoning, and Declarative Problem Solving. Cambridge University Press, New York (2003)
3. Bench-Capon, T., Prakken, H., Sartor, G.: Argumentation in legal reasoning. Argumentation in Artificial Intelligence, pp. 363–382. Springer, Heidelberg (2009)
4. Bordini, R.H., Wooldridge, M., Hübner, J.F.: Programming Multi-Agent Systems in AgentSpeak using Jason (Wiley Series in Agent Technology). John Wiley & Sons, New York (2007)
5. Clark, K.V.: Negation as failure. In: Minker, J. (ed.) Logic and Data Bases, vol. 1, pp. 293–322. Plenum Press, New York (1978)
6. Coste-Marquis, S., Devred, C., Marquis, P.: Prudent semantics for argumentation frameworks. In: 17th IEEE International Conference on Tools with Artificial Intelligence (ICTAI), pp. 568–572. IEEE Computer Society (2005)
7. Dung, P.M.: On the acceptability of arguments and its fundamental role in nonmonotonic reasoning, logic programming and n-person games. Artif. Intell. **77**(2), 321–358 (1995)
8. Dung, P.M., Thang, P.M.: A unified framework for representation and development of dialectical proof procedures in argumentation. In: Boutilier, C. (ed.) Proceedings of the 21st International Joint Conference on Artificial Intelligence (IJCAI), pp. 746–751 (2009)
9. Eshghi, K., Kowalski, R.A.: Abduction compared with negation by failure. In: ICLP, pp. 234–254 (1989)
10. Fan, X., Toni, F., Hussain, A.: Two-agent conflict resolution with assumption-based argumentation. In: Computational Models of ArgumentComputational Models of Argument (COMMA), pp. 231–242 (2010)
11. Gaggl, S.A.: Solving argumentation frameworks using answer set programming. Master's thesis, Technische Universitt, Wien (2009)
12. Gelfond, M., Lifschitz, V.: The stable model semantics for logic programming. pp. 1070–1080. MIT Press (1988)
13. Gordon, T.F., Prakken, H., Walton, D.: The carneades model of argument and burden of proof. Artif. Intell. **171**(10–15), 875–896 (2007)
14. Gordon, T.F., Walton, D.: Legal reasoning with argumentation schemes. In: International Conference on Artificial Intelligence and Law (ICAIL), pp. 137–146. ACM (2009)
15. Kakas, A.C.: Default reasoning via negation as failure. In: Lakemeyer, G., Nebel, B. (eds.) ECAI-WS 1992. LNCS, vol. 810, pp. 160–178. Springer, Heidelberg (1994)
16. Prakken, H.: Formalising ordinary legal disputes: a case study. Artif. Intell. Law **16**(4), 333–359 (2008)
17. Prakken, H., Sartor, G.: Formalising arguments about the burden of persuasion. In: Proceedings of the 11th international Conference on Artificial intelligence and law, ICAIL '07, pp. 97–106. ACM. New York (2007)
18. Prakken, H., Sartor, G.: More on presumptions and burdens of proof. In: Francesconi, E., Sartor, G., Tiscornia, D. (eds.) JURIX, volume 189 of Frontiers in Artificial Intelligence and Applications, pp. 176–185. IOS Press (2008)
19. Rao, A.S., Georgeff, M.P.: BDI agents: from theory to practice. In: Proceeding of the first International Conference on Multi-Agent Systems (ICMAS-95), pp. 312–319 (1995)
20. Satoh, K., Asai, K., Kogawa, T., Kubota, M., Nakamura, M., Nishigai, Y., Shirakawa, K., Takano, C.: PROLEG: an implementation of the presupposed ultimate fact theory of Japanese civil code by PROLOG technology. In: Bekki, D. (ed.) JSAI-isAI 2010. LNCS, vol. 6797, pp. 153–164. Springer, Heidelberg (2011)

21. Satoh, K., Kogawa, T., Okada, N., Omori, K., Omura, S., Tsuchiya, K.: On general-
    ity of PROLEG knowledgerepresentation. In: Proceedings of the 6th International
    Workshop on Juris-informatics (JURISIN 2012), Miyazaki, Japan, pp. 115–128
    (2012)
22. Satoh, K., Kubota, M., Nishigai, Y., Takano, C.: Translating the Japanese pre-
    supposed ultimate fact theory into logic programming. In: Proceedings of the 2009
    Conference on Legal Knowledge and Information Systems: JURIX 2009, Amster-
    dam, The Netherlands, pp. 162–171. IOS Press (2009)
23. Satoh, K.: Logic programming and burden of proof in legal reasoning. New Gener.
    Comput. **30**(4), 297–326 (2012)
24. Sergot, M.J., Sadri, F., Kowalski, R.A., Kriwaczek, F., Hammond, P., Cory, H.T.:
    The British nationality act as a logic program. Commun. ACM **29**(5), 370–386
    (1986)
25. Thang, P.M., Dung, P.M., Hung, N.D.: Towards a common framework for dialecti-
    cal proof procedures in abstract argumentation. J. Logic Comput. **19**(6), 1071–1109
    (2009)
26. López y López, F., Luck, M.: A model of normative multi-agent systems and
    dynamic relationships. In: Lindemann, G., Moldt, D., Paolucci, M. (eds.) RASTA
    2002. LNCS (LNAI), vol. 2934, pp. 259–280. Springer, Heidelberg (2004)
27. López y López, F., Luck, M., d'Inverno, M.: A normative framework for agent-
    based systems. Comput. Math. Organiz. Theor. **12**, 227–250 (2006)
28. Yoshino, H.: On the logical foundations of compound predicate formulae for legal
    knowledge representation. Artif. Intell. Law **5**(1–2), 77–96 (1997)

Answering Yes/No Questions
in Legal Bar Exams

Mi-Young Kim[1($\boxtimes$)], Ying Xu[1], Randy Goebel[1], and Ken Satoh[2]

[1] Department of Computing Science, University of Alberta, Edmonton, Canada
{miyoung2, yx2, rgoebel}@ualberta.ca
[2] National Institute of Informatics/Sokendai, Tokyo, Japan
ksatoh@nii.ac.jp

Abstract. The development of Question Answering (QA) systems has become important because it reveals research issues that require insight from a variety of disciplines, including Artificial Intelligence, Information Extraction, Natural Language Processing, and Psychology. Our goal here is to develop a QA approach to answer yes/no questions relevant to civil laws in legal bar exams. A bar examination is intended to determine whether a candidate is qualified to practice law in a given jurisdiction. We have found that the development of a QA system for this task provides insight into the challenges of formalizing reasoning about legal text, and about how to exploit advances in computational linguistics. We separate our QA approach into two steps. The first step is to identify legal documents relevant to the exam questions; the second step is to answer the questions by analyzing the relevant documents. In our initial approach described here, the first step has been already solved for us: the appropriate articles for each question have been identified by legal experts. So here, we focus on the second task, which can be considered as a form of Recognizing Textual Entailment (RTE), where input to the system is a question sentence and its corresponding civil law article(s), and the output is a binary answer: whether the question sentence is entailed from the article(s). We propose a hybrid method, which combines simple rules and an unsupervised learning model using deep linguistic features. We first construct a knowledge base for negation and antonym words for the legal domain. We then identify potential premise and conclusion components of input questions and documents, based on text patterns and separating commas. We further classify the questions into easy and difficult ones, and develop a two-phase method for answering yes/no questions. We answer easy questions by negation/antonym detection. For more difficult questions, we adapt an unsupervised machine learning method based on morphological, syntactic, and lexical semantic analysis on identified premises and conclusions. This provides the basis to compare the semantic correlation between a question and a legal article. Our experimental results show reasonable performance, which improves the baseline system, and outperforms an SVM-based supervised machine learning model.

Keywords: Legal text mining · Natural language processing · Question answering · Recognizing textual entailment

© Springer International Publishing Switzerland 2014
Y. Nakano et al. (Eds.): JSAI-isAI 2013, LNAI 8417, pp. 199–213, 2014.
DOI: 10.1007/978-3-319-10061-6_14

1 Introduction

The last decade has challenged many disciplines with a deluge of written information, typically in digital form. In the legal domain, this situation was anticipated and referred to as the "information crisis" in law, and served as the impetus for the development of legal full-text information extraction systems [1].

Our immediate goal is to automatically answer yes/no questions relevant to civil law in legal bar exams. Legal bar examinations are intended to determine whether a candidate is qualified to practice law in a given jurisdiction. The task can be conceived as the first of evaluating the semantic equivalence between input questions and relevant law articles. This task is related to Recognizing Textual Entailment (RTE), where the task is to confirm whether a question sentence is entailed by a corresponding civil law article; the output is a binary classification decision, "yes" or "no". Since the input questions and articles are all domain-specific, they share the same technical terms, and therefore detecting semantic relationship is easier than for open-domain questions.

Earlier studies have concluded that simple word overlap measures (e.g., bag of words, n-grams) have a surprising degree of utility [3], but they still need to be improved. A common problem identified in these earlier systems is the lack of understanding the semantic relation between words and phrases. Later systems that include more linguistic features extracted from resources such as WordNet showed better performance [4]. Previous studies have also shown that syntactic features from parse trees are also helpful in this task [5]. Even more recent studies gained further leverage from systematic exploration of the syntactic feature space through analysis of parse trees [6]. Our methods also extract some deep linguistic features, such as lexically semantic information from thesauri, and syntactic dependency information.

An interesting recent development in the area of recognizing textual entailment (RTE) has been the application of so-called natural logics [7]. Natural logics provide a form of meaning representations that are essentially phrase-structured natural language sentences; from these one can compute entailments as substitutions for constituents (words or phrases). Any implementation of a natural logic will require the specification of conditions for monotonicity, subsectiveness, subsumption and exclusion properties of the predicates and modifiers identified in the vocabulary of the text, as well as vocabulary-independent meta-axioms that support reasoning with these properties. In addition, the natural logic inference systems need to incorporate a lot of background domain-dependent subsumption facts (e.g., walking and running are both subsumed by some kind of human locomotion). In our study, since the questions and corresponding documents are all in a restricted legal domain, they share the same technical terms. Because it is easy to compare lexical terms in this domain, we do not implement a general logic which needs to supply general, vocabulary-independent meta-axioms, but instead use unsupervised learning by constructing a domain-specific knowledge base and extracting deep linguistic features.

Question answering system comprises the extraction of a relevant paragraph of a source text that somehow aligns with the information need expressed by a natural language question. In order to automatically answer the bar exam yes/no questions, we first have to find the corresponding articles based on the Q/A technologies, and then we

Table 1. 'No' question types in legal bar exam

'No' question types	Proportion	'No' question types	Proportion
Negation	0.32	Constraints in premise	0.2
Using (semi-) anonym word	0.176	Constraints in conclusion	0.04
Paraphrasing of a phrase	0.12	Etc.	0.08
Exceptional case written in article	0.064		

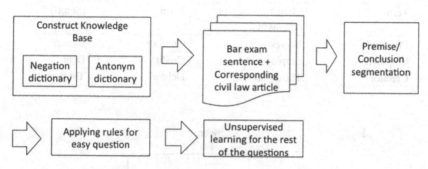

Fig. 1. Overall procedure of our method

have to compare the meaning of the input question and the corresponding article. We can then produce yes/no answers. In our case here, legal experts have already annotated corresponding civil law articles for questions. So for now, we focus only on comparing the meaning of the input questions and corresponding articles, to provide the basis for our yes/no assessment.

The rest of our paper is structured as follows. First, we explain the details of our method in Sect. 2, and we describe experimental setup, results, and error analysis in Sect. 3. Section 4 explains related work, and finally our future work and conclusions are described in Sect. 5.

2 Our Method

In order to answer yes/no questions according to the corresponding legal articles, we have to align structures and words embedded in the sentence pairs. These alignments are not given as inputs, and to determine them is a non-trivial task. This alignment-based approach has been shown effective by many RTE, QA, and MTE systems [6, 8]. But alignment is not the only approach. Other studies have successfully applied theorem proving and logical induction techniques, translating both sentences to more abstract knowledge representations and then doing inference on these representations [9]. In comparison to previous work that exploits various ad hoc or heuristic methods, we intend to build on more principled techniques.

Part of our method is to classify the yes/no questions into a spectrum from simple to difficult, according to the observations on the data. Table 1 shows "no" question types

Table 2. Examples of negation types

Negation type	Example
Negation affix	not, no, less…
Negation words	Unreasonable, block, withdraw, cancel, shrink, forbid, prohibit..
Negation concepts	n457, n444 …

Table 3. Examples of antonym dictionary

Term	(Semi) Antonym	Term	(Semi-) Antonym
Principal	Interest	Creditor	Debtor
Employer	Employee	Credit	Debt
Creditor	Third-Party	Debtor	Third-party

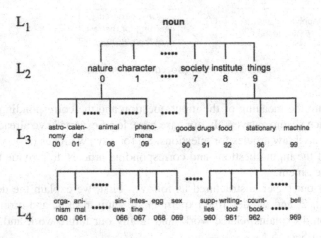

Fig. 2. Concept hierarchy of the Kadokawa thesaurus

in our data. "Negation" is the highest proportion with 32 %, and the second largest category is when there are different constraints in the premises between question and article. The third largest category arises when a word in the article is replaced with an "antonym" word or "different-meaning"-word. The fourth is a question with paraphrasing that cannot be resolved without expert knowledge, and the fifth are those where there is a difference of constraints in the conclusion between question and corresponding article. We notice that the categories of negation and antonym make up about 50 % of the total.

To address these two categories, we first construct two kinds of knowledge bases: a negation dictionary and an antonym dictionary. When an input pair (question, corresponding article) is given, we look for the premise and conclusion parts of the input question and corresponding civil law article. We then use a rule-based method to solve easy questions. Subsequent categories are addressed by exploiting some machine learning, based on exploiting some deeper linguistic information. We describe each process in detail in the following subsections. The overall workflow of our method is shown in Fig. 1.

2.1   Constructing Supportive Knowledge Bases

The most important features in determining semantic equivalence or near-equivalence, are accurate attribution of negation and confirming the use of antonyms. In our approach, we construct a negation knowledge base from the civil law articles. We identify two types of negation expressions: one is to note negation prefixes such as "not", "no", etc. The other is the case where the word itself conveys negative information. To extend our identification of negation words, we also use the Kadokawa thesaurus [10] which has a 4-level hierarchy of about 1,100 semantic classes, as shown in Fig. 2. Concept nodes in level L1, L2, and L3 are further divided into 10 subclasses. Table 2 shows examples of negation types.

We also manually collect the legal terms that can be used as antonyms or semi-antonyms having the same named entities from the civil law articles. Table 3 shows examples of the antonym dictionary. In a preprocessing phase, we add features "NEG" for the negation words in the questions and articles, and add features "ANT" for the (semi-) antonyms in the questions by comparing the words in the corresponding articles.

2.2   Premise and Conclusion Detection

The general idea of determining alignment between question and law article is easier if we first divide the question and articles into premise and conclusion. We compare the premise (conclusion) of a question and that of the corresponding article, and then examine negation or contradiction of intended meaning if it exists. Sentences in the legal law articles are usually long (average 21.25 words/sentence according to our data), and a comma is the most common delimiter between phrases.

Based on commas and keywords of a premise, we segment sentences. The keywords of premise are as follows: "in case(s)", "if", "unless", "with respect to", "when", and comma. After segmentation, the last segment is considered to be a conclusion, and the rest of the sentence is considered as a premise as follows:

$$conclusion := segment_{last}(sentence, keyword),$$

$$premise := \sum_{i \neq last} segment_i(sentence, keyword),$$

In our context, it is typical that a law article consists of two sentences. The first sentence is the main explanation of the law, and the second is for any exceptions. The second sentence only includes specific terms in the exceptional case and conclusion of the case. Consider the following example:

<Civil Law Article 295-1>: If a possessor of a Thing belonging to another person has a claim that has arisen with respect to that Thing, he/she may retain that thing until

*that claim is satisfied. **Provided, however, that this shall not apply if such claim has not yet fallen due**.*

Our central task here is to determine if a question belongs to the overall case or to the exceptional case. To do so, we cannot simply use the count of overlapped words as features, because typically the first article sentence, which represents the overall case, has more overlapping words with the question than the second article sentence, which represents the exceptional case. Since the second sentence typically includes only terms specific to describing exceptional cases and the conclusion in the case, we first compare the premise of the second sentence and that of the question: if the content words in the premise of the exceptional sentence are included in the question above a threshold, then we conclude that the question belongs to the exceptional case. Our measure for this is as follows:

$$if\left(\frac{n(w_i(article_{n\_exception}, premise) \in W(q_n, premise))}{n(W(article_{n\_exception}, premise))} > = threshold\right)$$

$$then, \; article_n := article_{n\_exception}$$

$$otherwise, \; article_n := article_{n\_main},$$

where we define $article_{n\_main}$ as the main sentence of the corresponding article of n-th question, and $article_{n\_exception}$ as the second sentence describing exceptional case. We use only the lemma of each word for comparison, and consider only contents words (verb, adjective, and noun). We currently set the threshold to 0.7 based on a 10 % of random sampling of all data in our experiments.

2.3 Applying Rules for Easy Questions

Because our language domain is restricted for both the input questions and law articles, there are some questions that can be answered easily using only negation and antonym information. If the question and article share the same word as the root in each syntactic tree, we consider the question as easy, which means it can be answered using only negation/antonym detection. Here is an example:

*Question: If person A sells owned land X to person B, but soon after, sells the same land X to person C then if the registration title is transferred to B, **then person B can assert against C in the acquisition of ownership of land X.***

---

if (neg_level(article_n, premise)+neg_level(article_n, conclusion)
   = neg_level(q_n, premise)+neg_level(q_n, conclusion)),
             Answer_n := yes,
otherwise,   *Answer_n := no,*
where   *neg_level() := 1 if negation and antonym occur odd number of*
                  times.
        neg_level() := 0 otherwise.

Fig. 3. Answering rule for easy questions

Article 177: **Acquisitions of, losses of and changes in real rights concerning immovable properties may not be asserted against third parties,** *unless the same are registered pursuant to the applicable provisions of the Real Estate Registration Act and other laws regarding registration.*

➜ *Conclusion of the question : then person B can assert against C in the acquisition of ownership of land X.*
  Premise of the question : : If person A sells owned land X to person B, but soon after, sells the same land X to person C then if the registration title is transferred to B,
  Conclusion of the article : Acquisitions of, losses of and changes in real rights concerning immovable properties may **not** *be asserted against third parties,*
  Premise of the article : **unless** *the same are registered pursuant to the applicable provisions of the Real Estate Registration Act and other laws regarding registration.*

In the above example, the conclusions in both the question and the article use the root word "assert" of the syntactic tree. So, this example can be answered using only the confirming negation and antonym information. If the sum of the negation levels of a question is the same with that of the corresponding article, then we determine the answer is "yes", and otherwise "no".

The negation level is computed as following: if [negation + antonym] occurs an odd number of times in a premise (conclusion), its negation level is "1". Otherwise if the [negation + antonym] occurs an even number of times, including zero, its negation level is "0". In the above example, the negation level of the premise of the question is zero, and that of the conclusion of the question is also zero. The negation level of a premise of the article is one, and that of a conclusion of the article is also one. Since the sum of the negation levels of the question is the same with that of the corresponding article, we determine the answer of the question is "yes".

Our precise description for this rule is shown in Fig. 3. The output of our rule-based system is also used below in an unsupervised learning model for assigning labels of premise (conclusion) clusters for non-easy questions.

2.4   Unsupervised Learning for the Non-easy Questions

For the questions not confirmed as easy, we need to construct deeper representations. Fully general solutions are extremely difficult, if not impossible; for our first approximation to the non-easy cases, we have developed a method using unsupervised learning with more detailed linguistic information. Since we do not know the impact each linguistic attribute has on our task, we run a machine learning algorithm that 'learns' what information is relevant in the text to achieve our goal.

The types of features we use are as follows:

Word matching Having the same lemma.
Tree structure features Considering only the dependents of a root.
Lexical semantic features Having the same Kadokawa thesaurus concept code.

We use our learning method on linguistic features to confirm the following semantic entailment features:

$Feature\,1$: $if\ w_{root}(q_n, premise) = w_{root}(article_n, premise)$

$Feature\,2$: $if\ w_{root}(q_n, conclusion) = w_{root}(article_n, conclusion)$

$Feature\,3$: $if\ w_{dep\_i}(q_n, conclusion) \in W_{dep}(article_n, conclusion)$

$Feature\,4$: $if\ c_{root}(q_n, premise) = c_{root}(article_n, premise)$

$Feature\,5$: $if\ c_{root}(q_n, conclusion) = c_{root}(article_n, conclusion)$

$Feature\,6$: $if\ neg\_level(q_n, premise) = neg\_level(article_n, premise)$

$Feature\,7$: $if\ neg\_level(q_n, conclusion) =$

$$neg\_level(article_n, conclusion)$$

Features 1, 2, 3 consider both lexical and syntactic information, and Features 4 and 5 consider semantic information. Features 6 and 7 incorporate negation and antonym information. Features 1 and 2 are used to check if premises (conclusions) of a question and corresponding article share the same root word in the syntactic tree. Feature 3 is to determine if each dependent of a root in the conclusion of a question appears in the article. We heuristically limit the number of dependents as those three nearest to the root. Features 4 and 5 confirm if the root words of premises (conclusions) of the question and corresponding article share the same concept code. We use some morphological and syntactic analysis to extract lemma and dependency information. Details of the morphological and syntactic analyzer are given in Sect. 3.

The inputs for our unsupervised learning model are all the questions and corresponding articles. The outputs are two clusters of the questions. The yes/no outputs of easy questions which have been already obtained are used as a key for assigning yes/no label of each cluster. The cluster which includes higher portion of "yes" of the easy questions is assigned the label "yes", and the other cluster is assigned "no". For the non-easy questions, we determine their yes/no answers following their clustering labels. For the easy questions, we use results of the rule of Fig. 3, regardless of the clustering labels of the questions, because the rule produces more accurate answers for easy questions than the clustering output.

3 Experiments

3.1 Experimental Setup

In the general formulation of the textual entailment problem, given an input text sentence and a hypothesis sentence, the task is to make predictions about whether or not the hypothesis is entailed by the input sentence. We report the accuracy of our method in answering yes/no questions of legal bar exams by predicting whether the questions can be entailed by the corresponding civil law articles.

There is a balanced positive-negative sample distribution in the dataset (49.8 % yes, and 50.2 % no), so we consider the baseline for true/false evaluation is the accuracy when returning always "no", which is 50.2 %. Note that other systems that give state-of-the-art performance on RTE use non-comparable techniques such as theorem-proving and logical induction, and often involve significant manual engineering specifically for RTE. It is thus difficult to make meaningful comparisons with the methods employed in our model.

Therefore the basis for our calibration is with the yes/no questions of legal bar exam sentences related to civil laws. The experts (law school students) annotated corresponding articles for each question. The correspondence type of (question, article) can be divided into three categories: The first is (one question, one article), the second is (one question, multiple articles), and the last is (one question, precedence which is not an article). The proportion of (one question, one article) is 25.63 % of overall questions, and we target only the first case, which is a one-to-one correspondence between question and article. Our data has 247 questions, with total 1044 civil law articles.

The original examinations are provided in Japanese, and our initial implementation used a Korean translation, provided by the Excite translation tool (http://excite.translation.jp/world/). Because most of our study team members are not proficient in Japanese, we translated the Japanese data into Korean. The reason that we chose Korean is that the characteristics of Korean and Japanese language are similar, and the translation quality between two languages ensures relatively stable performance. In addition, because our study team includes a Korean researcher, we can easily analyze the errors and intermediate rules in Korean. We used a Korean morphological analyzer and dependency parser [11], which extracts enriched information including the use of the Kadokawa thesaurus for lexical semantic information. We use a simple unsupervised learning method, since the data size is not big enough to separate it into training and test data.

We compare our method with SVM, a supervised learning model. Using the SVM tool included in the Weka [27] software, we performed cross-validation for the 247 questions using 7 features explained in Sect. 2.4. We used a linear kernel SVM because it is popular for real-time applications as they enjoy both faster training and classification speeds, with significantly less memory requirements than non-linear kernels because of the compact representation of the decision function.

3.2 Experimental Results

Evaluation of question answering systems is in general almost as complex as question-answering itself. So one must make the choice to consider several features of QA

Table 4. Performance of our system

Our method	Accuracy (%)
Baseline	50.20
Rule-based model for easy questions	68.36
Rule-based model for all questions	60.02
Unsupervised learning for difficult questions (K-means)	54.62
Unsupervised learning (K-means) for all questions	56.73
Rule for easy questions + unsupervised learning for difficult questions	61.13
Supervised learning (SVM) for all questions	58.01
Supervised learning (SVM) for difficult questions	55.78

Table 5. Error types

Error type	Accuracy (%)	Error type	Accuracy (%)
Specific example case	7.45	Paraphrasing	42.55
Exceptional case	8.09	Constraints in premise	28.09
Condition, conclusion mismatch	3.19	Reference to another article	3.19
Etc.	7.45		

systems in the evaluation process, e.g., query language difficulty, content language difficulty, question difficulty, usability, accuracy, confidence, speed and breadth of domain [12].

Table 4 shows our results. A rule-based model for easy questions showed accuracy of 68.36 %, and it covered 117 questions, which is 47.18 % of all questions. When we applied the rule-based method for all questions, the accuracy was decreased into 60.02 %. We use a K-means clustering algorithm with K = 2 for unsupervised learning for the rest of the questions, and it showed accuracy of 54.62 %. The overall performance when combining the use of rules and unsupervised learning showed 61.13 % of accuracy which outperformed unsupervised learning for all questions, and even SVM, the supervised learning model we use with a linear kernel. According to p-value measures between the baseline and each model in the true/false determination, all models significantly outperformed the baseline. Since previous methods use supervised learning with syntactic and lexical information, we consider the supervised learning experiment with SVM in Table 4 approximately represents the performance of previous methods.

3.3 Error Analysis

From unsuccessful instances, we classified the error types as shown in Table 5. The biggest error arises, of course, from the paraphrasing problem, which should be solved by expert knowledge and much larger corpora. The second biggest error is because of complex constraints in conditions. As with the other error types, there are cases where a question is an example case of the corresponding article, and the corresponding article embeds another article. In further work, we will need to complement our knowledge base with some kind of paraphrasing dictionary, perhaps with the help of experts. We also found cases that indicate the need to do more extensive temporal analysis.

It will be interesting if we compare our performance using Korean-translated sentences with that using original Japanese sentences. We would expect the system using original sentences will show better performance than ours, because there exist no translation errors. As future work, we will construct a Japanese system using paraphrase/synonym/antonym dictionaries for Japanese, and then analyze how the translation affects performance.

3.4 Using PROLEG

Of course the capture of legal concepts and their relationships is central to the improvement of systems such as ours, but the automatic construction of this kind of knowledge is equivalent to the general problem of open information extraction. However, in the legal domain, there are examples of legal representation systems that have already been used to capture some of this knowledge [24–26].

The one we know best is PROLEG [2], which is a PROLOG-based legal reasoning system. A PROLEG program is a general description of a legal reasoning case, which outputs a trace of derivation, and this trace is represented in the form of an argument between plaintiff and defendant. The main function of PROLEG is to capture and simulate the judge's decision process, and a derivation trace is a by-product of legal reasoning performed by a judge in the form of argument. We have constructed PROLEG logics for civil laws, and we intend to use PROLEG in our base system to improve performance by exploiting the deeper legal knowledge captured in PROLEG.

Here follows an example of PROLEG usage. We have a question <18-16-B> and the corresponding civil law article No. 333 as follows:

<Question 18-16-B>

In cases where movable X was delivered from person A to person B, and then from person B to person C based on a sale, the transfer of movable, person A can deter movable X as exercise of statutory liens for sale of movables.

<Civil law article 333>

Statutory liens may not be exercised with respect to the movables that are the subject matter of the same after the obligors have delivered those movables to third-party acquirers.

We have the following PROLEG rules and exceptions related with the article 333:

```
<PROLEG>
1. 'effect of statutory lien'(Obligee,Obligor,Third_Party,Object)<=
     'statutory lien over movables'(Obligee,Obligor,Cause).
2.'statutory lien over movables' (Obligee, Obligor,contract ('Sales',Obligee,
     Obligor, Object, T_contract))<=
     contract(Obligee,Obligor,contract('Sales',Obligee,Obligor,Object,T_contract)).
3. exception('effect of statutory lien'(Obligee, Obligor,Third_Party, Object),
     'exception of third party acquirers'(Obligor,Third_Party,Obligee,Object)).
4. 'exception of third party acquirers'(Obligor,Third_Party,Obligee,Object)<=
     contract(Obligor,Third_Party,contract('Sales',Obligor,Third_Party,Object,
     T_contract)),
     delivery(Obligor,Third_Party,contract('Sales',Obligor,Third_Party,Object,T_contract),
     T_delivery).
```

For readability, we express the above PROLEG rules and exceptions using the letters A, B, C, D, E and F.

(1) A <= B.
(2) B <= C.
(3) exception (A, D)
(4) D <= E, F.

In rule (3), we have the "exception" meta-predicate which takes two arguments. The former of the arguments is the head of default rule, and the latter is the head of exceptional rule. Then, "exception(A, D)" means "if D, then not A".

We can represent the above question into the following PROLEG:

'effect of statutory lien'(personA,personB,personC,movableX) <=
    contract(personA,personB, contract('Sales', personA,personB, movable,t_contract1)),
    contract(personB, personC, contract('Sales', personB, personC, movableX, t_contract2)),
    delivery(personB, personC, contract('Sales', personB, personC, movableX, t_contract2),
    t_delivery),

which means A <= C, E, F.

Since we have E and F in the premise of the question, we also have D according to rule 4). Therefore "not A" is derived according to rule 3). Since "not A" contradicts the conclusion of the question, which is "A", the answer of this question is "no". This kind of logical reasoning will likely improve performance.

However, to do this, we need to confirm correspondence between words in a question and predicate names and arguments in PROLEG. To find the corresponding PROLEG rules and fill the argument variables correctly, we need more extensive natural language processing techniques, including some general information extraction processes like co-reference resolution, query expansion, paraphrasing, synonym dictionary construction, and syntactic graph matching. As we augment our NLP tools, the PROLEG-based text entailment will provide a deeper level understanding of the questions/articles, and improve performance.

4   Related Work

W. Bdour et al. [13] developed a Yes/No Arabic Question Answering System. They used a kind of logical representation, which bridges the distinct representations of the functional structure obtained for questions and passages. This method is not appropriate for our task. If a false question sentence is constructed by replacing named entities with terms of different meaning in the legal article, a logic representation can be helpful. However, false questions are not simply constructed by substituting specific named entities, and any logical representation can make the problem more complex. Kouylekov and Magnini [14] experimented with various cost functions and found a combination scheme to work the best for RTE. Vanderwende et al. [15] used syntactic heuristic matching rules with a lexical-similarity back-off model. Nielsen et al. [16] extracted features from dependency paths, and combined them with word-alignment features in a mixture of experts classifier. Zanzotto et al. [17] proposed a syntactic cross-pair similarity measure for RTE. Harmeling [18] took a similar classification-based approach with transformation sequence features. Marsi et al. [19] described a system using dependency-based paraphrasing techniques. All previous systems uniformly conclude that syntactic information is helpful in RTE, and we also use syntactic information combined with lexical semantic information.

There are also many QA studies in the legal field. The first one is ResPubliQA 2009 [20]. It describes the first round of ResPubliQA, a Question Answering (QA) evaluation task over European legislation, proposed at the Cross Language Evaluation Forum (CLEF) 2009. The ResPubliQA 2009 exercise is aimed at retrieving answers to a set of 500 questions. The answer of a question is a paragraph of the test collection. The hypothetical user considered for this exercise is a person interested in making inquiries in the law domain, specifically on the European legislation. There is another system for QA of legal documents reported by Monroy et al. [21]. They experiment by using natural language techniques such as lemmatizing and using manual and automatic thesauri for improving question based document retrieval. In addition, there was a method based on syntactic tree matching [22], and knowledge-based method using a variety of thesaurus and dictionaries [23]. As further research, we can enrich our knowledge base with deeper analysis of data, and add paraphrasing dictionary getting help from experts.

5 Conclusion

We have proposed a method to answer yes/no questions from legal bar exams related to civil law. We construct our own knowledge base by analyzing negation patterns and antonyms in the civil law articles. To make the alignment easy, we first segment questions and articles into premise and conclusion. We then extract deep linguistic features with lexical, syntactic information based on morphological analysis and dependency trees, and lexical semantic information using the Kadokawa thesaurus. Our method consists of two phases. First, we apply our own simple rules for easy questions, and then adopt unsupervised learning for other questions. This achieved quite encouraging results in both true and false determination. To improve our approach in future work, we need to create deeper representations (e.g., to deal with embedded articles and paraphrase), and analyze the temporal aspects of legal sentences. In addition, we will complement our knowledge base with paraphrasing dictionary with the help of experts. We also have access to a logic-based reconstruction of legal rules in a system called PROLEG [2], which we believe can augment our unsupervised learning process with more precise legal information.

Acknowledgements. This research was supported by the Alberta Innovates Centre for Machine Learning (AICML) and the iCORE division of Alberta Innovates Technology Futures.

References

1. Merkl, D., Schweighofer, E.: En route to data mining in legal text corpora: clustering, neural computation, and international treaties. In: Proceedings of International Workshop on Database and Expert Systems Applications, pp. 465–470 (1997)
2. Satoh, K., Asai, K., Kogawa, T., Kubota, M., Nakamura, M., Nishigai, Y., Shirakawa, K., Takano, C.: PROLEG: an implementation of the presupposed ultimate fact theory of Japanese civil code by PROLOG Technology. In: Bekki, D. (ed.) JSAI-isAI 2010. LNCS (LNAI), vol. 6797, pp. 153–164. Springer, Heidelberg (2011)

3. Jikoun, V., de Rijke, M.: Recognizing textual entailment using lexical similarity. In: Proceedings of the PASCAL Challenges Workshop on RTE (2005)
4. MacCartney, B., Grenager, T., de Marneffe, M.-C., Cer, D., Manning, C.D.: Learning to recognize features of valid textual entailments. In: Proceedings of HLT-NAACL (2006)
5. Sno, R., Vanderwende, L., Menezes, A.: Effectively using syntax for recognizing false entailment. In: Proceedings of HLT-NAACL (2006)
6. Das, D., Smith, N.A.: Paraphrase identification as probabilistic quasi-synchronous recognition. In: Proceedings of ACL-IJCNLP (2009)
7. Schubert, L.K., Durme, B.V., Bazrafshan, M.: Entailment inference in a natural logic-like general reasoner. In: Proceedings of the AAAI 2010 Fall Symposium on Commonsense Knowledge (2010)
8. MacCartney, B., Galley, M., Manning, C.D.: A phrase-based alignment model for natural language inference. In: Proceedings of EMNLP (2008)
9. MacCartney, B., Manning, C.D.: Natural logic for textual inference. In: Proceedings of Workshop on Textual Entailment and Paraphrasing at ACL (2007)
10. Ohno, S., Hamanishi, M.: New Synonym Dictionary. Kadokawa Shoten, Tokyo (1981)
11. Kim, M.-Y., Kang, S.-J., Lee, J.-H.: Resolving ambiguity in inter-chunk dependency parsing. In: Proceedings of 6t^h Natural Language Processing Pacific Rim Symposium, pp. 263–270 (2001)
12. Walas, M.: How to answer yes/no spatial questions using qualitative reasoning? In: Proceedings of the International Conference on Computational Linguistics and Intelligent Text Processing, pp. 330–341 (2012)
13. Bdour, W.N., Gharaibeh, N.K.: Development of yes/no arabic question answering system. Int. J. Artif. Intell. Appl. 4(1), 51–63 (2013)
14. Kouylekov, M., Magnini, B.: Tree edit distance for recognizing textual entailment: estimating the cost of insertion. In: Proceedings of the Second PASCAL Challenges Workshop on RTE (2006)
15. Vanderwende, L., Menezes, A., Snow, R.: Microsoft research at RTE-2: syntactic contributions in the entailment task: an implementation. In: Proceedings of the Second PASCAL Challenges Workshop on RTE (2006)
16. Nielsen, R.D., Ward, W., Martin, J.H.: Toward dependency path based entailment. In: Proceedings of the Second PASCAL Challenges Workshop on RTE (2006)
17. Zanzotto, F.M., Moschitti, A., Pennacchiotti, M., Pazienza, M.T.: Learning textual entailment from examples. In: Proceedings of the Second PASCAL Challenges Workshop on RTE (2006)
18. Harmeling, S.: An extensible probabilistic transformation-based approach to the third recognizing textual entailment challenge. In: Proceedings of ACL PASCAL Workshop on Textual Entailment and Paraphrasing (2007)
19. Marsi, E., Krahmer, E., Bosma, W.: Dependency-based paraphrasing for recognizing textual entailment. In: Proceedings of ACL PASCAL Workshop on Textual Entailment and Paraphrasing (2007)
20. Penas, A., Forner, P., Sutcliffe, R., Rodrigo, A., Forascu, C., Alegria, I., Giampiccolo, D., Moreau, N., Osenova, P.: Overview of ResPubliQA 2009: question answering evaluation over european legislation. In: Proceedings of the Cross-Language Evaluation Forum Conference on Multilingual Information Access Evaluation: Text Retrieval Experiments, pp. 174–196 (2009)
21. Monroy, A., Calvo, H., Gelbukh, A.: NLP for shallow question answering of legal documents using graphs. In: Gelbukh, A. (ed.) CICLing 2009. LNCS, vol. 5449, pp. 498–508. Springer, Heidelberg (2009)

22. Mai, Z., Zhang, Y., Ji, D.: Recognizing text entailment via syntactic tree matching. In: Proceedings of NTCIR-9 Workshop Meeting (2011)
23. Arya, D.A., Yaligar, V., Prabhu, R.D., Reddy, R., Acharaya, R.: A knowledge based approach for recognizing textual entailment for natural language inference using data mining. Int. J. Comput. Sci. Eng. **2**(6), 2133–2140 (2010)
24. Bench-capon, T.: What makes a system a legal expert? In: Legal Knowledge and Information Systems: JURIX 2012, pp. 11–20 (2012)
25. Alberti, M., Gomes, A.S., Gonçalves, R., Leite, J., Slota, M.: Normative systems represented as hybrid knowledge bases. In: Leite, J., Torroni, P., Ågotnes, T., Boella, G., van der Torre, L. (eds.) CLIMA XII 2011. LNCS (LNAI), vol. 6814, pp. 330–346. Springer, Heidelberg (2011)
26. Lundstrom, J.E., Aceto, G., Hamfelt, A.: Towards a dynamic metalogic implementation of legal argumentation. In: Proceedings of ICAIL, pp. 91–95 (2011)
27. Hall, M., Frank, E., Holmes, G., Pfahringer, B., Reutemann, P., Witten, I.H.: The WEKA data mining software: an update. SIGKDD Explor. **11**(1), 10–18 (2009)

Answering Legal Questions
by Mining Reference Information

Oanh Thi Tran[✉], Bach Xuan Ngo, Minh Le Nguyen, and Akira Shimazu

School of Information Science, Japan Advanced Institute of Science and Technology,
1-1 Asahidai, Nomi, Ishikawa 923-1292, Japan
{oanhtt,bachnx,nguyenml,shimazu}@jaist.ac.jp

Abstract. This paper presents a study on exploiting reference information to build a question answering system restricted to the legal domain. Most previous research focuses on answering legal questions whose answers can be found in one document (The term *'documents'* corresponds to articles, paragraphs, items, or sub-items according to the naming rules used in the legal domain.) without using reference information. However, there are many legal questions whose answers could not be found without linking information from multiple documents. This connection is represented by explicit or implicit references. To the best of our knowledge, this type of questions is not adequately considered in previous work. To cope with them, we propose a novel approach which allow us to exploit the reference information among legal documents to find answers. This approach also uses requisite-effectuation structures of legal sentences and some effective similarity measures to support finding correct answers without training data. The experimental results showed that the proposed method is quite effective and outperform a traditional QA method, which does not use reference information.

1 Introduction

A question answering (QA) system is a system that is able to automatically respond answers to questions posed by human in a natural language by retrieving information from a collection of documents. This is an important task and has drawn much attention in natural language processing research. Particularly, there are several top conferences which have organized special tracks for the topic of QA such as Text Retrieval Conference (TREC[1]) and Cross Language Evaluation Forum (CLEF[2]).

When considering an application of QAs in a specific domain, especially the legal domain, we saw that there is little work particularly devoted to this kind of research, despite its wide uses and applications. In the legal domain, QAs could be applied to help citizens and law-makers have easier access to legal information. Previous works [1,5,12] showed that a common problem is that traditional QAs

---

[1] http://trect.nist.gov
[2] http://www.clef-campaign.org

© Springer International Publishing Switzerland 2014
Y. Nakano et al. (Eds.): JSAI-isAI 2013, LNAI 8417, pp. 214–229, 2014.
DOI: 10.1007/978-3-319-10061-6_15

are not adequate to find the correct answers to legal questions. This was mostly caused by special structures, specific terms and long sentences.

In many laws, it is common that specific terms, legal objects, provisions, etc. are first defined in one document and then referred many times in other documents by using briefer expressions. This, as a result, helps to guarantee the soundness as well as the consistency in a law system. In order to link documents, references are used to identify the position as well as the fragment of texts referred to. These references bring precious information and their resolution helps interpreting the law. This is a useful characteristic of legal texts which can benefit the process of finding correct answers to many legal questions. Until now, however, there is no research on using this advantage of references to help finding correct answers. Of several work dedicated to legal QAs [1,5,12,18], they mostly focus on legal questions whose answers can be found from merely one document. However, the fact is that many legal questions requiring answers are combined from multiple documents which are linked based on references (as we can see later in Sect. 3). This type of questions is not adequately considered in previous research [1,5,12,18]. Building a good reference resolver is not a trivial task. Therefore, for some languages, authors usually find an alternative solution to help indirectly representing the relationship between documents [9] rather than using the reference information.

In this paper, we investigate an application of reference resolution to a legal QA system. We focus on one type of questions which can be benefited from the reference information such as the example in Fig. 1. In this example, to answer the question, it is necessary to link the information from two documents to find the answer. The linking information is expressed via the relation of the reference-referent in the colored italic texts. The italic texts bounded in angle brackets of the document 'Article 12, Paragraph 4' is a reference, which refers to the italic texts bounded in square brackets of the document 'Article 12, Paragraph 1'. Such questions are quite popular in the legal domain.

Sometimes, users are not only interested in obtaining just an answer, but also want to know its evidence. In this paper, we also provide proofs of the answer. The main contribution of our work can be concluded in the following points:

- Building a legal QA system by adequately considering one type of legal questions that can be benefited from reference information.
- Testing the proposed system on several legal questions yields promising results, and it also outperforms a traditional QA system.

This paper is organized as follows. Section 2 presents related work. Section 3 describes important characteristics of legal texts exploited in this research, i.e. references-referents structures between documents. In this section, we also describe in more details the type of legal questions considered in this work. Section 4 presents a proposed framework, which exploits the characteristics of legal texts, especially the reference information shown in Sect. 3. Section 5 presents experiments to compare the proposed system with a traditional QA system. Finally, Sect. 6 concludes the paper and discusses future research.

Question	市町村長は、被保険者の氏名及び住所の変更に関する事項の届出を受理したときは、何をしなければなりませんか？ "What should a mayor of a municipality do if he/she receives the notification on changing the address or name of an insured person?"
Clues	*Article 12, paragraph 1* 被保険者（第三号被保険者を除く。次項において同じ。）は、厚生労働省令の定めるところにより、[*その資格の取得及び喪失並びに種別の変更に関する事項並びに氏名及び住所の変更に関する事項を市町村長に届け出なければならない*]。 Pursuant to the provisions of the Ministry of Health, Labour and Welfare, the insured person (except for the third type. The same for the next para), must [*notify to the mayor of a municipality matters relating to the change of name, address, as well as matters related to change of type and loss and acquisition of the qualification*]. ∘ *Article 12, paragraph 4* "市町村長は、<*第一項又は第二項の規定による届出*>を受理したときは、厚生労働省令の定めるところにより、厚生労働大臣にこれを報告しなければならない。" When a mayor of a municipality receives <*the notification in the provision of Para 1 or Para 2*>, pursuant to the provisions of the Ministry of Health, Labour and Welfare, s/he must report it to the Minister of Health, Labour and Welfare.
Answer	厚生労働省令の定めるところにより、厚生労働大臣にこれを報告しなければならない。 pursuant to the provisions of the Ministry of Health, Labour and Welfare, s/he must report it to the Minister of Health, Labour and Welfare.

Fig. 1. A question is solved in this paper. In this figure, references are bounded in angle brackets (⟨⟩) while their referents are bounded in square brackets ([ ]).

2 Related Work

In the legal domain, there is not much research dedicated to QAs. To the best of our knowledge, there exists no work on legal QA, which focuses directly on making use of reference information between legal documents.

In [12], Paulo et al. present a QA system for Portuguese juridical documents. The proposed approach is based on computational linguistic theories: syntactical analysis followed by semantic analysis; and finally, a semantic/pragmatic interpretation using ontology and logical inference. The QA system was applied to the complete set of decisions from several Portuguese juridical institutions. It uses very expensive sources. This work was applied to the judge texts rather than the statute texts as in our work. Therefore, it cannot use the characteristic of the statute texts.

Monroy et al. [9] focus on building a QA system for Spanish at the shallow level by using graphs. The system gives answers which consist of a set of articles related to the question and also the relevant articles related with them to complement the answer. This method represents the link between documents by using the similarity (i.e. TF.IDF) between them via terms in documents. They also limit questions that mainly ask if it is possible to perform certain action.

Recently, Tomura et al. [18] present a study on building a QA system for Japanese legal texts. In this work, they deal with 5 types of questions whose answers can be found from one document using the requisite-effectuation structures of law sentences. This work shares the same type of law with our work - the Japanese National Pension Law (JNPL).

3   A Type of Legal Questions Raised from Characteristics of Legal Texts

Firstly, we introduce some important characteristics of legal texts. Then, we describe a type of legal questions, which is mostly raised from these characteristics.

3.1   The Characteristics of Legal Texts

One important characteristics of legal texts is that they usually have some specific structures at both sentence and paragraph levels. At sentence levels, law sentences usually have some specific structures [3]. At paragraph levels, sentences in the same paragraph usually have close relations. Another important characteristic of legal texts is that, at the discourse level, legal documents contain many reference phenomena which need solving in order to understand their contents.

Reference Phenomena in Legal Texts. Legal texts contain many reference phenomena within them. Legal references relate to terms, definitions, provisions, etc. For example, when law-makers describe conditions of a law in *Article 12, Paragraph 4* of the JNPL, they recall the definition of a type of notification by using a reference *'the notification in the provisions of Para 1 or Para 2'*. If this reference is resolved, we can fully understand which notification (explained in *Article 12, Paragraph 1*) is actually referred to in this document.

References (Mentions) [19] in legal texts have their own structures, which are different from mentions in general texts. A mention usually consists of two main parts: a position part and a content part. The later part may be a noun or a noun phrase, which determines the referred object. The former part conforms to some regular expressions which locate the position of the referred object. Referents (Antecedents) are definitions or explanations of related terms or provisions. They can be nouns, noun phrases, sentences, paragraphs of articles or even whole articles in some cases. They help readers fully comprehend the law, and also help lawmakers create concise and easy-to-understand legal texts.

Logical Parts and Logical Structures of Legal Texts. At the sentence level, a law sentence can roughly be divided into two *high-level*[3] logical parts: *requisite part* and *effectuation part* [3, 4, 17] in the form of:

requisite part $\Rightarrow$ *effectuation part*

Each *requisite part* or *effectuation part* consists of several logical parts. A logical part is a clause or a phrase in a law sentence that conveys a part

---

[3] The reason why they call them high-level is that each *requisite part* or *effectuation part* consists of several logical parts.

Fig. 2. An example of law sentences and their logical parts (A: Antecedent part;
C: Consequent part; T: Topic part).

of the meaning of legal texts. Each logical part contains a specific kind of information according to its type. Three main types of logical parts are *antecedent part*, *consequent part*, and *topic part*. A logical part in consequent type describes a law provision; a logical part in antecedent type indicates cases (or the context) the law provision can be applied; and a logical part in topic type describes subjects related to the law provision. In a simple case[4], the *requisite part* only consists of a *topic part* or an *antecedent part*; and the *effectuation part* only consists of one *consequent part*.

Figure 2 shows four cases of law sentences and their logical parts. Logical structures in four cases can be seen more in the paper of [4].

At the paragraph level, a paragraph usually contains a main sentence and one or more subordinate sentences [16]. To be concrete, in a paragraph, the first sentence presents a law provision, and the other sentences describe cases in which the law provision can be applied.

---

[4] To understand more about four cases of legal sentences and their logical parts, please check the paper of Bach et al. [2].

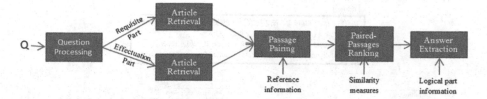

Fig. 3. A proposed framework to extract answers for this type of legal questions.

3.2 A Type of Question Raised from the Characteristics of Legal Texts

Generally speaking, to find correct answers, a QA system should have the ability to interpret content of documents. At the discourse level, legal documents are highly related by references, which usually bring precious information. A law cannot be correctly interpreted without reading some of the referenced items within it. These references can be placed on requisite parts or effectuation parts of sentences. This means that if a sentence contains a reference, the real content of its requisite or effectuation parts actually lie in a different document. For example, the sentence in '*article 12, paragraph 4*' of the tracing example has its requisite part lying in '*article 12, paragraph 1*' because it imports the definition of the notification in that document. The challenge for us is to be able to identify references and to jointly interpret them. Therefore, a good QA system should have the ability to follow these connections, which are represented via the relation of the references and their referents.

There are many legal questions falling under this type of questions because many users tend to ask about the beneficial conditions of laws, or the beneficiaries that can be achieved if some conditions of laws are satisfied[5].

4   A Proposed Framework for a Legal QA System

Based on the above analyses, we propose a novel framework to solve the problem as presented in Fig. 3. This framework includes five steps. In the first step, each input question is split into two parts, i.e. a requisite part and an effectuation part. In the next step - Article Retrieval, two collections of relevant articles are retrieved by using the content words and their synonyms of two parts respectively. Next, in the passage pairing step, a passage of articles in the first collection is aligned to a passage of articles in the second one by using the reference-referent if available. The result is a set of paired-passages which are likely to contain evidence for finding the correct answer. To determine the best pair, we rank pairs by using effective similarity measures derived from previous work [7]. The best pair will be used to extract the correct answer by using logical structures of legal texts.

---

[5] We can use these characteristics for a QA system as shown in Tomura, K., A study on a question answering system for laws, Master thesis, JAIST, 2013. The system answers to a question based on only one document.

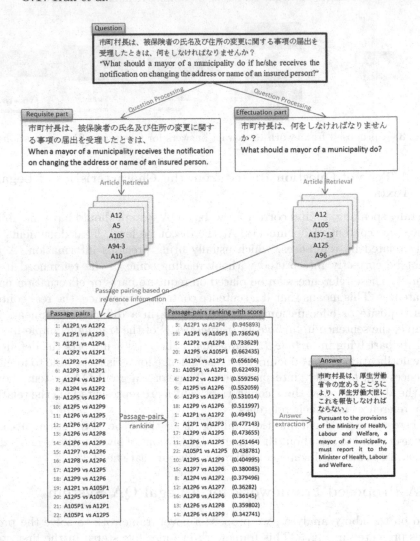

Fig. 4. A true example of the proposed system.

Figure 4 illustrates a running example of the proposed system. The question was first processed via the first step to divide it into two parts, i.e. requisite part and effectuation part. The next step retrieved 2 article collections corresponding to two logical parts. With N equals to 5, C_r includes 5 articles (A12, A5, A105, A94-3, A10) and C_e includes 5 articles (A12, A105, A137-13, A125, A96). Passages in the articles of C_r are paired with passages in the articles of C_e if they contain at least one reference which refers to the other passages or vice verse. This step leaded to the results including 22 passage pairs. The next step measured the similarity scores between these pairs and the question. The pair, A12P1 and A12P4, with the highest score is chosen as containing the answer. The question word lies in the effectuation part and the reference lies in A12P4,

question q	市町村長は、被保険者の氏名及び住所の変更に関する事項を受理したときは、何をしなければなりませんか？ *What should a mayor of a municipality do if he/she receives the notification on changing the address or name of an insured person?*
q after analyzing logical parts	<T3>市町村長は、</T3><A>被保険者の氏名及び住所の変更に関する事項を受理したときは、</A><C>何をしなければなりますか？</C>
Requisite part q_r	*(T3, A)* = when a mayor of a municipality receives the notification on changing the address or name of an insured person.
Effectuation part q_e	*(T3, C)* = What should a mayor of a municipality do?

Fig. 5. An example of the question processing step (A: Antecedent part; C: consequent part; T: Topic part).

so the system extracted the effectuation part of A12P4 as the correct answer. In the following sub-sections, we will present in more details about these steps.

4.1 Question Processing

The goal of this part is to split the question into two parts, i.e. the requisite part q_r and the effectuation part q_e. We exploit an implementation of Bach et al. [2] to recognize these parts. An example is given in Fig. 5. Each part is then preprocessed by word segmentation, POS tagging, and dependency parsing using Cabocha tool[6]. We keep content words and remove stop-words by using a list downloaded from this website[7]. Removing stop-words helps the model ignore function words and high-frequency, but low-content words.

In fact, the forms and words in user's questions might be different from real laws 's. Therefore, the exact wording of the answers might look nothing like the questions. Thus, it is necessary to expand the question by adding terms in hopes of matching the particular from of the answer as it appears. In other words, to increase the number of relevant articles, we also use the synonyms of each keyword in the question by using a Japanese synonym list[8].

4.2 Article Retrieval

Based on the content words extracted from the previous step, we retrieved relevant articles from the corpus using Boolean *AND* and *OR* queries. The information retrieval system selects a set of potentially relevant articles that are likely to contain the evidence for finding correct answers. To retrieve, we implemented cosine similarity between the question and an article. In the vector space model [6], articles and questions are represented as vectors of features representing the terms (keywords) that occur within the collection. The value of each feature is

---

[6] http://code.google.com/p/cabocha/

[7] http://www.ranks.nl/stopwords/japanese.html

[8] We used synonym list extracted from Japanese WordNet Copyright 2009, 2012 by National Institute of Information and Communications Technology (NiCT).

called the *term weight*. Here, we use conventional tf-idf [8] term weighting which is very useful and popular in many information retrieval tasks [10,11,20].

This step returns two article sets C_r and C_e. C_r contains a set of relevant articles for the requisite part of the question, q_r. C_e contains a set of relevant articles for the effectuation part of the question, q_e.

4.3 Passage Pairing

The purpose of this step is to link passages of articles in two sets C_r and C_e using the reference-referent information. Two passages in two sets are linked if one passage contains at least one reference, which refers to a referent in the other passage. In more detail, we pair each paragraph p_r in an article of the set C_r to a paragraph p_e of an article in the set C_e if there exists one reference in p_r referring to a fragment of texts in the paragraph p_e and vice verse.

4.4 Paired-Passages Ranking

In this step, all pairs in the form of (p_r, p_e) are ranked using a ranking function. The ranking function is a linear combination of some similarity scores between the passage pair and the question. The similarity score of each passage pair (p_r, p_e) with the question (q_r, q_e) is calculated using the following equation:

$$TotalScore((p_r, p_e), (q_r, q_e)) = TotalScore(p_r, q_r) + TotalScore(p_e, q_e) \quad (1)$$

Each $TotalScore(,)$ between an answer passage p_x and a question part q_x is calculated using the following equation:

$$TotalScore(p_x, q_x) = \sum_{i=1}^{n} \lambda_i \times score_i(p_x, q_x) \quad (2)$$

where λ_i is the weight of $score_i$; each $score_i(p_x, q_x)$ corresponds to one score in the following sets of scores derived from the work of Surdeanu et al. [7]. Accordingly, n is the total number of scores derived from that work. For the sake of simplicity, we set all λ_i equal to 1.

– *Similarities*
  The similarity between an part of a question q and the passage p is measured using the length-normalized *BM25* formula [14,15]. For completeness, we also include the value of the $tf.idf$ measure as presented in the article retrieval step. To understand the contribution of the syntactic and semantic processors, we compute the above similarity measures using three different representations of the question and passage content as follows:
  • *Words (W)* - the text is considered as a bag of words.
  • *Dependencies (D)* - the text is represented as a bag of binary syntactic dependencies. We extract dependency paths of length 1, i.e., direct head-modifier relations.

<T3>市町村長は、</T3><A>第一項又は第二項の規定による届出を受理したときは、</A><C>厚生労働省令の定めるところにより、厚生労働大臣にこれを報告しなければならない。</C>
<T3> a mayor of a municipality, </T3> <A>when receiving *the notification in the provision of Para 1 or Para 2* </A>, <C>pursuant to the provisions of the Ministry of Health, Labour and Welfare, s/he must report it to the Minister of Health, Labour and Welfare. <C>

Fig. 6. An example of the answer extraction step (A: Antecedent part; C: consequent part; T: Topic part).

- *Bigrams (B)* - the text is represented as a bag of bi-grams. This view is added to ensuring a fair analysis of the above syntactic views.
- *Density and frequency scores*
  These scores measure the density and frequency of question terms in the passage text.
  - *Same word sequence* - computes the number of non-stop question words that are recognized in the same order in the passage.
  - *Answer span* - the largest distance (in words) between two non-stop question words in the passage.
  - *Same sentence match* - number of non-stop question terms matched in a single sentence in the passage.
  - *Overall match* - number of non-stop question terms matched in the complete passage. These scores are normalized into [0,1]. These last two scores are computed also for the two remaining text representations.
  - *Informativeness* - models the amount of information contained in the answer passage by counting the number of non-stop nouns, verbs, and adjectives in the passage that do not appear in the question.

4.5 Answer Extraction

At the paragraph level, a paragraph usually contains a main sentence and one or more subordinate sentences. Here, we used an implementation of Bach et al. [2,3] to recognize the logical structures of paragraphs to extract the answer.

- If the question part lies in the effectuation part of the question, we extract the effectuation part of the main paragraph[9] as the answer and vice versa. To determine this, we use clues of question words such as *dare, itsu, doko, desuka, masenka, masuka, dono, nani, etc.*
- If the answer contains references, we also extract their referents to help people fully understand it.

Figure 6 shows an example of this step for the input question. Because the question asks about the consequence of an action, we extract its consequence part (including the topic part and the consequent part) as the final answer.

---

[9] The paragraph which contains the reference referring to the referenced paragraph.

5   Experimental Results of the QA System

In this section, we first present experimental setups including testing data, evaluation measure and a traditional QA system. The purpose of implementing this traditional QA system is to compare the performance of our QA system using reference information and not using them. Then, we present experimental results of our QA system using the proposed method and a traditional QA system.

5.1   Experimental Setups

This sub-section presents the testing legal questions and evaluation measure to estimate the system's performance. We also briefly present a traditional QA system to prove that using reference information in answering this special type of questions yields better results.

Data. We tested our system using 51 legal questions on the Japanese National Pension Law. To help us understand more about the behavior of the systems, we categorize these questions into two main classes based on how they use the reference information in determining the answers.

1. Obligatory references-resolving questions: Relevant sentences, which provide evidence to answers, lie in different documents. These sentences are linked through the reference information. This class is sub-divided as follows:
   (a) Bi-document-linking questions: Only two documents are linked using reference information to provide evidence to answers. A majority of legal questions falls into this case.
   (b) Multi-document-linking questions: more than two documents are linked to find answers using reference information.
2. Optional references-resolving questions: In this case, the referenced documents play the role of explaining more about the terms/phrases in the users questions. Therefore, QA systems can still find the answers without using reference information.

The class of each question is given in Table 1.

Evaluation Measure. To evaluate the performance, we use the evaluation measure of ResPubliQA 2009 [13] which is a QA evaluation task over European Legislation, proposed at the CLEF 2009. Because the two systems always output answers to all questions, the evaluation measure $c@1$ [13] becomes the accuracy measure calculated as follows:

$$Accuracy = \frac{\#CorrectlyAnsweredQuestions}{\#Questions} \tag{3}$$

A Traditional QA System. The traditional QA system consists of four steps as presented in Fig. 7. In the question processing step, the question is processed as

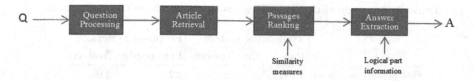

Fig. 7. The framework of the traditional QA system.

same as in the proposed method except for dividing it into its logical parts. The second step, article retrieval, retrieves top N relevant articles as in our method by using all question words and their synonyms. The third step, the passages ranking, ranks all passages in each relevant articles based on their similarity scores with the question. In this step, we use all two sets of scores as in our methods. The last step is as same as the answer extraction step of the proposed method.

5.2   Experimental Results Using the Traditional QA System and the Proposed System

Table 1 presents the experimental results using the traditional QA system and the proposed system on 51 legal questions. Table 2 presents the experimental results of the two systems using two evaluation criteria. The first criterion is to measure the performance based on the extracted paragraphs. This means that if the systems correctly determine the main paragraphs (which contain the answers). The second criterion is to measure the performance based on the

Table 1. Experimental results of two QA systems using the traditional method and the proposed method on 51 legal questions.

No.	Traditional Method	Proposed Method	Ques.' Class	No.	Traditional Method	Proposed Method	Ques.' Class
0	Wrong (referenced para)	True	1a	26	Wrong (True main para)	True	1a
1	True	True	2	27	Wrong (referenced para)	True	1a
2	Wrong (referenced para)	Wrong	1a	28	True	True	1a
3	Wrong (referenced para)	True	1a	29	Wrong (referenced para)	Wrong	1a
4	Wrong (referenced para)	True	1a	30	Wrong	True	1a
5	Wrong (referenced para)	Wrong	1b	31	True	Wrong	1b
6	True	True	1a	32	True	True	1a
7	Wrong (referenced para)	True	1a	33	Wrong (referenced para)	True	1a
8	Wrong	True	1a	34	Wrong (referenced para)	True	1a
9	Wrong (referenced para)	True	1b	35	Wrong (referenced para)	Wrong	1b
10	Wrong	Wrong	1b	36	Wrong (referenced para)	True	1a
11	Wrong	Wrong	1a	37	Wrong (referenced para)	True	1b
12	Wrong	Wrong	1a	38	Wrong (referenced para)	Wrong	1b
13	True	True	1b	39	True	True	1a
14	Wrong (referenced para)	True	1b	40	Wrong	Wrong	1b
15	Wrong (referenced para)	True	1a	41	Wrong	Wrong	1b
16	Wrong	Wrong	1b	42	True	True	1a
17	Wrong (referenced para)	True	1a	43	Wrong (referenced para)	Wrong	1a
18	Wrong	Wrong	1a	44	Wrong (referenced para)	True	1a
19	Wrong (referenced para)	Wrong	1a	45	True	Wrong (True main para)	1a
20	Wrong (True main para)	True	1a	46	Wrong	True	1a
21	True	True	1a	47	Wrong (referenced para)	Wrong	1a
22	Wrong (referenced para)	Wrong	1a	48	Wrong	True	1a
23	Wrong	Wrong	1a	49	True	True	1a
24	Wrong (referenced para)	Wrong	1a	50	Wrong (True main para)	True	1b
25	True	True	1a				

Table 2. Accuracy of the QA system using two methods on 51 questions.

	Traditional system		Proposed system	
	Paragraphs	Answers	Paragraphs	Answers
#CorrectQuestions	15	12	**32**	**31**
Accuracy(%)	29.4	23.5	**62.7**	**60.8**

extracted answers. This means that if the systems correctly find the answers. Next, we describe the performance of each QA system in more details.

The Traditional QA System. Although, the traditional QA system did not use the reference information, it still correctly found the main paragraphs for 15 questions and correctly extracted answers for 12 questions. The reasons are high word overlaps between the questions and the main paragraphs. Another reason is that in some questions, the usage of reference information is optional to the process of finding their answers (i.e. in Question 2 in Fig. 8, it is not necessary to link the information from the main para A92-4P2 to the referenced para A92-4P1. But, the information in A92-4P1 helps us understand more about the situation). There are 24 questions that the traditional system finds out the referenced paragraphs instead of the main paragraphs (i.e. the tracing example question). The reason is that their majorities of question words contained in the referenced paragraphs (as in the example question). Because their answers are not contained in these paragraphs, the system is impossible to extract their correct answer.

For the remaining questions, the traditional system could not determine relevant paragraphs. Hence, it is unable to find their answers.

In questions 20, 26, and 50, the system correctly finds the main paragraphs. However, because the correct answers lie in the referents of the references contained in these main paragraphs, it cannot extract their correct answers. The accuracy of the traditional system, therefore, is 23.5 %.

The Proposed QA System. There are 20 questions whose answers could not be found. The reason may be that the similarity measures could not capture the entire context between questions and the paragraphs, which contain the answers; or the errors of the processing tools. For examples, in question 48, the system correctly determines the paragraph pair, which contains the answer. However, it extracts the wrong answer because of the error of the requisite-effectuation tool[10] (in Q.48, the answer is '*14.6 % per year*' instead of the extracted answer bounded in tags ⟨C⟩ and ⟨/C⟩.

There are 13 questions, in which finding their answers requires that the main paragraph must be linked to more than one document to provide the contexts for the correct answers (i.e. to find the answer of Question 5 in Fig. 4, it is necessary to link the information from the document A96P3 to the document

---

[10] This tool got the accuracy of ∼ 90 %.

Fig. 8. Some typical examples of the systems.

A96P1 via the document A96P2). Although the proposed framework does not allow us to process on more than two documents, it can still find the correct main paragraph containing the correct answers (in 5 questions). In these cases, the system correctly determines one linking pair between the main paragraph and one of the referenced paragraphs. Because the main paragraphs are correctly determined, the proposed method can extract the correct answers. This method also provides concrete evidences of the answers by showing the paragraph pairs, which contain the answers.

The accuracy of the proposed system, therefore, is 60.8%. It can be seen that the proposed system outperformed the traditional system, which did not exploit the reference information. Even if the traditional system can find the main paragraph because of high word overlaps, it cannot provide the evidence to help users believe in the systems output. However, our method can do this.

6   Conclusion and Future Work

This paper presented an application of reference information to build a legal QA system. We focused on one type of questions whose answers can not be found from merely one document. To find their correct answers, it is necessary to link documents via the relation of reference-referent. To achieve the goal, we first built a reference resolver. Based on that, we proposed a novel framework which allows us to exploit the reference information between legal documents to find answers. This approach also uses the requisite-effectuation structures of legal sentences and some effective similarity measures based on legal terms to support finding correct answers without training data. The experimental results showed that the proposed method was quite promising and outperformed a traditional method which did not use reference information.

In our framework, there is an assumption that questions and related paragraphs can be divided into two parts. Therefore, the proposed system is restricted to legal questions asking about the requisite and the effectuation problems. In fact, there are many questions falling under this category because users tend to ask about the beneficial conditions of laws, or the beneficiaries that can be achieved if some conditions are satisfied. As an initial step, we selected these questions manually. In the future, we aim at building a question classifier, which can automatically filter this type of questions. In addition, the assumption about dividing a paragraph into its logical structure is also quite reasonable. We counted the frequency of paragraphs having requisite-effectuation structures in the JPL corpus which are not definitions, and got 537 paragraphs among 547 paragraphs[11]. Hence, the ratio of paragraphs having the requisite-effectuation structure is very high (98.2 %). In our corpus, definition sentences are also marked using requisite and effectuation tags where a defined term is an effectuation and an explanation part a requisite. Therefore, our method is also applied to definition paragraphs. Another aspect is that we focused on providing the QA system with questions which are more easier to find the answers. In fact, natural questions are usually ambiguous, therefore they need complicated preprocessing techniques. These two problems will be further considered in the future work. Moreover, we also plan to extend the framework so that the system can handle more than two-linked documents.

References

1. Anne, R.D., Yilmazel, O., Liddy, E.D.: Evaluation of restricted domain question-answering systems. In: Proceedings of Workshop on Question Answering in Restricted Domains, ACL, pp. 2–7 (2004)
2. Bach, N.X., Minh, N.L., Oanh, T.T., Shimazu, A.: A two-phase framework for learning logical structures of paragraphs in legal articles. J. ACM TALIP **12**((1, article no 3)), 1–32 (2013)

---

[11] We did not count the number of definition in parentheses and only count paragraph main sentences.

3. Bach, N.X., Minh, N.L., Shimazu, A.: Recognition of requisite part and effectuation part in law sentences. In: Proceedings of ICCPOL, pp. 29–34 (2010)
4. Bach, N.X., Minh, N.L., Shimazu, A.: Rre task: The task of recognition of requisite part and effectuation part in law sentences. J. IJCPOL **23**(2), 109–130 (2010)
5. Doan-Nguyen, H., Kosseim, L.: The problem of precision in restricted-domain question-answering. In: Proceedings of Workshop on Question Answering in Restricted Domains, ACL, pp. 8–15 (2004)
6. Jurafsky, D., Martin, J.H.: Speech and Language Processing: An Introduction to Natural Language Processing, Computational Linguistics and Speech Recognition. Prentice Hall Series in Artificial Intelligence, 2nd edn. Prentice Hall, Upper Saddle River (2009)
7. Ciaramita, M., Surdeanu, M., Zaragoza, H.: Learning to rank answers on large online qa collections. In: Proceedings of the ACL-HLT, pp. 719–727 (2008)
8. Manning, C.D., Raghana, P., Schutze, H.: Introduction to Information Retrieval. Cambridge University Press, Cambridge (2008)
9. Monroy, A., Calvo, H., Gelbukh, A.: NLP for shallow question answering of legal documents using graphs. In: Gelbukh, A. (ed.) CICLing 2009. LNCS, vol. 5449, pp. 498–508. Springer, Heidelberg (2009)
10. Paik, J.H.: A novel tf-idf weighting scheme for effective ranking. In: Proceedings of ACM SIGIR, pp. 343–352 (2013)
11. Paltoglo, G., Thelwall, M.: A study of information retrieval weighting schemes for sentiment analysis. In: Proceedings of ACL, pp. 1386–1395 (2010)
12. Paulo, Q., Rodrigues, I.P.: A question-answering system for portuguese juridical documents. In: Proceedings of the ICAIL, pp. 256–257 (2005)
13. Peñas, A.: Overview of ResPubliQA 2009: question answering evaluation over european legislation. In: Peters, C., Di Nunzio, G.M., Kurimo, M., Mandl, T., Mostefa, D., Peñas, A., Roda, G. (eds.) CLEF 2009 Workshop, Part I. LNCS, vol. 6241, pp. 174–196. Springer, Heidelberg (2010)
14. Robertson, S., Walker, S.: On relevance weights with little relevance information. In: Proceedings of ACM SIGIR Conference on Research and Development in Information Retrieval, pp. 16–24 (1997)
15. Robertson, S., Zaragoza, S.: The probabilistic relevance framework: Bm25 and beyond. J. Found. Trends Inf. Retrieval **3**, 333–389 (2009)
16. Takano, K., Nakamura, M., Oyama, Y., Shimazu, A.: Semantic analysis of paragraphs consisting of multiple sentences - towards development of a logical formulation system. In: Proceedings of JURIX, pp. 117–126 (2010)
17. Tanaka, K., Kawazoe, I., Narita, H.: Standard structure of legal provisions for the legal knowledge processing by natural language. In Research report on Natural Language Processing, IPSJ, pp. 79–86 (1993) (in Japanese)
18. Tomura, K.: Study on question answering system for laws. Technical report, School of Information Science, Japan Advanced Institute of Science and Technology (JAIST) (2013)
19. Tran, O.T., Ngo, B.X., Le Nguyen, M., Shimazu, A.: Reference resolution in legal texts. In: Proceedings of ICAIL, pp. 101–110 (2013)
20. Zhang, W., Yoshida, T., Tang, X.: A comparative study of TF*IDF, LSI and multi-words for text classification. Expert Syst. Appl. Int. J. **38**, 2758–2765 (2011)

Belief Re-Revision in Chivalry Case

Pimolluck Jirakunkanok$^{(\boxtimes)}$, Shinya Hirose, Katsuhiko Sano, and Satoshi Tojo

School of Information Science, Japan Advanced Institute of Science and Technology,
1-1 Asahidai, Nomi, Ishikawa 923-1292, Japan
{pimolluck.jira,s.hirose525,v-sano,tojo}@jaist.ac.jp

Abstract. We propose a formalization of legal judgment revision in terms of dynamic epistemic logic, with two dynamic operators; commitment and permission. Each of these operations changes the accessibility to possible worlds, restricting to personal belief as local announcement. The commitment operator removes some accessible links for an agent to come to believe an announced proposition, while the permission operator restores them to tolerate former belief state. In order to demonstrate our formalization, we analyze judge's belief change in *Chivalry Case* in which a self-defense causes a misconception. Furthermore, we show an implementation of our logical formalization to demonstrate that it can be used in a practical way.

Keywords: Belief revision · Belief change · Legal case · Dynamic epistemic logic · Logic of belief

1 Introduction

Many studies [1–3] described the use of logical approaches in the legal systems. Recently, there are several theoretical and technical developments in modal logic, and among them, dynamic epistemic logic (DEL) [4–6] is significant as a logical tool to study belief change. In the real world, the knowledge or belief changes through the time, e.g., observations by agents, communication between agents, and so on. This paper focuses on belief change of a judge in a court.

Thus far, the previous work [7] introduced a dynamic operator for formalizing belief change of a judge, where the dynamic operator represented an agent's commitment. In that work, the formalization provided only the process of removing links but did not include the process of restoring new links. In other words, that work only dealt with monotonic changes of agent's belief but this work can also cover *non-monotonic* changes of them. For this reason, we employ the notion of belief *re-revision*. In the conventional settings, belief revision simply abandons former belief states and we cannot revive those former states in the later stage. In our proposal, however, we intend to get back to the former state; that is, the belief *re-revision* is not only a sequence of multiple belief revisions but also a restoration of former belief.

In this paper, we consider *Chivalry case*, which concerns a self-defense, i.e., an act of defending oneself or any other person from attacking by others. If an act

© Springer International Publishing Switzerland 2014
Y. Nakano et al. (Eds.): JSAI-isAI 2013, LNAI 8417, pp. 230–245, 2014.
DOI: 10.1007/978-3-319-10061-6_16

is considered to be the self-defense, Article 36-1 of Penal Code will be applied, however, if the act is considered to be an excessive defense, Article 36-2 of Penal Code will be applied (see Sect. 2 for more details). In this case, the judgment consists of three trials with different judges. However, in this paper, the judges of all trails are considered as a single agent.

The remainder of the paper is organized as follows. Section 2 describes the details of *Chivalry case*. Then, a logical formal tool for analyzing *Chivalry case* is presented in Sect. 3. In Sect. 4, we propose a dynamic logical analysis of *Chivalry case*. After that, Sect. 5 provides an implementation of logical formalization. Finally, our conclusion and future works are stated in Sect. 6.

2   Target Legal Case

2.1   Outline of Chivalry Case

First, we summarize an outline of Chivalry case [8, pp.58–59] as follows.

> One day, while o was drunken, her friend f was helping her. Then, d, a passer-by, accidentally met them. Since they looked wrestling, d misunderstood that f was assaulting o, d jumped in them, and tried to help o. Then, d came to near to f with both hands open. When f looked at d, f crossed his hands with fists in front of him to protect himself from d. On the other hand, d misunderstood that f posed to fight, d, who happened to be a *karate* (Japanese martial arts) master, quickly tried to kick f's face with an art of karate, also to protect himself. However, his left leg strongly kicked f's face, which made f fall down on the ground, and as a result, f's skull was crushed. Eight days later, f was dead by breeding cerebral dura mater and its crushed wound.

This story can be further summarized as the following sequences.

(1) f came to help o who was drunken.
(2) d misunderstood that f was assaulting o.
(3) d came to help o.
(4) f posed to protect himself.
(5) d misunderstood that f posed to fight.
(6) d attacked f.
(7) f was dead.

From the above summary of the story, the defendant has two misconceptions with f as follows.

– d misunderstood that f was assaulting o.
– d misunderstood that f posed to fight.

The story shows that the defendant believed that f would attack him, so he kicked f to protect himself. Thus, an act of the defendant might be thought as a self-defense in terms of the following Article 36 of Penal Code:[1]

> Article 36 (self-defense)
> (1) An act unavoidably performed to protect the rights of oneself or any other person against imminent and unlawful infringement is not punishable.
> (2) An act exceeding the limits of self-defense may lead to the punishment being reduced or may exculpate the offender in light of the circumstances.

Nevertheless, the defendant's attacking is not an actual self-defense or called as a *virtual* self-defense because the defendant attacked f by misconception that f posed to fight; in fact, f just posed to protect himself. That is, f did not really intend to attack the defendant. To judge this case, there were three trials as follows.

The first trial was conducted at Chiba regional court on February 7, 1984. The court judged the defendant to be innocent by the following reasons.

- Based on the supposed misconception of the defendant, his act belongs to a category of tolerable self-defense. Although the result is significant, this does not affect the adequacy of the defensive act.
- As the defendant is English, such a misconception cannot be his fault.
- Since his defensive attack by misconception is not intended, and it is not his fault, the defendant is innocent by the above Article 36-1 of Penal Code.

Since the prosecutor appealed the court ruling, the judge of the second court re-interpreted claims and evidences of the first trial. In principle, an appeal court may admit new claims and evidences than those of the first trial. In the second trial, Tokyo High Court judged the defendant to be guilty on November 11, 1984. The defendant was sentenced to be imprisonment for 18 months with parole of three years. The reasons of the judgment were as follows.

- The defendant possessed other alternative methods to protect himself. Notwithstanding this, the act of kicking by the defendant is such dangerous that the attack would be lethal.
- The act of the defendant is comparable to an excessive defense resulting in death. Therefore, the following Article 205 of Penal Code(see Footnote 1) is applied to this case. However, because of Penal Code Article 36-2, the penalty is reduced.

> Article 205 (Injury Causing Death)
> A person who causes another to suffer injury resulting in death shall be punished by imprisonment with work for a definite term of not less than 3 years.

---

[1] An English translation of the article can be referenced from http://www.japaneselawtranslation.go.jp/.

At last, in the final trial, Japan Supreme Court adopted the result of the second trail on March 26, 1987; it is obvious that the defendant's act of kicking f is the excessive self-defense, by misconception of f's intended attack, and the case is accidental mortality. However, the Penal Code Article 36-2 is also applied and the penalty is reduced, based on the preceding case of Jury 7, 1966.

In short, the judge first believed that the act of the defendant was a reasonable self-defense, so the defendant was innocent by applying Article 36-1 of Penal Code. After that, the judge changed his/her evaluation about the defendant's act. Since the kicking of the defendant was such dangerous enough to kill f, the defendant's act was not the reasonable self-defense or could be called as an excessive self-defense. Thus, Article 205 of Penal Code was applied to judge that the defendant was guilty because his act caused f's death. Nevertheless, the penalty was reduced as a result of Article 36-2 of Penal Code.

2.2 Our Perspective of Analyzing Chivalry Case

In order to analyze the judge's belief change in Chivalry case, it is clear that we need a logical tool satisfying the following two requirements at least:

(R1) To represent the misconception of the defendant.
(R2) To represent the judge's belief on the defendant's belief, i.e., iterated beliefs between agents.

We also need to analyze the *changes* of the judge's belief. Let us extract two belief changes from Chivalry case. In the decision of the first trial, the judge believed that the defendant's act belonged to the reasonable self-defense. In the beginning of the trial, however, the judge did not have such belief, i.e., he/she should not have any biases for the defendant. Let us symbolize 'the defendant's act belonged to the reasonable self-defense' by q. Then, we can summarize this belief change of the judge as follows: q was not in the belief of the judge but later q was in the belief of the judge. Let us say that the proposition φ is *monotonic* with respect to an agent j's belief if, once φ becomes a j's belief, then φ continues to be in the j's belief.

However, the above proposition q is not monotonic with respect to the judge j in Chivalry case, if we regard the judge of the first trial and the judge of the second trial as a single agent j. This is because the second trial decided that the defendant's act *did not* belong to the reasonable self-defense, i.e., $\neg q$ is in the belief of the judge j. In this sense, j has a different belief $\neg q$ from the previous belief q. If q is monotonic with respect to the j's belief, then the decision of the second trial implies that j believes both q and $\neg q$, in other words, j comes to have a contradictory belief. In practice, the most of the propositions are non-monotonic in the legal context.

Now, we can specify our final requirement of analyzing the judge's belief change.

(R3) To represent the judge's belief change on non-monotonic propositions.

3   Formal Tool for Analyzing Target Legal Case

3.1   Static Logic for Agents' Belief

In order to analyze the previous Chivalry case from logical point of view, we introduce a modal language which enables us to formalize the agent's belief, which satisfies the requirements (R1) and (R2) in the previous section.

Let G be a fixed *finite* set of agents. Our syntax $\mathcal{L}$ consists of the following vocabulary: (i) a set $\mathsf{Prop} = \{p, q, r, \dots\}$ of propositional letters, (ii) Boolean connectives: $\neg$, $\wedge$, (iii) belief operators B_i ($i \in G$), as well as (iv) the global modality E and (v) the modal constant n, denoting the actual state. A set of formulas of $\mathcal{L}$ is inductively defined as follows:

$$\varphi ::= p \mid \mathsf{n} \mid \neg\varphi \mid \varphi \wedge \psi \mid \mathsf{B}_i\,\varphi \mid \mathsf{E}\varphi,$$

where $p \in \mathsf{Prop}$ and $i \in G$.

We define $\vee$, $\rightarrow$, $\leftrightarrow$ as ordinary abbreviations and use $\mathsf{A}\,\varphi$ to mean $\neg\mathsf{E}\neg\varphi$.

As before, let f, d, and j be agents of a friend of the observed drunken, a defendant who is a highly ranked karate (Japanese martial arts) master, and the judge of Chivalry Case, respectively. Let us also denote 'f posed to fight against d' by p and 'the kick of d was beyond the self-defense' by q, respectively. We can provide some formalization which is relevant to our legal example as in Table 1. Note that we can regard '$\neg p \wedge \mathsf{B}_d\,p$' as a formalization of the sentence 'd misunderstood that f posed to fight against d.'

Table 1. Examples of static logical formalization for Chivalry case

$\mathsf{B}_d\,p$	: d believes that f posed to fight against d
$\neg p \wedge \mathsf{B}_d\,p$	: f did not pose to fight against d,
	but d believes that f posed to fight against d.
$\mathsf{B}_j(\neg p \wedge \mathsf{B}_d\,p)$	: j believes that d misunderstood that f posed to fight against d.
$\mathsf{B}_j\,\neg q$	: j believes that d's kick was not beyond the self-defense.
$\mathsf{B}_j\,q$	: j believes that d's kick was beyond the self-defense.

Let us provide Kripke semantics with our syntax. A *model* $\mathfrak{M}$ is a tuple

$$\mathfrak{M} = (W, (R_i)_{i \in G}, @, V),$$

where W is a non-empty set of states, called *domain*, $R_i \subseteq W \times W$, $@ \in W$ is a distinguished element called the *actual state*, and $V : \mathsf{Prop} \rightarrow \mathcal{P}(W)$ is a valuation. Given any model $\mathfrak{M}$, any world $w \in W$, and any formula φ, we define the *satisfaction relation* $\mathfrak{M}, w \models \varphi$ inductively as follows.[2]

---

[2] In this paper, we treat the notion of belief as an ordinary model operator. However, we note that there is also another approach by Baltag and Smets [9]. They proposed a notion of plausibility models to deal with agents' beliefs.

$$\mathfrak{M}, w \models p \qquad \text{iff } w \in V(p)$$
$$\mathfrak{M}, w \models \mathsf{n} \qquad \text{iff } w = @$$
$$\mathfrak{M}, w \models \neg\varphi \qquad \text{iff } \mathfrak{M}, w \not\models \varphi$$
$$\mathfrak{M}, w \models \varphi \wedge \psi \quad \text{iff } \mathfrak{M}, w \models \varphi \text{ and } \mathfrak{M}, w \models \psi$$
$$\mathfrak{M}, w \models \mathsf{B}_i \varphi \quad \text{iff } R_i(w) \subseteq [\![\varphi]\!]_{\mathfrak{M}}$$
$$\mathfrak{M}, w \models \mathsf{E}\varphi \qquad \text{iff } \mathfrak{M}, v \models \varphi \text{ for some } v \in W,$$

where $R_i(w) := \{ w' \in W \mid wR_iw' \}$ and $[\![\varphi]\!]_{\mathfrak{M}} = \{ w \in W \mid \mathfrak{M}, w \models \varphi \}$ (we drop the subscript $\mathfrak{M}$ from $[\![\varphi]\!]_{\mathfrak{M}}$, if it is clear from the context). We say that φ is *valid* on $\mathfrak{M}$ if $\mathfrak{M}, w \models \varphi$ for all $w \in W$. As for $\mathsf{A}\,\varphi$, we have the following derived semantic clause:

$$\mathfrak{M}, w \models \mathsf{A}\,\varphi \text{ iff } \mathfrak{M}, v \models \varphi \text{ for all } v \in W.$$

3.2 Dynamic Operators for Belief Change of Judge

In the first trial of Chivalry case, a judge should be open to several possibilities of the defendant's belief. In the process of the trials, the judge receives some new evidences and/or rejects some confirmed evidences, and then, he/she removes and/or sometimes adds some possibilities to reach his/her own decision. We want to simulate the effect of this process by introducing two dynamic operators for action to our static syntax. These two operators allow us to satisfy the requirement (R3) of Sect. 2.

We introduce two kinds of dynamic operators $[\mathsf{com}_j(\varphi)]$ and $[\mathsf{per}_j(\varphi)]$.[3] Our intended reading of $[\mathsf{com}_j(\varphi)]\psi$ is 'after the agent j commits him/herself to φ, ψ', and we read $[\mathsf{per}_j(\varphi)]\psi$ as 'after the agent j permitted φ to be the case, ψ'. Semantically speaking, $[\mathsf{com}_j(\varphi)]$ restricts j's attention to the φ's states, and $[\mathsf{per}_j(\varphi)]$ enlarges j's attention to cover *all* the φ's states. We denote the expanded syntax with all $[\mathsf{com}_j(\varphi)]$ and $[\mathsf{per}_j(\varphi)]$ by $\mathcal{L}^+$. Table 2 demonstrates the dynamic logical formalization of the judge's belief, where we keep the same reading of agents and propositions as in Table 1.

Table 2. Examples of dynamic logical formalization for Chivalry case

$[\mathsf{com}_j(\neg p \wedge \mathsf{B}_d\,p)]\,\mathsf{B}_j\,\mathsf{B}_d\,p$: After j's commitment of d's misunderstanding of p, j believes that d believes that p.
$\mathsf{B}_j\,q \wedge [\mathsf{per}_j(\neg q)]\neg\,\mathsf{B}_j\,q$: j first believes that q, but after j's permission of $\neg q$, j does not believe that q.

Let us fix a Kripke model $\mathfrak{M} = (W, (R_i)_{i \in G}, @, V)$. A semantic clause for $[\mathsf{com}_j(\varphi)]\psi$ on $\mathfrak{M}$ and $w \in W$ is defined as follows.

$$\mathfrak{M}, w \models [\mathsf{com}_j(\varphi)]\psi \text{ iff } \mathfrak{M}^{\mathsf{com}_j(\varphi)}, w \models \psi,$$

---

[3] This study assumes that, when an agent receives a piece of information φ, the agent has already decided if he/she uses the commitment or the permission operator for φ. Thus, we will not analyze a process of the decision in this paper.

where $\mathfrak{M}^{\mathsf{com}_j(\varphi)} = (W, (R_i)_{i \in G \setminus \{j\}}, S_j, @, V)$ and S_j is defined as: for all $x \in W$,

$$S_j(x) := \begin{cases} R_j(x) \cap \llbracket \varphi \rrbracket_{\mathfrak{M}} & \text{if } x = @, \\ R_j(x) & \text{otherwise.} \end{cases}$$

Let us move to a semantic clause for $[\mathsf{per}_j(\varphi)]$ on $\mathfrak{M}$ and $w \in W$.

$$\mathfrak{M}, w \models [\mathsf{per}_j(\varphi)]\psi \text{ iff } \mathfrak{M}^{\mathsf{per}_j(\varphi)}, w \models \psi,$$

where $\mathfrak{M}^{\mathsf{per}_j(\varphi)} = (W, (R_i)_{i \in G \setminus \{j\}}, S'_j, @, V)$ and S'_j is defined as: for all $x \in W$,

$$S'_j(x) := \begin{cases} R_j(x) \cup \llbracket \varphi \rrbracket_{\mathfrak{M}} & \text{if } x = @, \\ R_j(x) & \text{otherwise.} \end{cases}$$

Note that the effects of the agent j's commitment and permission of φ are restricted only at the distinguished element $@$. While $[\mathsf{com}_j(\varphi)]$ is compatible with the *monotonicity* of j's belief (i.e., once a formula $\mathsf{B}_j \psi$ holds, it continues to be true), $[\mathsf{per}_j(\varphi)]$ may break this monotonicity. This is because an addition of new j's links allows the belief change from $\mathsf{B}_j \psi$ into $\neg \mathsf{B}_j \psi$. In this sense, $[\mathsf{per}_j(\varphi)]$ can capture the non-monotonic change of agent's belief.

The formulas of Table 3 are called *reduction axioms*, which can be regarded as a necessary criteria for reducing the semantic completeness result with dynamic operators to the one without them. With the help of the necessitation rules for $[\mathsf{com}_i(\varphi)]$ and $[\mathsf{per}_i(\varphi)]$: from ψ, we may infer $[\mathsf{com}_i(\varphi)]\psi$ and $[\mathsf{per}_i(\varphi)]\psi$, reduction axioms allow us to rewrite a formula of $\mathcal{L}^+$ to an equivalent formula in $\mathcal{L}$ with no occurrence of $[\mathsf{com}_i(\varphi)]$ and $[\mathsf{per}_i(\varphi)]$.[4]

Table 3. Reduction axioms for $[\mathsf{com}_i(\varphi)]$ and $[\mathsf{per}_i(\varphi)]$

$[\mathsf{com}_i(\varphi)]p$	$\leftrightarrow p$	$[\mathsf{per}_i(\varphi)]p$	$\leftrightarrow p$
$[\mathsf{com}_i(\varphi)]\mathsf{n}$	$\leftrightarrow \mathsf{n}$	$[\mathsf{per}_i(\varphi)]\mathsf{n}$	$\leftrightarrow \mathsf{n}$
$[\mathsf{com}_i(\varphi)]\neg\psi$	$\leftrightarrow \neg[\mathsf{com}_i(\varphi)]\psi$	$[\mathsf{per}_i(\varphi)]\neg\psi$	$\leftrightarrow \neg[\mathsf{per}_i(\varphi)]\psi$
$[\mathsf{com}_i(\varphi)](\psi \wedge \chi)$	$\leftrightarrow [\mathsf{com}_i(\varphi)]\psi \wedge [\mathsf{com}_i(\varphi)]\chi$	$[\mathsf{per}_i(\varphi)](\psi \wedge \chi)$	$\leftrightarrow [\mathsf{per}_i(\varphi)]\psi \wedge [\mathsf{per}_i(\varphi)]\chi$
$[\mathsf{com}_i(\varphi)]\,\mathsf{E}\,\psi$	$\leftrightarrow \mathsf{E}[\mathsf{com}_i(\varphi)]\psi$	$[\mathsf{per}_i(\varphi)]\,\mathsf{E}\,\psi$	$\leftrightarrow \mathsf{E}[\mathsf{per}_i(\varphi)]\psi$
$[\mathsf{com}_i(\varphi)]\,\mathsf{B}_k\,\psi$	$\leftrightarrow \mathsf{B}_k[\mathsf{com}_i(\varphi)]\psi$	$[\mathsf{per}_i(\varphi)]\,\mathsf{B}_k\,\psi$	$\leftrightarrow \mathsf{B}_k[\mathsf{per}_i(\varphi)]\psi \quad (i \neq k)$
$[\mathsf{com}_i(\varphi)]\,\mathsf{B}_i\,\psi$	$\leftrightarrow (\mathsf{n} \rightarrow \mathsf{B}_i(\varphi \rightarrow [\mathsf{com}_i(\varphi)]\psi)) \wedge (\neg\mathsf{n} \rightarrow \mathsf{B}_i[\mathsf{com}_i(\varphi)]\psi)$		
$[\mathsf{per}_i(\varphi)]\,\mathsf{B}_i\,\psi$	$\leftrightarrow (\mathsf{n} \rightarrow (\mathsf{B}_i[\mathsf{per}_i(\varphi)]\psi \wedge \mathsf{A}(\varphi \rightarrow [\mathsf{per}_i(\varphi)]\psi))) \wedge (\neg\mathsf{n} \rightarrow \mathsf{B}_i[\mathsf{per}_i(\varphi)]\psi)$		

Proposition 1. *All the formulas of Table 3 are valid on all models.*

---

[4] Remark that we cannot obtain the reduction axioms for iterated commitments $[\mathsf{com}_i(\varphi)][\mathsf{com}_j(\theta)]\psi$ or iterated permissions $[\mathsf{per}_i(\varphi)][\mathsf{per}_j(\theta)]\psi$ for different agents i and j. This is one of the main differences of our operators from the public announcement operator (see, e.g., [10]).

Proof. We only establish the equivalence for $[\text{per}_i(\varphi)]\,B_i\,\psi$ here. Let us fix any model $\mathfrak{M}$ and any state w of $\mathfrak{M}$. We can proceed as follows: $\mathfrak{M}, w \models [\text{per}_i(\varphi)]\,B_i\,\psi$ iff $\mathfrak{M}^{\text{per}_i(\varphi)}, w \models B_i\,\psi$ iff $S_i'(w) \subseteq [\![\psi]\!]_{\mathfrak{M}^{\text{per}_i(\varphi)}}$ iff

$$\begin{cases} R_i(w) \cup [\![\varphi]\!]_{\mathfrak{M}} \subseteq [\![\psi]\!]_{\mathfrak{M}^{\text{per}_i(\varphi)}} & \text{if } w = @ \\ R_i(w) \subseteq [\![\psi]\!]_{\mathfrak{M}^{\text{per}_i(\varphi)}} & \text{if } w \neq @ \end{cases}$$

$$\text{iff} \quad \begin{cases} R_i(w) \cup [\![\varphi]\!]_{\mathfrak{M}} \subseteq [\![[\text{per}_i(\varphi)]\psi]\!]_{\mathfrak{M}} & \text{if } w = @ \\ R_i(w) \subseteq [\![[\text{per}_i(\varphi)]\psi]\!]_{\mathfrak{M}} & \text{if } w \neq @ \end{cases}$$

$$\text{iff} \quad \begin{cases} R_i(w) \subseteq [\![[\text{per}_i(\varphi)]\psi]\!]_{\mathfrak{M}} \text{ and } [\![\varphi]\!]_{\mathfrak{M}} \subseteq [\![[\text{per}_i(\varphi)]\psi]\!]_{\mathfrak{M}} & \text{if } w = @ \\ \mathfrak{M}, w \models B_i[\text{per}_i(\varphi)]\psi & \text{if } w \neq @ \end{cases}$$

$$\text{iff} \quad \begin{cases} \mathfrak{M}, w \models B_i[\text{per}_i(\varphi)]\psi \text{ and } \mathfrak{M}, w \models A(\varphi \rightarrow [\text{per}_i(\varphi)]\psi) & \text{if } w = @ \\ \mathfrak{M}, w \models B_i[\text{per}_i(\varphi)]\psi & \text{if } w \neq @ \end{cases}$$

The final line can be easily shown to be equivalent with the right-hand side of the equivalent axiom for $[\text{per}_i(\varphi)]\,B_i\,\psi$ in Table 3. $\square$

For a complete axiomatization of our dynamic logic with two operators, the readers can refer to Appendix A.

4 Dynamic Logical Analysis of Target Legal Case

Let us describe our semantic idea to construct a model for analyzing *Chivalry case*. Firstly, we assume that all possibilities are represented by a square W as in Fig. 1(a). Secondly, when the judge obtains a piece of information 'f posed to fight against d' (p), W is divided into two equal horizontal parts, i.e., p and $\neg p$ as shown in Fig. 1(b). From *Chivalry case*, in fact, f did not pose to fight against d ($\neg p$), so we focus on the possibility of $\neg p$. Thirdly, the judge considers a piece of information 'd believes that f posed to fight against d' ($B_d\,p$) by dividing the possibility of $\neg p$ into two equal horizontal parts, i.e., $B_d\,p$ and $\neg B_d\,p$ as shown in Fig. 1(c). Finally, the judge considers a piece of information 'the kick of d was beyond the self-defense' (q) by dividing W into two equal vertical parts, i.e., q and $\neg q$ as shown in Fig. 1(d), where we assume that each of the partitions of the $\neg p$-states is non-empty.

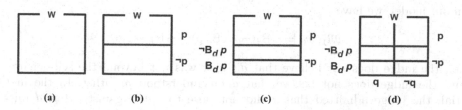

Fig. 1. A concept to construct a model for analyzing *Chivalry case*

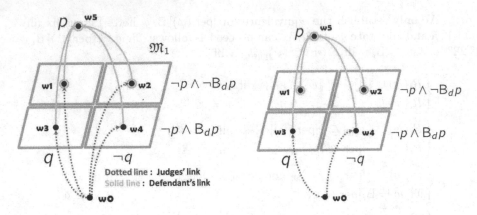

Fig. 2. A model $\mathfrak{M}_1$ of Definition 1   **Fig. 3.** Update $\mathfrak{M}_1$ by $[\text{com}_j(\neg p \wedge \mathsf{B}_d\, p)]$

In this section, we will not analyze the result of the final trial because the final trail only adopted the result of the second trial, i.e., the result of the final trail is the same as that of the second trial.

4.1   First Trial

In order to analyze the result of the first trial, we simplify the model from the above idea and introduce the following specific model $\mathfrak{M}_1$ (for the graphical representation, see Fig. 2, where the solid circles around the states w_1, w_2 and w_5 mean that they are reflexive states with respect to R_d-relation).

Definition 1. *Let $G = \{\, d, j\,\}$, where recalls that d and j mean the defendant and the judge, respectively. Define $\mathfrak{M}_1$ as follows:*

$W = \{\, w_0, w_1, w_2, w_3, w_4, w_5\,\}$.
$R_j = \{\, (w_0, w_1), (w_0, w_2), (w_0, w_3), (w_0, w_4)\,\}$.
$R_d = \{\, (w_1, w_1), (w_2, w_2), (w_1, w_5), (w_2, w_5), (w_3, w_5), (w_4, w_5), (w_5, w_5)\,\}$.
@ := w_0.
$V(p) = \{\, w_5\,\}$ *and* $V(q) = \{\, w_1, w_3\,\}$.

We regard the state w_0 as j's viewpoint at the initial stage of the first trial. In our model, we have

$$\mathfrak{M}_1, w_0 \models \neg\, \mathsf{B}_j(\neg p \wedge \mathsf{B}_d\, p) \wedge \neg\, \mathsf{B}_j\, \neg q,$$

i.e., the judge does not believe that d's kick was not beyond the self-defense, and the judge does not believe that d misunderstood p, either. In the first trial, the judge admitted that f did not pose to fight against d but d misunderstood that f posed to fight against d (i.e., p). Based on this, the judge also committed him/herself to $\neg q$, i.e., d's kick was not beyond the self-defense.

Those processes can be formalized by the successive updates of the model $\mathfrak{M}_1$ by $[\mathsf{com}_j(\neg p \wedge \mathsf{B}_d p)]$ and $[\mathsf{com}_j(\neg q)]$. Let us see the effects of the updates step by step (Figs. 3 and 4). By the update of $[\mathsf{com}_j(\neg p \wedge \mathsf{B}_d p)]$, we delete all the links from w_0 into the states where $\neg p \wedge \mathsf{B}_d p$ is false. That is, we eliminate (w_0, w_1) and (w_0, w_2) from R_j (see Fig. 3). The result becomes $\{(w_0, w_3), (w_0, w_4)\}$. After this, the update of $[\mathsf{com}_j(\neg q)]$ requires us to delete all the links from w_0 into the states where q is true. Therefore, we furthermore need to delete (w_0, w_3). Then, the final accessibility relation for j becomes $\{(w_0, w_4)\}$ and the judge j now believes both $\neg p \wedge \mathsf{B}_d p$ and $\neg q$ (see Fig. 4). Therefore,

$$\mathfrak{M}_1, w_0 \models [\mathsf{com}_j(\neg p \wedge \mathsf{B}_d p)][\mathsf{com}_j(\neg q)]\, \mathsf{B}_j((\neg p \wedge \mathsf{B}_d p) \wedge \neg q).$$

4.2 Second Trial

Let us denote by $\mathfrak{M}_2$ the updated model of $\mathfrak{M}_1$ by $[\mathsf{com}_j(\neg p \wedge \mathsf{B}_d p)]$ and $[\mathsf{com}_j(\neg q)]$, i.e., the model of Fig. 4. Precisely, $\mathfrak{M}_2$ is defined as follows.

Definition 2. *Define $\mathfrak{M}_2$ is the same model as $\mathfrak{M}_1$ except $R_j = \{(w_0, w_4)\}$.*

We can regard $\mathfrak{M}_2$ as a model of the beginning of the second trial. The judge of the second trial also committed him/herself that d misunderstood p but there is a difference between the first and the second trials, i.e., whether the judge accepts q or not. As we have seen in the previous section, the judge of the first trial accepts $\neg q$. On the other hand, the judge of the second trial rejects $\neg q$ but accepts q instead. In order to overturn the decision, the judge of the second trial first needs to permit the possibility of q, i.e., the possibility that d's kick was beyond the self-defense. This was done by the operator $[\mathsf{per}_j(q)]$ in our logical framework. $[\mathsf{per}_j(q)]$ allows us to 'revive' the older links from w_0 to the states where q is true. By the update of $[\mathsf{per}_j(q)]$, R_j becomes $\{(w_0, w_1), (w_0, w_3), (w_0, w_4)\}$ (see Fig. 5). Then, the judge becomes undetermined on q, i.e., $\neg \mathsf{B}_j q$ and $\neg \mathsf{B}_j \neg q$ hold at w_0, i.e.,

$$\mathfrak{M}_2, w_0 \models \mathsf{B}_j \neg q \wedge [\mathsf{per}_j(q)](\neg \mathsf{B}_j q \wedge \neg \mathsf{B}_j \neg q).$$

After this reviving the links into the q-states, the judge successively commits him/herself to $\neg p \wedge \mathsf{B}_d p$ (the same as in the first trial) and then q (instead of $\neg q$ in the first trial) (Figs. 6 and 7, respectively). By the update of $[\mathsf{com}_j(\neg p \wedge \mathsf{B}_d p)]$, j's accessibility relation is changed into $\{(w_0, w_3), (w_0, w_4)\}$ (see Fig. 6). Furthermore, the update of $[\mathsf{com}_j(q)]$ deletes j's accessibility relation into $\{(w_0, w_3)\}$, which implies that j now believes that q at w_0 (see Fig. 7). To sum up, we obtain the following *non-monotonic change* of j's belief from $\mathsf{B}_j(\neg q)$ into $\mathsf{B}_j q$:

$$\mathfrak{M}_2, w_0 \models \mathsf{B}_j \neg q \wedge [\mathsf{per}_j(q)][\mathsf{com}_j(p \wedge \neg \mathsf{B}_j p)][\mathsf{com}_j(q)]\, \mathsf{B}_j q.$$

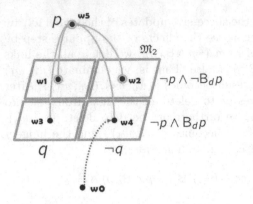

Fig. 4. Update Fig. 3 by $[\mathsf{com}_j(\neg q)]$

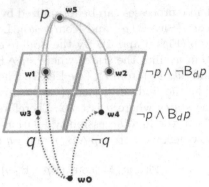

Fig. 5. Update $\mathfrak{M}_2$ by $[\mathsf{per}_j(q)]$

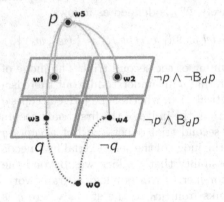

Fig. 6. Update Fig. 5 by $[\mathsf{com}_j(\neg p \wedge \mathsf{B}_d p)]$

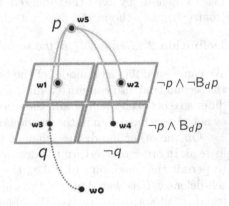

Fig. 7. Update Fig. 6 by $[\mathsf{com}_j(q)]$

5 Implementation

This section introduces three algorithms calculating the truth value of the formulas $\mathsf{B}_j \psi$ and the updates by $[\mathsf{com}_j(\varphi)]$ and $[\mathsf{per}_j(\varphi)]$ (Algorithm 1, Algorithm 2 and Algorithm 3, respectively, see below), and demonstrates the algorithms in our example of Chivalry case.

First, let us explain Algorithm 1. findall finds all the accessible worlds from w, i.e. findall includes those w' such that $wR_j w'$ into X. Then Algorithm 1 investigates if ψ holds for all $w' \in X$, to judge if $\mathsf{B}_j \psi$ holds or not.

Second, Algorithm 2 concerns an agent's commitment operator $[\mathsf{com}_j(\varphi)]$. findall collects all the accessible worlds w' from the actual world (@), to build set X. Next, findall finds all those in W', in which φ holds, to build set Y. Then, the accessibility for j is renewed to be a set S_j of accessibility, in which all the accessible worlds from '@' are restricted to those where φ holds. Finally, we output a new model for evaluating $(W, S_j, R_d, @, V), w \models \psi$ by using Algorithm 1.

Third, Algorithm 3 concerns an agent's permission operator $[\mathsf{per}_j(\varphi)]$. findall collects all the accessible worlds w' in W, in which φ holds, to build set Y. Then, the accessibility for j is renewed to be a set S_j of accessibility, in which all the accessible worlds from '@' are enlarged to those where φ holds. Finally, we output a new model for evaluating $(W, S_j, R_d, @, V), w \models \psi$ by using Algorithm 1.

We have implemented the above algorithms in SWI-Prolog^{TM} 6.5.2. This program outputs the veridicality of propositions, together with world accessibility relations in dot format. Thus, we can visualize the dot file by Graphviz^{TM}.

Algorithm 1. Evaluation of the truth of $\mathsf{B}_j\,\psi$

---

input $\mathfrak{M} = (W, R_j, R_d, @, V)$, $\mathsf{B}_j\,\psi$, w
findall $(w' \in W,\ wR_jw')$
  add w' **to** X
end findall
forall $(w' \in X,\ \mathfrak{M}, w' \models \psi)$
  $\mathfrak{M}, w \models \mathsf{B}_j\,\psi$
else
  $\mathfrak{M}, w \not\models \mathsf{B}_j\,\psi$
end forall

Algorithm 2. Calculation of $[\mathsf{com}_j(\varphi)]$

---

input $\mathfrak{M}_1 = (W, R_j, R_d, @, V)$, $[\mathsf{com}_j(\varphi)]$, w
findall $(w' \in W,\ @R_jw')$
  add w' **to** X
end findall
findall $(w' \in X,\ \mathfrak{M}_1, w' \models \varphi)$
  add w' **to** Y
end findall
$S_j := (R_j \setminus \{(@, w') | w' \in X\}) \cup \{(@, w') | w' \in Y\}$
output $\mathfrak{M}_2 = (W, S_j, R_d, @, V)$

Algorithm 3. Calculation of $[\mathsf{per}_j(\varphi)]$

---

input $\mathfrak{M}_1 = (W, R_j, R_d, @, V)$, $[\mathsf{per}_j(\varphi)]$, w
findall $(w' \in W,\ \mathfrak{M}_1, w' \models \varphi)$
  add w' **to** Y
end findall
$S_j := R_j \cup \{(@, w') | w' \in Y\}$
output $\mathfrak{M}_2 = (W, S_j, R_d, @, V)$

Let $\mathfrak{N}_1$ be a model which is defined as follows.

$W = \{\, w_0, w_1, w_2, w_3, w_4, w_5 \,\}$
$R_j = \{\, (w_0, w_1), (w_0, w_3) \,\}$
$R_d = \{\, (w_1, w_1), (w_2, w_2), (w_1, w_5), (w_2, w_5), (w_3, w_5), (w_4, w_5), (w_5, w_5) \,\}$
$@ := w_0$
$V(p) = \{\, w_5 \,\}$ and $V(q) = \{\, w_1, w_3 \,\}$

Example 1. This example shows a process for performing the permission operator composing of three steps (see Fig. 8) as follows. First, we input a model $\mathfrak{N}_1$ by `load`. Second, our permission operator $[\text{per}_j(q)]$ is represented by `per q`. In this example, $[\text{per}_j(\neg q)]$ is calculated by `input(per ~q)`. In Fig. 8, we can find that three links $\{\,$ `[w0,w0]`, `[w0,w2]`, `[w0,w4]` $\,\}$ are added during the execution. After calculating the permission operator, we have a new model $\mathfrak{N}_2$ which is the same model as

Fig. 8. Execution Log 1

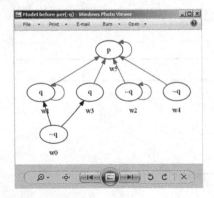

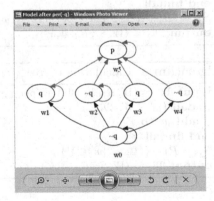

Fig. 9. Kripke model before $[\text{per}_j(\neg q)]$

Fig. 10. Kripke model after $[\text{per}_j(\neg q)]$

$\mathfrak{N}_1$ except $R_j = \{ (w_0, w_0), (w_0, w_1), (w_0, w_2), (w_0, w_3), (w_0, w_4) \}$. Third, we are to verify if $\mathfrak{N}_2, w_0 \models B_j q$ represented by input($q, w0$) in Prolog syntax. As a result, the system can find five links $\{$ [w0,w1], [w0,w3], [w0,w0], [w0,w2], [w0,w4] $\}$, that is $w_0 R_j w_1, w_0 R_j w_3, w_0 R_j w_0, w_0 R_j w_2, w_0 R_j w_4$. Then, the system recursively verifies $\{w_1, w_3, w_0, w_2, w_4\} \subseteq V(q)$, i.e., input(q,w1), input(q,w3), input(q, w0), input(q,w2) and input(q,w4), and finally answers false. Moreover, the system can automatically visualize the resultant states, as in Fig. 9 and in Fig. 10, each of which corresponds to the models before and after the permission $[per_j(\neg q)]$ in Fig. 8.

6   Conclusion

In this research, we have provided the permission operator $[per_j(\varphi)]$, together with the commitment operator $[com_j(\varphi)]$; the former operator tolerates a proposition, restoring the accessibility to those worlds in which the proposition holds. With these equipments, we have shown an example of *Chivalry* case, where an action might look the spirit of chivalry though the action was actually done by misconception. In the process of the trials, a court gave firstly a decision, but after then, with a different interpretation of penal code, the second judgment overturned the first. We could successfully depict this process by revising belief state in multiple times.

Our contribution in this paper is two-fold. One is that we have formalized the accessibility restoration; that is, we can re-revise one's belief, in terms of dynamic epistemic logic. The other is that we have actually implemented the process in a computer system to show its adequacy.

In order to obtain more fine-grained theory of belief re-revision, one may employ the idea of *Evidential Logic* by van Benthem and Pacuit [11]. Their basic idea is to provide a set of sets of states, regarded as evidences, with each state, and propose a rich variety of dynamic operations for evidences. In their paper, we can find update operations of public announcement $[!\varphi]$ and evidence addition $[+\varphi]$ for neighborhood models, which are similar to commitment and permission operators, respectively. However, the above update operations change all the agents' knowledge and belief. In this sense, their operations are public one. Also, their update operations change a frame structure (i.e., neighborhood structure) not only of the current state but also all of the other states. Unlike our commitment and permission operations, the above operations are not local in this sense. This means that the framework in [11] does not fit well with our purpose.

Note that our model aims at the formalization of belief change in a court, and not at that of norms; thus, the objective of our paper does not directly concern a reasoning system. Instead, we contribute to clarify agents' belief states, with which we obtain an accountability of judgment. In this paper, the commitment and permission operators could capture belief change of only one agent. Thus, the change of judgment was not triggered by agent communication, e.g., with an appearance of new evidence; in our case the second adjudication was merely a reinterpretation of the first judgment. In order for us to develop our formalization furthermore, our next target should be to employ the notion of agent communication as an announcement to other agents.

A Complete Axiomatization of Dynamic Logic with Commitment and Permission Operators

In this section, we give an axiomatization of our dynamic logic with two operators and sketch its completeness proof.

Table 4. Axiomatization of static logic in $\mathcal{L}$

Axioms	
(CT) All propositional Tautologies	(Incl) $A\varphi \rightarrow B_i\varphi$
(K) $B_i(\varphi \rightarrow \psi) \rightarrow (B_i\varphi \rightarrow B_i\psi)$ $(i \in G)$.	(Incl$_n$) En
(K$_A$) $A(\varphi \rightarrow \psi) \rightarrow (A\varphi \rightarrow A\psi)$	(Nom$_E$) $E(n \wedge \varphi) \rightarrow A(n \rightarrow \varphi)$
(Ref$_E$) $\varphi \rightarrow E\varphi$	**Inference Rules**
(Tran$_E$) $EE\varphi \rightarrow E\varphi$	(MP) Modus Ponens
(Sym$_E$) $\varphi \rightarrow AE\varphi$	(Nec) From φ, we may infer $B_i\varphi$ $(i \in G)$.
	(Nec$_A$) From φ, we may infer $A\varphi$.

Theorem 1. *The set of all the valid formulas of $\mathcal{L}$ is completely axiomatized by all the axioms and the rules of Table 4.*

Proof. (Sketch) Soundness is easy to establish. So, let us concentrate on the completeness proof. We can regard our syntax $\mathcal{L}$ as a syntax $\mathcal{H}(\mathsf{E})$ of hybrid logic [12, p.72] by the following identification: n as a nominal and E as the global modality, where a 'nominal' means a new sort of propositional variable which is true at exactly one state. Our axiomatization of Table 4 is the equivalent axiomatization of hybrid logic with nominals and the global modality in [12, p.72], provided the set of nominals consists of n alone. Then, [12, Theorem 5.4.2] implies that our axiomatization is strongly complete with respect to the class of all models, since the proof of [12, Theorem 5.4.2] does not depend on the number of nominals in the syntax (in fact, we could obtain the stronger result to cover the additional axioms for B_i called *Sahlqvist axioms* [12, p.87, Theorem 5.4.2]).   □

Theorem 2. *The set of all the valid formulas of $\mathcal{L}^+$ is completely axiomatized by all the axioms and the rules of Table 4 and 3 as well as the necessitation rules for $[\mathsf{com}_i(\varphi)]$ and $[\mathsf{per}_i(\varphi)]$: from ψ, we may infer $[\mathsf{com}_i(\varphi)]\psi$ and $[\mathsf{com}_i(\varphi)]\psi$.*

Proof. (Sketch) By $\vdash \psi$ (or $\vdash^+ \psi$), we mean that ψ is a theorem of the axiomatization for $\mathcal{L}$ (or, $\mathcal{L}^+$, respectively.) The soundness part is mainly due to Proposition 1. One can also check that the necessitation rules for $[\mathsf{com}_i(\varphi)]$ and $[\mathsf{per}_i(\varphi)]$ preserve the validity on the class of all models. As for the completeness part, we can reduce the completeness of our dynamic extension to the static counterpart (i.e., Theorem 1) as follows. With the help of the reduction axioms of Table 3, we can define a mapping t sending a formula ψ of $\mathcal{L}^+$ to a formula $t(\psi)$ of $\mathcal{L}$, where we start rewriting the *innermost occurrences* of $[\mathsf{com}_i(\varphi)]$ and $[\mathsf{per}_i(\varphi)]$. We can define this mapping t such that $\psi \leftrightarrow t(\psi)$ is valid on all models and $\vdash^+ \psi \leftrightarrow t(\psi)$. Then, we can proceed as follows. Fix any formula ψ of $\mathcal{L}^+$ such that ψ is valid on all models. By the validity of $\psi \leftrightarrow t(\psi)$ on all models, we obtain $t(\psi)$ is valid on all models. By Theorem 1, $\vdash t(\psi)$, which implies $\vdash^+ t(\psi)$. Finally, it follows from $\vdash^+ \psi \leftrightarrow t(\psi)$ that $\vdash^+ \psi$, as desired. □

References

1. Prakken, H., Sartor, G.: The role of logic in computational models of legal argument: a critical survey. In: Kakas, A.C., Sadri, F. (eds.) Computational Logic: Logic Programming and Beyond. LNCS (LNAI), vol. 2408, p. 342. Springer, Heidelberg (2002)
2. Bench-Capon, T.J.M., Prakken, H.: Introducing the logic and law corner. J. Log. Comput. **18**(1), 1–12 (2008)
3. Grossi, D., Rotolo, A.: Logic in the law: a concise overview. Log. Philos. Today Stud. Log. **30**, 251–274 (2011)
4. van Ditmarsch, H., van der Hoek, W., Kooi, B.: Dynamic Epistemic Logic. Springer, Heidelberg (2008)
5. van Benthem, J.: Dynamic logic for belief revision. J. Appl. Non-Class. Log. **14**(2), 129–155 (2004)
6. van Ditmarsch, H.P., van der Hoek, W., Kooi, B.P.: Public announcements and belief expansion. In: Schmidt, R., Pratt-Hartmann, I., Reynolds, M., Wansing, H. (eds.) Advances in Modal Logic, pp. 335–346. King's College Publications, London (2005)
7. Sano, K., Hatano, R., Tojo, S.: Misconception in legal cases from dynamic logical viewpoints. In: Proceedings of the Sixth International Workshop of Juris-Informatics, pp. 101–113 (2012)
8. Nishida, N., Yamaguchi, A., Saeki, H.: Keihou Hanrei Hyakusen 1 Souron. 6 edn. Yuuhikaku (in Japanese) (2008)
9. Baltag, A., Smets, S.: Dynamic belief revision over multi-agent plausibility models. In: Proceedings LOFT 2006, pp. 11–24 (2006)
10. Plaza, J.A.: Logics of public communications. In: Emrich, M.L., Pfeifer, M.S., Hadzikadic, M., Ras, Z.W. (eds.) Proceedings of the 4th International Symposium on Methodologies for Intelligent Systems, pp. 201–216 (1989)
11. van Benthem, J., Pacuit, E.: Dynamic logics of evidence-based beliefs. Stud. Logica **99**(1–3), 61–92 (2011)
12. ten Cate, B.: Model theory for extended modal languages. Ph.D. Thesis, University of Amsterdam, Institute for Logic, Language and Computation (2005)

MiMI2013

How Do We Talk in Table Cooking?: Overlaps and Silence Appearing in Embodied Interaction

Rui Sakaida[1(✉)], Fumitoshi Kato[2], and Masaki Suwa[2]

[1] Graduate School of Media and Governance, Keio University,
5322 Endo, Fujisawa-shi, Kanagawa 252-0882, Japan
lui@sfc.keio.ac.jp
[2] Faculty of Environment and Information Studies, Keio University,
5322 Endo, Fujisawa-shi, Kanagawa 252-0882, Japan
{fk,suwa}@sfc.keio.ac.jp

Abstract. Cooking and eating on a table is known as a Japanese dining style. As we cook "monja-yaki" on a table, how do we communicate with others? This paper indicates that cooking acts cause utterances to overlap and generate silence more frequently than when not cooking. The order of overlaps in table cooking is shown in two aspects: (1) accidental overlaps are not always repaired in cooking, and (2) co-telling of how to cook sometimes allows utterances to overlap. Besides, while cooking, there occur some kinds of sequence organization with bodily actions: (1) adjacency pairs are organized not only by language but also bodily actions, and (2) even if adjacency pairs are not sufficiently organized with language, bodily actions could complement the absence or insufficiency. Such orders of sequence organization of actions may make silence occur more frequently. Repeated occurrences of overlaps and silence in cooking may result from embodied interaction.

Keywords: Table cooking · Overlap · Silence · Repair · Co-telling · Embodied sequence organization · Adjacency pair

1 Introduction

1.1 Japanese Table Cooking Style

It is often the case that we participate in conversations with our body engaged in some activities. A table talk is one of the most frequent examples of interaction accompanied by bodily motions in our daily life. In a table talk, each participant has to coordinate one's own utterances and eating acts, as well as the others' utterances and one's own utterances [1]. That is why a table talk is a very complicated and intelligent activity.

In Japan, we often cook and eat dishes on a table (not in a kitchen), such as *nabe-ryori* (one-pot meal), *yakiniku* (grilled meat), *okonomi-yaki* (Japanese-style pancake with various ingredients), or *monja-yaki* (Japanese-style pancake thinner and laxer than okonomi-yaki). Cooking and eating on a table is known as a traditional Japanese dining style. In this research, we call a dining style of this sort *table cooking*. The Japanese often say that a table cooking such as nabe-ryori enhances social relationships among the participants. However, how we can coordinate cooking and talking simultaneously has not been studied yet.

© Springer International Publishing Switzerland 2014
Y. Nakano et al. (Eds.): JSAI-isAI 2013, LNAI 8417, pp. 249–266, 2014.
DOI: 10.1007/978-3-319-10061-6_17

In coordinating cooking and talking, there seems to occur a little different order of interaction from when eating and talking. In eating, we have to use one mouth both to eat dishes and to talk with the other participants. Mukawa [2] and Tokunaga [3] found that we tend to try to speak even when our mouth is filled with food, so that eating acts may not hamper our conversation. Such preference for coordinating conversations may be because eating is an individual activity while talking is an interactive one. On the other hand, in table cooking, we have to coordinate both cooking and talking, with the other participants. In addition, not only the conversation but also co-cooking seems to have some kinds of sequence organization as interaction.

In this paper, we investigate situatedness of interaction in cooking acts by examining the relationship between overlaps of utterances and silence and cooking acts, and clarify the "order of interactions" brought about by a table cooking.

1.2   Multiparty Interaction in Cooking Monja-Yaki

A table cooking of monja-yaki (often called "monja") is very interesting to observe. In analyzing interaction in table cooking, there are unlimited variables to observe, e.g. the kind of dish cooked on the table or cooking tools used that influence the way and the process of cooking. For instance, in cooking nabe-ryori (one-pot meal), it is likely that one of the participants monopolizes cooking. That is because only one pair of chopsticks and/or one ladle is often used in cooking nabe-ryori, and one of the participants becomes a "chair person" (called "nabe-bugyo" in Japanese) of cooking.

On the other hand, monja-yaki needs to be cooked by several participants, because the process of cooking monja-yaki is complicated. While one is pouring the ingredients into a hot plate, another has to hash them up so that they can be cooked well. Monja-yaki is more difficult to cook than nabe-ryori, and that difficulty could make troubles or encourage participants to teach each other how to cook. The "order of interactions" generated by a table cooking of monja-yaki depends on the number of participants and/or the relationships among them. It is possible that general features of multimodal interaction, e.g. the order of turn-taking, influence the way we cook and talk. In this paper, we observe a table cooking of monja-yaki by the three participants close to each other. Those who are not necessarily close to each other would show another order of interactions in table cooking.

1.3   Overlaps and Silence in Cooking

The first author invited two friends to a restaurant and conducted an experiment of cooking monja-yaki. Observing the interaction as a participant, the first author got an impression that a table cooking may cause overlap of utterances and cause silence to occur more frequently than when not cooking.

While cooking, we have to engage in both cooking acts and a conversation. Never can we cook without gazing at cooking tools or ingredients of the dishes. Therefore, we look at the others' faces less frequently than while not cooking. Generally, in Japanese conversations, a hearer gazed at by the current speaker is likely to be the next speaker [4], and it is indicated that the participants' gaze exchanges can realize smooth turn-taking. Cooking on a table could hamper smooth turn-taking, which may result in the increase of overlaps and silence.

There seems to be another reason, too, for which our utterances tend to overlap in a table cooking. When several participants are engaged in cooking, all the participants do not always have equal amount of knowledge about how to cook. In a multiparty interaction, where more than two participants are involved, two or more advanced participants sometimes tell their knowledge collaboratively to less advanced one(s). This type of tutoring is called "co-telling". In co-telling, it is known that two participants frequently co-create one sentence, repairing each other's utterances, and that is why overlapping utterances are often produced (e.g. [5]).

In addition, while cooking and talking, there seems to be some kinds of sequence organization by both language and bodily actions. For instance, when asked a question or offered something by others while cooking, instead of answering them with language, the participants often responded to them with some actions. Schegloff [6] pointed out that sequence of actions could be dealt with by conversation analysts. In such cases, we may speak less frequently than when not cooking, which leads to the increase of longer gaps or silence.

Despite many overlapping utterances and silence for a long time, we do not feel that cooking acts disturb conversations. Although turn-taking rules are designed to prevent too many overlaps and too long gaps or silence [7], there are likely to be lots of overlaps and silence in conversations, especially when we are engaged in cooking.

In this paper, first, we analyze how often overlaps and silence occur in three phases, e.g. when people choose the dishes from the menu, wait for arrival of dishes, and cook monja-yaki. Second, referring to transcripts, we analyze some interesting cases of sequence organization with overlaps or gaps in cooking.

2 Method

2.1 The Data

The first author (called S) invited two friends (called U and H) to a monja-yaki restaurant in Kanagawa, Japan. We recorded our conversations on the table with two ultra-small digital video cameras and three voice recorders (Fig. 1). Three-party conversation is appropriate for observation of a table talk, for conversations by three participants are not likely to be split into more than one group [2]. In order to generate daily life conversations and not to put pressure on the participants to talk without any silence, we did not tell the participants what topics to talk about.

Fig. 1. Capture image of video data and top view of the table.

2.2   Excerption and Annotation

The conversation data was excerpted and divided into three phases: (1) seeing the menu and deciding what to eat, (2) waiting for the dishes to arrive, and (3) cooking monja-yaki. In phases (1) and (3), the participants talked with their bodies engaged in seeing the menu or cooking monja-yaki. In phase (2), the participants could focus on talking without any bodily acts, except non-verbal communications, such as gestures or exchanging glances. We compare the phases with bodily acts and the other from a viewpoint of the frequency of overlaps and silence.

Using annotation software ELAN[1], we made annotations of utterances and cooking acts for each participant. We also composed Japanese transcripts [8] of some suggestive examples. Overlapping utterances are put in [ ] in the transcripts[2].

2.3   Combination of Quantitative Analysis and Conversation Analysis

In this research, we analyze the interactions by means of both quantitative analysis and qualitative analysis, i.e., conversation analysis (CA). As for quantitative analysis, the length and the number of overlaps and silence are calculated and compared among the three phases mentioned above. Regarding qualitative analysis, some noteworthy examples concerning overlaps and silence are transcribed in detail, following the traditional method of transcription in CA.

It is notable that in this research both quantitative and qualitative analyses contribute to each other, indicating remarkable perspectives of analysis. For instance, calculating the frequency of overlapping utterances, we can see whose utterances are likely to overlap more frequently than the others', or which combination of the participants generates more overlaps. If two skilled participants' utterances overlap more frequently in the cooking phase, co-telling of how to cook may be occurring many times there. Or, in transcribing fine-grained interactions, similar phenomena are observed several times in the same phase and that tendency may be represented in the quantitative patterns as well. We try to combine the results of quantitative analysis with the CA findings suitably.

3   Overlaps

3.1   Quantitative Analysis

In this section, calculating the total hours and the number of overlapping utterances about each phase (Table 1), we analyze the frequency of overlaps. Overlaps are the time when more than one participant is talking for 100 ms or more[3].

---

[1] http://tla.mpi.nl/tools/tla-tools/elan/

[2] Transcript symbols are explained at the end of this paper.

[3] Although overlapping of back-channeling expressions is ruled out as examples of overlapping utterances in general, we call all the overlaps including back-channeling expressions "overlapping utterances" in this paper.

Table 1. Length and number of overlaps in each phase.

(1) Deciding what to eat	S&H	S&U	H&U	S, H&U	Total
Length of overlaps (sec.)	7.73	3.18	7.05	0	17.96
Length of overlaps among total length of utterances by the concerned participants (%)	7.02	3.37	6.34	0.00	12.89
Number of overlaps (time(s))	16	7	17	0	40
Number of overlaps among total number of utterances by the concerned participants (%)	14.16	7.61	15.32	0.00	25.95
Average length of overlaps (sec.)	0.48	0.45	0.41	-	0.45

(2) Waiting for the dishes	S&H	S&U	H&U	S, H&U	Total
Length of overlaps (sec.)	3.11	2.62	4.04	0.33	9.43
Length of overlaps among total length of utterances by the concerned participants (%)	4.63	3.04	6.16	0.00	9.44
Number of overlaps (time(s))	8	9	5	2	20
Number of overlaps among total number of utterances by the concerned participants (%)	12.12	10.84	9.09	0.02	19.61
Average length of overlaps (sec.)	0.39	0.29	0.81	0.16	0.47

(3) Cooking monja-yaki	S&H	S&U	H&U	S, H&U	Total
Length of overlaps (sec.)	5.92	12.42	3.91	0.87	21.38
Length of overlaps among total length of utterances by the concerned participants (%)	6.08	10.28	3.01	0.00	14.08
Number of overlaps (time(s))	13	19	10	3	39
Number of overlaps among total number of utterances by the concerned participants (%)	13.13	16.96	8.55	0.02	25.61
Average length of overlaps (sec.)	0.46	0.65	0.39	0.29	0.55

First, we calculated the length and the number of overlaps among the total length and the number of utterances by all the participants (Fig. 2). In phases (1) and (3), the percentages of the length and the number of overlaps were respectively higher than those in phase (2). It is possible that the participants were forced to turn their gaze on the menu or the dishes being cooked and they had difficulty in exchanging glances and coordinating their utterances.

Second, the length and the number of overlaps of each participant were calculated (Fig. 3). In phase (1), the percentages of the length and the number of all the participants were almost the same. In phase (2), all the percentages except the length of participant H were lower than phase (1). In particular, the number of overlaps of S in phase (2) is much less than that of the previous phase, and it is the same as that of U. In phase (3), the length and the number of S, and the number of U were especially high. While only the percentage of the length of H was lower than the previous phase, that of the number of hers was as high as in phase (1).

In general, our hypothesis that overlaps are more frequent while cooking than when not cooking was mostly supported. Then, why are overlapping utterances more likely to occur while cooking or looking at a menu? In the following sections, with several

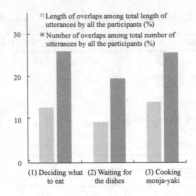

Fig. 2. Length and number of overlaps among total utterances.

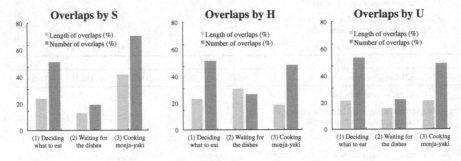

Fig. 3. Length and number of overlaps by each participant.

transcripts, we will indicate the order of overlaps in a table cooking in two aspects: (1) accidental overlaps do not always need to be repaired in cooking, and (2) co-telling of how to cook sometimes causes utterances to overlap.

3.2 Accidental Overlaps and Insufficient Repair

Although the frequencies and lengths of overlapping utterances in phases (1) and (3) were somewhat similar, the qualitative features of the overlaps were different between the two phases.

Overlapping utterances can be classified into five types (Fig. 4), from a viewpoint of when the latter utterance starts and stops overlapping with the former[4]: (a) Simultaneous Start (two utterances are started simultaneously, and either of them is completed before the other), (b) Included in the Other (the latter is started after the former is started, and the latter is completed before the former is completed), (c) Turn-taking

---

[4] In this paper, "simultaneously" means that the latter utterance is started less than 100 ms after the former utterance is started, and "One utterance is started after (completed before) the other" means that one utterance is started (completed) 100 ms or more after (before) the other.

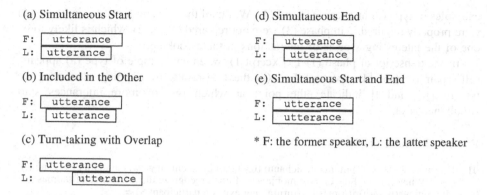

(a) Simultaneous Start

F: | utterance |
L: | utterance |

(b) Included in the Other

F: | utterance |
L: | utterance |

(c) Turn-taking with Overlap

F: | utterance |
L: | utterance |

(d) Simultaneous End

F: | utterance |
L: | utterance |

(e) Simultaneous Start and End

F: | utterance |
L: | utterance |

* F: the former speaker, L: the latter speaker

Fig. 4. Five types of overlapping utterances.

with Overlap (the latter is started after the former is started, and the former is completed before the latter is completed), (d) Simultaneous End (the latter is started after the former is started, and the two utterances are completed simultaneously), (e) Simultaneous Start and End (two utterances are started and completed simultaneously) (modified after [9]).

Among the five types, type (a) and (e) have a different feature from the others. In general, when a hearer starts to overlap with the speaker's utterance, the latter speaker, more or less, intends or expects to make his/her own utterance overlap with the former's utterance. However, as for type (a) and (e), two utterances consequently "accidentally" overlap, for neither of the two speakers can anticipate the beginning of the other's utterance. When two participants start to speak at the same time, one of them or both of them may not be heard or understood completely. In such cases, the speaker him/herself or the others often "repair" the insufficiently understood utterances [10].

Nevertheless, in table cooking, it may not be frequent that an accidental overlap of utterances (type (a) or (e)) is repaired either by the participant who made the trouble or by the other participant(s). In fact, in our experiments of conversations with monja-yaki, all the accidentally overlapped utterances were not repaired. Of all the overlaps, 3 examples in phase (1) and 4 examples in phase (3) were type (a). There were no

Table 2. Number of accidental overlaps in phases (1) and (3).

	Total of overlaps (time (s))	Accidental overlaps (time (s))	Repaired accidental overlaps (time (s))	Not repaired accidental overlaps (time (s))
(1) Deciding what to eat	40	3	3	0
(3) Cooking monja-yaki	39	4	2	2

examples of type (e) in phases (1) and (3). While all the 3 examples of (a) in phase (1) were properly repaired, 2 in phase (3) were not repaired (Table 2), which is likely to be one of the interesting aspects of interactions in table cooking.

In the transcript of phase (1)-1 (Excerpt 1), when an example of type (a) appears, self-repair is smoothly accomplished (in the transcripts, overlapping utterances are put in [ ], and [[indicates the point at which two or more utterances start simultaneously).

Excerpt 1 (Phase (1)-1).

```
01   H:   watashi mon- (.) monja ni sichauto (0.4) monja no chigai ga wakaranain desuyone (.)
          "When I,      I try to have monja,      I don't recognize the difference of monjas."
02   H:   imi wa(h)ka(h)ri(h)masu? ((turning her eyes on participant S))=
          "Do you understand?"
03   U:   =[[ e: douiu koto ]::?(0.4)
          "Well, what does it mean?"
04   S:   =[[nan no chigai?]
          "Difference of what?"
05   U:   aji no [chigai tte kanji?]
          "Is it the difference of the taste?"
06   H:          [ zenbu monja   ] tte monja (.) ni naru
          "All monjas will be monjas."
```

Answering the question "Imi wakarimasu? (Do you understand?)" by H in the second line, U in the third line and S in the fourth line started to speak simultaneously. Judging from her eyes on S and the polite expression "wakarimasu", H in the second line seems to have addressed S[5]. However, soon after H in the second line, U in the third line and S in the fourth line began to ask questions, in order to clarify H's question. U said, "E, douiu koto? (Well, what does it mean?)" and S said, "Nan no chigai? (Difference of what?)" These two utterances overlapped accidentally, and their utterances may not have been properly heard by H. 400 ms after U in the third line, U in the fifth line tried to repair the trouble for herself, saying, "Aji no chigai tte kanji? (Is it the difference of the taste?)" Since the expression of U in the third line was more abstract than S in the fourth line, U in the fifth line may have combined her previous question with more specific question of S's. The question by U in the fifth line, which was a yes-no question, seems to have been easier for H to answer than U in the third line and S in the fourth line, which were wh-questions. Realizing it was necessary to repair incomprehensibility due to the overlap of utterances and to make the question easier to answer for H, U succeeded in repairing for herself.

On the other hand, in the transcript of phase (3)-1 (Excerpt 2), the trouble caused by an overlap was left without being repaired by any participants. The trouble was due to overlapping utterances of type (a).

---

[5] Participant S is one year older than H and U, and H usually uses polite expressions to S, not to U. Though U is also younger than S, U does not use polite expressions to S so often.

Excerpt 2 (Phase (3)-1).

```
01  U:   ruisan wa (.) e (.) monja tte kansai? (.) kanto? (.)
         "Rui*,      well, are monjas from the Kansai ((region))? Or the Kanto ((region))?"
Gaze :   (1                    )(2 )(3                )(2        )

02  S:   kanto [dayo]
         "((It is from)) the Kanto."
03  U:         [ a:  ]: (0.6)
               "Oh."
Gaze :   (2                )(1    )

04  S:   [[ kansai  na ] i
         "((There are)) not ((any monjas)) in Kansai."
05  H:   [[tsukishima?]
         "((Is it from)) Tsukishima?"
Gaze :   ( 1 )( 4 )(  2  )(1)

06   :   (8.4) ((No one answered H or repaired the trouble.))
```

* Rui is participant S.

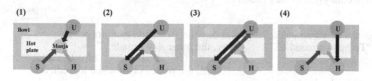

Fig. 5. Gaze direction (all the participants) in the transcript of phase (3)-1.

In this transcript, the participants were engaged in turning their eyes on the monja-yaki cooked on the hot plate. Caused to look at the monja, the participants do not seem to have focused on smooth turn-taking, exchanging glances with each other. First, all the participants were looking at the monja ((1) in Fig. 5). U in the first line addressed S and asked a question to him, "Monja tte Kansai? Kanto? (Are monjas from Kansai? Or Kanto?)" At the same time as U started to ask the question, U turned her eyes on S ((2) in Fig. 5), and immediately, S also turned his eyes on U ((3) in Fig. 5). U seems to have looked at S in order to have S answer her question, and S was preparing to answer it, with his eyes on U. S, however, turned his eyes on the monja again, before U completed the question ((2) in Fig. 5). S in the second line answered U, saying, "Kanto dayo. (It is from the Kanto.)" While S was answering it, S's eyes remained on the monja. As soon as S finished answering, U also turned her eyes on the monja ((1) in Fig. 5). 0.6 ms after U in the third line responded to S, S in the fourth line and H in the fifth line started to speak simultaneously. S told an additional answer to U, "Kansai nai (There are not any monjas in Kansai)" and H was to tell new information about the birthplace of monja, "Tsukishima? (Are monjas from Tsukishima[6]?)". This was an accidental overlap, but no one answered H or repaired the trouble. S was speaking while H in the fifth line was speaking, so S may not have heard H clearly and answered

---

[6] Tsukishima is the place in Tokyo (in the Kanto region), which is said to be the birthplace of monja-yaki.

her. On the other hand, U seems to have noticed H saying something. Just after H started to speak, U looked at H in a moment (less than 300 ms) ((4) in Fig. 5). Nevertheless, U neither answered H nor asked H to repeat the utterance. S and H were looking at the monja while they were speaking. Hearing S and H, U turned her eyes on H, S, and finally the monja in a short time. It is possible that U's intention to realize smooth turn-takings was diminished on account of S and H concentrating on cooking, and U gave up talking about it with S and H. This kind of closure of topics may be typical of interaction in table cooking.

In phase (3), a different interaction of overlaps of type (a) was also observed. In the transcript of phase (3)-2 (Excerpt 3), two overlapping troubles in the sixth, seventh, and eighth lines were not repaired. One of the troubles is due to overlapping utterances of type (a).

Excerpt 3 (Phase (3)-2).

```
01  H:  monja tte saki guzai nose- (0.2) nanka (0.4) gusha tte yarun deshita k[ke?]
        "In ((cooking)) monja, first, the ingredients are put...well...
        ((Are they)) to be in a muddle?"
02  S:                                                              [sou]ssuyo(.)
                                                                    "That's right."

03  U:  gusha tte yarun dakke (0.6) a[re dayone(.)konaida tsukutta no-]
        "To be in a muddle?"          "Say, the other day I made..."
04  S:                              [ ano::   (.)   shita  ni::   shiru ] ga arunde: (0.6)
                                     "Well, the paste is below the ingredients,

05  S:  sore wo [nokoshi-]
        so leave it ((in the bowl))..."
06  U:          [ toriaezu ] (0.6) [[  dasun   dayone  ]
                "For the present, ((we)) have to put ((them on the hot plate))."
07  S:                             [[sore wo nokoshite da]su- a chotto matta](.)
                                   "Leave it and put ((them)), oh, wait a moment."
08  H:                             [[ ° a sumimasen  (.)  ya-  (.)  yaru ° ]
                                   "Oh, sorry, d, ((I)) will do ((it))."
09  S:  are ga iru (.)
        "((We)) need that."
10  U:  a:[abura da]
        "Ah, oil."
11  H:    [abura [ abura abura (.) abu- ]
          "Oil, oil, oil, oi..."
12  S:           [abura wo (.) abura ga]
                 "Oil, oil..."
```

In this transcript, the participants were about to start cooking monja-yaki, confirming and deciding how they should cook it. First, H, not so skilled, asked a question about what to do first in cooking monja, "Gusha tte yarun deshitakke? (Are they to be in a muddle?)" S in the second line, a little more skilled, answered H, saying, "Soussuyo. (That's right.)" While S agreed to H's remark, U in the third line doubted whether H was really correct, and raised a question, "Gusha tte yarun dakke? (To be in a muddle?)" 600 ms after, U continued to tell her opinions, by telling her recent experience of eating monja, "Are dayone, konaida tsukuttano (Say, the other day I made...)" However, right after U began to tell the story, S in the fourth line started to tell the information about how to cook monja, without hesitating to overlap with U. As S did not seem to stop speaking, U gave up telling her experience. Why was S allowed to override U's story telling? While U's utterance in the third line is merely intended to tell her previous experience, S's utterance in the fourth line is a directive or an instruction, which is a talk-in-the-service-of-cooking. A talk regarding what to do next in the process of cooking is more or less urgent. If S had waited for a transition relevance place [7] to come until U finishing telling her experience, S might have missed the exact timing to give the directive. Similarly, in endodontic instructions with video broadcast, it is shown that detailed questions posed by the students take precedence over general lessons and they are allowed to break the flow of the instructor's talk [11].

U in the sixth line started to help S in the fifth line to tell what to do for the present[7], and the two utterances were partly overlapped. The latter part of U in the sixth line, S in the seventh line and H in the eighth line started to speak simultaneously and were also overlapped. U and S were trying to negotiate what to do at the present, U saying, "Dasun dayone. (We have to put them on the hot plate.)" and S saying, "Sore wo nokoshite dasu... (Leave it and put them...)" Just after the utterance, S found that the hot plate had not been oiled yet and said, "A chotto matta. (Oh, wait a moment.)" At the same time, H was trying to suggest to S that H should participate in cooking instead of S. However, because of the trouble of oil, H in the eighth line was not heard properly and all the participants were forced to begin solving the trouble (S in the ninth line, U in the tenth line, H in the eleventh line and S in the twelfth line). As a result, the suggestion by H was not shared with the others, and no one tried to repair the trouble.

In phase (3), among all the overlaps of type (a) (4 examples), 2 examples were not properly repaired by anyone. On the other hand, in phase (1), all the 3 examples were adequately repaired (all of them were self-repaired). While table cooking, if two participants start to speak simultaneously, the trouble of overlapping may not always be repaired properly. That may be because dealing with cooking acts is regarded as preferable to coordinating all the utterances. Especially when an urgent utterance such as a directive overlaps in the process of cooking, the overlap seems to remain unrepaired.

3.3 Overlaps Accompanied with Co-telling

Another reason why overlaps occur more frequently while cooking may be that more skilled participants tell how to cook monja-yaki to the less skilled. While cooking, the participants taught how to cook to each other several times. In this section, we show a case that two more skilled participants (S and U) told how to cook to the other (H), and then the utterances of the former two overlapped. This type of tutoring is called "co-telling". In a three-party conversation, when two speakers co-tell something to the third person, their utterances seem to overlap frequently [5].

In the transcript of phase (3)-3 (Excerpt 4), S and U co-told H how to cook monja.

---

[7] At this point, the utterances of S in the fifth line, U in the sixth line, and S in the seventh line are overlapped and this overlapping is regarded as co-telling of how to cook. As for "co-telling" in table cooking, we will mention in detail in the next section.

260   R. Sakaida et al.

Excerpt 4 (Phase (3)-3).

```
01  S:              kore (.) [ ano : : : (.) dote wo tsukura ] naito
                    "This, say...        ((you)) have to make a 'dote'*."
02  U:              [dote tsukutte (maru tsukutte)]
                    "Please make a 'dote', (make a circle)."
Gaze : (1 )(2 )(1 )(3    )(4   )(3                        )(5    )
```

* A "dote" means a bank in Japanese.

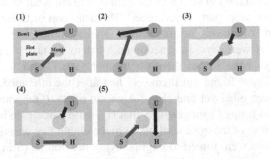

Fig. 6. Gaze direction (S and U) in the transcript of phase (3)-3.

First, U (the most skilled) found that it was time to make a "dote" (a bank) and pour the ingredients, and reached her hand to the bowl of the ingredients. Seeing a series of her cooking acts ((1) and (2) in Fig. 6), S (intermediately skilled) tried to tell H (not so skilled) to make a "dote". However, S in the first line was not able to vocalize the word "dote" quickly. He started a "word search (e.g. [12])", trying to express it with a gesture and saying, "Kore anoo... (This, say...)" Then, U in the second line moved her gaze from the bowl to the monja ((3) in Fig. 6), and said, "Dote tsukutte. (Please make a 'dote'.)" This utterance of U was meant to be a collaborative instruction to H. As a result, the two utterances overlapped by 1500 ms. In the situation that the more skilled had to tell H how to go on cooking as soon as possible, S and U realized "co-tellership" of instruction. That is why their utterances were allowed to overlap here and not repaired by anyone.

In phase (3), among all the overlapping utterances (39 examples), 4 examples were regarded as co-telling. On the other hand, in phase (1), there were no examples of co-telling. This result indicates that, in a three-party table cooking of monja-yaki, which is difficult to cook, overlapping while cooking may be partly responsible for two more skilled participants' co-telling about how to cook.

In fact, as was mentioned in Sect. 3.1 (Table 1), in phase (3), the length and the number of overlaps between S and U, who are more skilled than H, were much larger than those of the other combinations. In phases (1) and (2), on the contrary, the overlaps between S and U were not so frequent, compared to the other combinations (though the number of overlaps between them in phase (2) was more than the others). Further analysis is needed that investigates how many examples of overlaps by S and U resulted from collaborative instructions.

4 Silence

4.1 Quantitative Analysis

In this section, we analyze the frequency of silence in the three phases. We define "silence" as the time when none for 100 ms or more. The total length of silence in each phase (Table 3), the percentages of silence among total length of each phase (Fig. 7), and the average length of the silence in each phase (Fig. 7) were calculated.

Table 3. Total length of silences in each phase.

	(1) Deciding what to eat	(2) Waiting for the dishes	(3) Cooking monja-yaki
Total length of silence (s)	93.33	59.91	256.9
Length of silence among total length of each phase (%)	39.74	37.72	62.21

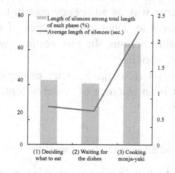

Fig. 7. Total length and average length of silences.

In phase (3), the percentage of silence was higher than in the other phases. Similarly, the average of silence in phase (3) was longer than in the other phases. It is indicated that silence occurs more often while cooking monja-yaki than when not cooking.

Although in both phases (1) and (3) the participants talked with their bodies engaged in some acts, silence in phase (1) was not so often, as well as in phase (2). In phase (1), the participants had to look through the menu, talk about what to eat, and decide it in a short time. It is possible that silence for a long time was not allowed because of the urgent task of decision-making.

On the other hand, silence in phase (3) occurred more frequently than the other phases. Our hypothesis that cooking acts make us silent was roughly supported. Even when long time silence occurred in cooking, we were not necessarily embarrassed. In phase (3), we observed several interesting examples of sequence organization related to cooking actions. In the next section, we will analyze two examples of sequence organization with bodily actions "adjacency pair".

4.2   Embodied Sequence Organization in Cooking

In the cooking phase, we observed several cases of sequence organization that consist of both utterances and bodily actions. Goodwin (e.g. [13]) has investigated how the visibility of our body accomplishes sequential embodied interactions. Mondada [14] stated that "sequentiality is a general principle governing not only talk but also action."

Adjacency pair [15] is a typical concept of sequence organization. Clark [16] proposed "projective pair", in order to expand adjacency pair into the wider concept including not only language but also actions. Schegloff [6] also discussed "sequences of actions", in which talk is accompanied by embodied action organized sequentially like adjacency pairs. In order to discuss sequences of actions, Schegloff offered an example in which one participant said, "Butter, please." and another participant passed butter to him. Mondada [14] transcribed embodied interactions called "multiactivity" in a surgical operating room, and revealed how "coagulation" in the operation is "collectively achieved" by a surgeon and an assistant, sequentially organized as "a paired action". Enomoto [17] analyzed passing and receiving interactions by a shopkeeper and a customer in a convenience store. She confirmed that, in bodily sequence organization of passing and receiving, each phase of first pair parts[8] appropriately occurs before that of second pair parts in the same way as adjacency pair in talk.

In table cooking, there seem to occur several types of sequence organization by both oral utterances and bodily actions. In this section, we analyze two examples typical of interactions in table cooking.

The transcript of phase (3)-4 (Excerpt 5) is an example where a projective pair involving bodily actions that is an instance of "offer-acceptance" occurred.

Excerpt 5 (Phase (3)-4).

01   S: ((S finished spreading oil all over the hot plate.)) hai
                                                                    "OK."
02   H: ((Putting the oil bottle back and not having anything to do, H stretched out her hand
             to receive the turner from S, but H put her hand back for a while without receiving
             it.))
03   S: ((Noticing H trying to receive the turner, S began to pass it to H.)) °hai°
                                                                    "Here you are."
((Right after this, H received the turner from S.))

In this scene, participant H offered to receive a turner from participant S, and S accepted the offer and passed it to her. First, H poured oil onto the hot plate from the bottle, and S was spreading it all over the hot plate with the turner. Finishing it, S said, "Hai. (OK.)", meaning that it was time to move on to the next cooking phase. At the same time, putting the oil bottle back and having nothing else to do, H stretched out her hand to receive the turner from S, without saying anything. S was about to put the

---

8 "First pair part (i.e., the first part of a pair)" and "second pair part" are concepts suggested by Schegloff et al. [13] that compose an "adjacency pair". The first pair part is an utterance produced by a speaker and should be followed by the second pair part, an utterance by another speaker.

turner back onto the table, but noticing that H was ready to receive the turner, S began to pass it to H. Just after saying, "Hai. (Here you are.)" quietly, S passed it to H.

In this interaction, there occurred an adjacency pair of offer-acceptance by not oral utterances but bodily acts of the two participants. In fact, in the third line, S said "Hai." and this utterance may have been meant to be acceptance by language. However, even if S had passed the turner to H without saying anything, it would not have been regarded as the absence of the second pair part. The utterance of S in the third line was not so loud and seems to have been an additional one. Fundamentally, this interaction was organized with two kinds of bodily actions, H's offering to receive the turner and S's passing it to H.

H's outstretched hand seems to be interpreted as an "offer" to receive the turner, rather than as a "request" to give her it here. In the previous scene, H had already tried to receive the turner from S and the attempt was to fail. H is a year younger than S, and she seems to have been embarrassed to have elderly S conduct cooking acts, so H "offered" to take turns to spread the oil.

The next example is a little more complicated than the previous one. In the transcript of phase (3)-5 (Excerpt 6), H asked S a question, and S did not answer it at all.

Excerpt 6 (Phase (3)-5).

((While U was speaking, H began to pour oil onto the hot plate. Just after it, in response to H, S started to spread the oil all over the hot plate with a turner.))
01 H: konna mon desuka (ne)
          "Is the oil enough?"
          ((This is a polite expression, with which H seems to have been addressed S.))
          ((On asking the question, H turned her eyes on the rack of seasoning in a moment, and gazed at the hotplate again.))
02 S: ((Without saying anything, S continued spreading the oil.))
03 H: ((H is holding the oil bottle.))
04 H: ((After a while, H stopped holding the bottle and put it back.))

In this scene, H was pouring oil onto the hot plate, and S was spreading it all over the hot plate. H asked a question "Konnamon desukane (Is oil enough?)", with a polite expression, which is interpreted to have been addressed to S (in the first line in Excerpt 6). However, instead of answering it, S continued spreading the oil for a while. This would be regarded as an absence of an answer from the viewpoint of adjacency pair of "question-answer". Nevertheless, H did not repeat the question to S, but instead kept on watching S to spread oil silently.

Why did H not repeat the question to S? When H asked the question, S was not able to answer it immediately, because S was not sure about whether the amount of oil was enough or not. Holding the oil bottle, H expected S to answer it before long, but H watched S spreading oil and became sure that the oil was sufficient, and then H stopped holding the bottle and gazing at the hot plate. In this interaction, because H assumed that oil was enough, watching the condition of the hot plate, the absence of S's answer to H's question was not regarded as a trouble.

In table cooking, (1) adjacency pairs are organized not only by language but also with bodily actions, and (2) even if adjacency pairs are not sufficiently organized with language, it is often the case that bodily actions complement the absence or insufficiency. Such orders of sequence organization of actions may make gaps or silence occur more frequently than in the other phases.

5 Conclusion and Future Work

In this paper, we analyzed how a table cooking influences the order of interaction. We indicated that cooking acts cause overlaps of utterances and generate silence more frequently than when not cooking. Showing several transcripts, we analyzed the order of overlaps in two aspects: (1) accidental overlaps are not always repaired in cooking, and (2) co-telling of how to cook sometimes allows utterances to overlap. In addition, we indicated some kinds of sequence organization with bodily actions: (1) adjacency pairs which are organized not only by language but also with bodily actions, and (2) even if adjacency pairs are not sufficiently organized with language, bodily actions could complement the absence or insufficiency. Repeated occurrences of overlaps and silence in cooking may result from the order of embodied interaction.

Our experiment has presented evidence suggesting that interactions in table cooking are situated in cooking acts. We conjecture that bodily motions irrelevant to the contents of a conversation generate an order of interaction different from a normal conversation; we are not necessarily supposed to exchange glances with each other, which would be a "social rule" in normal conversations, because of the obligation to engage in cooking acts. In addition, since monja-yaki is not so easy to cook and need to be cooked by more than one participant, we cannot help instructing each other or confirming how we should cook, instead of the most skilled one monopolizing cooking. In a three-party table cooking of monja-yaki, each participant making a commitment to cooking acts, there seems to be a kind of interactions in which a goal is achieved by all the participants.

As for sequence organization with bodily actions, there are a lot of subjects we have yet to take up and discuss. For example, how many types of sequence of actions could be regarded as an adjacency pair? Adjacency pair is the concept with which we have analyzed sequence organization in verbal conversations. Then, when dealing with sequence organization of bodily actions, we may have to revise the previous concept of adjacency pair or generate new concepts we could use appropriately. In verbal communications, we hear the others' utterances and gaze at them in order to realize smooth turn-takings. On the other hand, in bodily interaction, we must see the others' hand or body parts constantly (cf. sign language), and then it will matter where the participants are seated. In conversations, overlaps of utterances are regarded as inadequate, but in spatial interaction, how many participants can simultaneously act and interact with each other? For instance, Mondada [14] showed that a second pair part of actions of "coagulation" in an operation room can be "collectively accomplished" by a surgeon and an assistant, either "simultaneously" or "slightly dissociated". We will have to examine a lot of issues and form new theories different from those of conversations.

We will continue the study of multiparty interactions in table cooking and accumulate fundamental knowledge for analyzing dining table environment.

Acknowledgments. The authors thank Prof. Y. Den from Chiba University, Prof. K. Takanashi from Kyoto University, Prof. M. Enomoto from Tokyo University of Technology, Prof. H. Shirai from Keio University, and the two anonymous referees for their giving us a lot of beneficial advice.

Appendix: Transcript Symbols

[	The point of overlap onset.
[[	The point at which two or more utterances start simultaneously.
]	The point at which two overlapping utterances end.
=	No break or gap.
(0.0)	Elapsed time by tenths of seconds.
(.)	A brief interval within or between utterances.
::	Prolongation of the immediately prior sound.
°word°	The sounds softer than the surrounding talk.
(h)	Plosiveness with laughter.
(word)	Dubious utterances or words.
(( ))	Transcriber's descriptions.

References

1. Den, Y., Kowaki. T.: Annotation and preliminary analysis of eating activity in multi-party table talk. In: Proceedings of the 8th Workshop on Multimodal Corpora: How Should Multimodal Corpora Deal with the Situation?, pp. 30–33 (2012)
2. Mukawa, N., Tokunaga, H., Yuasa, M., Tsuda, Y., Tateyama, K., Kasamatsu, C.: Analysis on utterance behaviors embedded in eating actions: how are conversations and hand-mouth-motions controlled in three-party table talk? (in Japanese). IEICE Trans. Fundam. Electron. Commun. Comput. Sci. (Japanese edition) **J94-A**(7), 500–508 (2011)
3. Tokunaga, H., Mukawa, N., Kimura, A., Yuasa, M.: An analysis of interaction structures among table-talk participants based on gaze behaviors and dialog-acts (in Japanese). IEICE Trans. Fundam. Electron. Commun. Comput. Sci. (Japanese edition) **J96-D**(1), 3–14 (2013)
4. Enomoto, M., Den, Y.: Will the participant gazed at by the current speaker be the next speaker? (in Japanese). Jpn. J. Lang. Soc. (Japanese edition) **14**(1), 97–109 (2011)
5. Toyama, E., Kikuchi, K., Bono, M.: Joint construction of narrative space: coordination of gesture and sequence in Japanese three-party conversation. In: Proceedings of International Workshop on Multimodality in Multispace Interaction (MiMI 2011), pp. 49–60 (2011)
6. Schegloff, E.A.: Sequence Organization in Interaction: A Primer in Conversation Analysis, vol. 1. Cambridge University Press, Cambridge (2007)
7. Sacks, H., Schegloff, E., Jefferson, G.: A simplest systematics for the organization of turn-taking in conversation. Language **50**(4), 696–735 (1974)
8. Nishizaka, A., Kushida, S., Kumagai, T.: Introduction (in Japanese). Spec. Issue: Lang. Use Interact., Jpn. J. Lang. Soc. (Japanese edition) **10**(2), 13–15 (2008)
9. Enomoto, M.: When does the hearer start his turn? the turn-taking rules in Japanese conversation apply retrospectively after a possible completion point has passed (in Japanese). Cogn. Stud. **10**(2), 291–303 (2003)

10. Schegloff, E.A., Jefferson, G., Sacks, H.: The preference for self-correction in the organization of repair in conversation. Language **53**(2), 361–382 (1977)
11. Lindwall, O., Lymer, G.: Inquiries of the body: novice questions and the instructable observability of endodontic scenes. Discourse Stud. **16**(2), 271–294 (2014)
12. Hayashi, M.: Language and the Body as Resources for Collaborative Action: A Study of Word Searches in Japanese Conversation. Res. Lang. Soc. Interact. **36**(2), 109–141 (2003)
13. Goodwin, C.: Action and embodiment within situated human interaction. J. Pragmat. **32**, 1489–1522 (2000)
14. Mondada, L.: The organization of concurrent courses of action in surgical demonstrations. In: Streeck, J., Goodwin, C., LeBaron, C. (eds.) Embodied Interaction: Language and Body in the Material World, pp. 207–226. Cambridge University Press, Cambridge (2011)
15. Schegloff, E.A., Sacks, H.: Opening up Closings. Semiotica **8**, 289–327 (1973)
16. Clark, H.H.: Pragmatics of language performance. In: Horn, L.R., Ward, G. (eds.) Handbook of Pragmatics, 365–382. Blackwell, Oxford (2004)
17. Enomoto, M.: A description about adjacency pairs of social actions using non-verbal channels (in Japanese). In: Proceedings of the 52th Conference of Special Interest Group on Spoken Language Understanding and Dialogue Processing (SIG-SLUD), pp. 87–92 (2008)

Grounding a Sociable Robot's Movements in Multimodal, Situational Engagements

Morana Alač[⊠], Javier Movellan, and Mohsen Malmir

Department of Communication, University of California San Diego,
9500 Gilman Drive #0503, La Jolla, CA 91093, USA
alac@ucsd.edu

Abstract. To deal with the question of what a sociable robot is, we describe how an educational robot is encountered by children, teachers and designers in a preschool. We consider the importance of the robot's body by focusing on how its movements are contingently embedded in interactional situations. We point out that the effects of agency that these movements generate are inseparable from their grounding in locally coordinated, multimodal actions and interactions.

To define a *sociable robot* (e.g., [1]) one could discuss its computational architecture, the mechanics of its body, or its expected function. We engage this task by going beyond the robot's body: we think about the robot by attending to its relation to humans and the setting in which it dwells (see also [2]). As our interest is in describing how the robot is experienced in moment-by-moment practical encounters, we draw from *ethnomethodology* [3] and *conversation analysis* [4, 5], and turn to everyday, local, embodied practices that comprise actual situations in robotics research. This allows us to consider multiparty engagements in the local environment of *situated* practice [6] (rather than focus on the typically assumed unit of analysis where a single user interacts with technology). But even though we look beyond the robot's physical body, our intention is to highlight its relevance in interaction. We show how this relevance is achieved in particular instances of design practice and is inseparable from its situational grounding.

Sherry Turkle [7] reported that—when the gaze of the sociable robot Cog followed her as she walked around—she "had to fight [her] instinct to react to 'him' as a person" (85). Talking about her encounter with the same robot, Lucy Suchman [6] pointed out that its agency is not simply inherent to its body; it concerns an "extended network of human labor and affiliated technologies" (246). Here we look at effects that a movement of a robot's body generates by focusing on interaction. As we describe details of an everyday encounter with a sociable robot, we point out how the agential effects of the robot's movement are supported through an interactional coordination of people, objects and technologies in a preschool setting.

Our descriptions are thus aimed at accounting for the multimodal and sensory interactional organization of everyday practice. We consider how the interacting participants use talk, gesture, gaze, prosody, facial expressions, body orientation and spatial positioning as they engage the material aspects of the setting in which their action is lodged (e.g., [8, 9]).

© Springer International Publishing Switzerland 2014
Y. Nakano et al. (Eds.): JSAI-isAI 2013, LNAI 8417, pp. 267–281, 2014.
DOI: 10.1007/978-3-319-10061-6_18

In the tradition of *laboratory studies* (e.g., [10]), we ground these descriptions in a long-term participant observation of a machine-learning laboratory. The observational study was conducted by the first author. The other two authors are roboticists, who also engage in observational activities. They observed events at a university preschool located in the Western United States, where they – between 2004 and 2013 – immersed a robot in classroom activities. Their observations were part of the so called "iterative design cycle" method, where they continually updated the current version of the robot according to what they saw at the preschool. The first author has ethnographically participated in this project since 2005, and by following the robotics team, has often found herself at the preschool as well.[1] There, in Classroom 1, a group of 12–24-month-old children, together with their teachers and the robot's designers, engaged the robot. The robot is called RUBI, which stands for "Robot Using Bayesian Inference" [11], and is a low-cost, child-sized robot designed to function indoors as educational technology.

This long-term design endeavor[2] has seen the robot change through various instantiations [12]. The humanoid robot featured in this paper is equipped with a computer screen and two cameras that stand in for its eyes. The cameras are used to track people who interact with the robot. When involved in tracking human faces, the robot's head moves. The robot also has a radio-frequency identification (RFID) reader implanted in its right hand to recognize objects handed to it. The interaction with the robot, however, is expected to mainly revolve around its touch-sensitive screen. When in "running" mode, the screen displays educational games or a real-time video of the robot's surroundings, captured through its cameras. By displaying what the cameras record, the robot's screen allows its interlocutors to see what the robot "sees." Here we are concerned with what the robot sees/senses inasmuch as this seeing/sensing indicates how humans see the robot and its seeing of them. We track how "looking" and "sensing" is interactionally organized, and how it plays a part in the enactment of the robot's aliveness. Thus, we propose that the robot is an actor when it is treated as such in a specific interactional setting.

We focus on an excerpt from interaction that illustrates some of the complexities of the robot's agency. First, the toddlers' actions indicate the impact of the robot's movements for the experience of the robot's agency. At the same time, the excerpt also shows the situational grounding of this movement. The movement—grounded in a spatial organization of bodies and technologies and the dynamics of the multimodal semiotic interaction—participates in enacting the technological object as an actor.

The excerpt comes from the roboticists' third visit to the preschool. The toddlers had already become familiar with the robot during the two previous visits. During the

---

[1] Since this process of design and construction is meant to respond, at least in part, to the preschool visits, its contingencies do not only concern the work of the roboticists but also the classroom's interactions between the children and their teachers.

[2] Each of the RUBI robots can be mapped on a "project" ([10], pp. 53–80) as it participates in organizing laboratory practices within a temporal context. The appearance of the project's unity and sequential organization is achieved through local production and in situ activities of obtaining funding, responding to grant cycles, writing up results, as well as designing and building physical instantiations of the robotic machine.

third visit, the robot is accompanied by the laboratory's director or the *principal investigator* (PI), two *graduate students* (GS1 and GS2), and the *ethnographer* (Et). As the team enters the preschool, the ethnographer turns on her camcorder, aiming to capture a complex web of gazes and gestures that articulate and are articulated by the activities in Classroom 1.

Classroom 1 is a space for two-year-olds to play and learn while their parents are not present. Yet, because the preschool is part of the university, Classroom 1 is also predisposed as a research space. As depicted in Fig. 1, the classroom is divided into three areas where the two main areas—Area A and Area B—are connected by a door and a big window. The window allows for a direct monitoring between the two spaces. In addition, there is a small room—Area C—which has a one-way screen opening into Area B. This multifunctional space—both a classroom and an experimental space—is further organized as a *setting* [13][3] for research sessions. The presence and arrangement of multiple pieces of technology (e.g., the ethnographer's video camera, the roboticists' computers and the robot itself), their mingling with the objects that exist in Classroom 1, and the interactional engagements between the research team and the preschool inhabitants articulate the space as a *laboratory* [10].

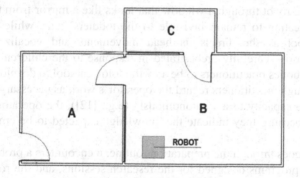

Fig. 1. The layout of Classroom 1

Following the team's arrival to the preschool, the roboticists engage in a ritual preparation of Area B for a research session. The PI asks one of the educators to keep the children busy elsewhere so that they can join the activity only once the scene has been set up. To stage the scene for the session, first the robot's computers have to be turned on and connected to an external laptop, then, the room's furniture needs to be appropriately arranged. Since the robot is not expected to perform any locomotion during the session, the PI places the robot in a corner of Area B, just in front of Area C. As he plugs the robot's computer into the wall socket, the PI arranges large, colorful cushions around the robot to cover any visual access to the wires, while allowing the children to be comfortably positioned in front of the robot. By arranging the barriers to perception, the PI's work is part of the "backstage" and "frontstage" ([14], pp. 22, 106,

---

[3] Setting for activity is repeatedly experienced, personally ordered and edited version of a more durable and physically, economically, socially organized space-in-time ([13], p. 71).

112; see also [6], p. 246) preparation. The arrangement in which the robot and the other classroom props are positioned sets the stage for certain kinds of actions. For example, it is almost impossible to stand behind the robot to observe its computers and wires (which would suggest that the robot is a piece of technology rather than a social actor.) Because these elements of the set up may "discredit the impression" of being in a presence of a live social actor, they are "suppressed" ([14], p. 111). When toddlers enter the room, they find the familiar environment, namely the playroom that grew into a research setting. There, they are expected to engage with the researchers and the robot as interlocutors. The organization of the cushions, the clearance in front of the robot, and the positioning of other actors who also face the robot make them not only notice the robot, but direct them to its face, hands, and the computer screen. This front region —where the robot is enacted as an interlocutor—is also managed through the use of gesture, talk, gaze, expressions of emotion and body orientation on the part of the roboticists and other preschool's inhabitants.

Concurrently, some of the members of the robotics team—together with the computers and wires—serve as back regions for the performance of the robot's agency. One of them is GS1, a graduate student, who, while the director organizes the space around the robot, positions herself in Area C. GS1's location allows her to observe the events around the robot through a window that looks like a mirror from the other side. She uses this location to remain invisible to the toddlers' gaze while being able to observe the robot as she directs its head movements and vocalizations from a laptop. GS1's work—carefully orchestrated in response to the children's conduct—is considered by robotics practitioners to be a methodological tool to develop autonomous robots. Even though practitioners regard the operator's work as necessary only until the robot regains the capacity to act autonomously (e.g., [12]), the operator's actions are also significant because they indicate the knowledge expected to be embedded in the robot's design.

As the team goes through the preparation routine, it encounters a problem: the robot will not run the programs designed for the research sessions, and the robot's operator has to reboot her laptop a couple of times. To diagnose the issue and coordinate the entire set up, the principal investigator swiftly moves between the operator's computer (in Area C) and the robot's body (in Area B) while the ethnographer continues to videotape the scene. As the roboticists struggle to turn on the robot's program, the toddlers—in particular *Perry* (P) and *Tansy* (T)—are visibly intrigued by the event. Even though one of the team's members—GS2—tries to prevent the toddlers from entering, they don't give up. Finally, they enter Area B. As the activity around the robot increases, and the adults have to manage the robot's appearance while they deal with the inopportune presence of the toddlers, it becomes clearer that the transformation of Area B is not under the absolute control of the adults.

At the beginning of the excerpt, the principal investigator closely monitors the robot's screen. As he turns around, he notices Tansy seated on the floor in front of the robot. By initiating a conversation with the toddler, the principal investigator accepts the arrangement of bodies and technology suggested by Tansy's presence (Fig. 2). He seats himself on the floor next to Tansy (as is usually done during the research sessions at the preschool), allowing this transformation of the scene to change the mode in which he engages the robot. He no longer treats the robot as a nonfunctioning thing, but also as an actor. At the same time, the PI and his team continue to prepare the robot for its "proper" functioning. In what follows, we see how this ad hoc management of the front- and back-stage regions is accomplished.

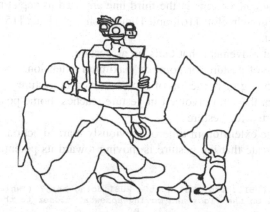

Fig. 2. The principal investigator, the robot and Tansy (from left, clockwise)

We follow the PI, the toddlers, and the rest of the team as they witness the moment in which the robot starts to "function properly." The situational grounding makes the robot's movements publicly available as relevant. They can then be recognized as meaningful events to be acted upon. Even though the toddlers do not yet exhibit a full linguistic mastery, their interactional capacities are remarkable. Their vocalizations, gestures and facial expressions indicate their co-participation in the robot's enactment as an actor. In describing the scene, we pay specific attention to how the robot achieves its agency through its movements grounded in the preschool interaction.

Each line of the transcript (marked by Arabic numerals - 1, 2, 3, ...) is divided by the participant contributions—human and non-human: PI (Principal Investigator), R (Robot), and the three toddlers—T (Tansy), P (Perry), and J (Joy). The contribution of each participant is further divided in a line of talk which follows the name of the participant, line of gaze—g, and line of hand gesture, where "rh" stands for "right hand" and "lh" stands for "left hand."

The line of talk is transcribed following Jefferson's [4] conventions:

= Equal signs indicate no interval between the end of a prior and start of a next piece of talk.

(0.0) Numbers in brackets indicate elapsed time in tenths of seconds.

(.) A dot in parentheses indicates a brief interval within or between utterances.

::: Colons indicate that the prior syllable is prolonged. The longer the colon row, the longer the prolongation.

- A dash indicates the sharp cut-off of the prior word or sound.

(guess) Parentheses indicate that transcriber is not sure about the words contained therein.

(( )) Double parentheses contain transcriber's descriptions.

.,? Punctuation markers are used to indicate 'the usual' intonation.

To transcribe the dynamics of the gaze (the second line), we adopted transcription conventions from Hindmarsh and Heath [15]:

PI, R, T, P, Te Initials stand for the target of the gaze.

_____ Continuous line indicates the continuity of the gaze direction.

The transcription conventions in the third line are used to depict the hand gesture, and are adopted from Schegloff [16], and Hindmarsh and Heath [15]:

p indicates point.

o indicates onset movement that ends up as gesture.

a indicates acme of gesture, or point of maximum extension.

r indicates beginning or retraction of limb involved in gesture.

hm indicates that the limb involved in gesture reaches 'home position' or position from which it departed for gesture.

.... Dots indicate extension in time of previously marked action.

,,, Commas indicate that the gesture is moving toward its potential target.

```
1
    PI   What's RUBI doing? Eh? What's RUBI doing here? ((while adjusting
         himself on the floor and moving somewhat closer to the robot))
    g    T_____R_____

    T    ((adjusting herself on the floor))
    g

2
    PI
    g    _____
    rh             ,,,,,,,,, p r,,hm ((touches the robot's screen))
    T
    g                              R_____

3
    PI
    g    _____T_____

    T                                (Ah)
    g    _____
    rh             ,,,,,,,,,,p....... ((points to the robot))

4
    PI   Yeah RUBI ((nods))
    g    _____

    T
    g    _____

5
    PI              Hi RUBI
    g    R_____T_____
    rh          ,,,,,,p.....r,,,hm ((briefly touches with the open hand
         the robot's face))

    T
    g    _____PI_____
```

```
6
    PI                  Hi RUBI
    g       __R_____
    lh         ,,,,,,p················r,,,hm ((points to and shakes the robot's left
           hand))

    T
    g              R_____
```

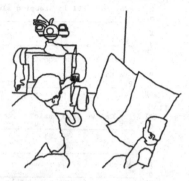

Fig. 3.

```
7
    PI              (She) is sleeping (.) RUBI is sleeping ((nods)) Yes
    g       T_____R__

    T                           (Eh)
    g       _____PI_____

8
    PI
    g       _____P___R_____

    T
    g       _____P___R_____

    P       To- toui:! ((running into the room))
```

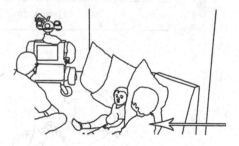

Fig. 4.

```
9
   PI
   g      _____

   T
   g      ___P_____R

   P    Toui:! Toui!
   g    R_____

10
   PI   RUBI is- RUBI is sleeping sleeping (.)     RUBI is sleeping
   g    P_____     R_____
   T
   g    _PI_____  R_____
   rh                                      ,p.....................r hm
        ((points to the robot))

   P      Ta ta
   g    PI_____R_____PI_____
   rh          ,,,,,,,,,,,,,,p__r hm ((points to the robot))

11
   PI      Can we have music?
   g      __P_____R_____

   T
   g      _____

   P    Ah-uh
   g    _____
   rh        ,,,,,,,,,,,p.................r hm ((points to the robot while slightly
        jumping up and down))

12
   PI
   g    _____P_____

   T
   g    _____

   P    ((steps toward the robot and back)) Bo:!=
   g    _____
   rh                                ,,p......((points to the
        robot))

13
   PI   =Bo:,
   g    _____
                                              R
   2h        o   a   r hm ((iconic gesture around his neck and face as if
        taking the head off, while making funny faces))

   T
   g    _____

   P                          ((jumps four times and laughs))
   g    _____PI_____R_____
```

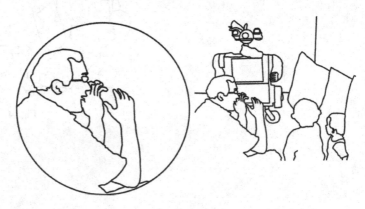

Fig. 5.

```
14
PI   Oh,                Hi RUBI,              Hi RUBI,
g    _____P_R_____
lh   ,,p..r hm ((points to the robot))
g    _____
T
g    _____
P                        ((retracts backwards and leaves the room))
g    _____PI_____R_____
R    ((video appears on the screen))
g    _____T_____
```

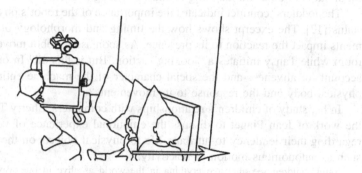

Fig. 6.

```
15
PI              ((changes seating position to engage in face-
     recognition activity with the robot))
g    _____
T
g    __ R_____Te_____
rh                        ,,p   r........hm ((points toward the robot))
R                        ((displays facial expressions in coordination
     with the PI))
g    _____PI_____
16
PI                      RUBI is there.
g    _____T___Te_____R_((one of the teachers entered
     the room and observes the scene))
rh                      ,,,,p.....r hm ((points toward the robot))
T    Uh      Uh Uh
g    _____PI____R_____
rh   ,,p...............p................................r hm ((points and stretches her
     arm two times toward the robot))
17
PI
g    _____
T
g    _____
R    Ha ha ha    Ha ha ha((laughs))
J                Ha               Ha ((entered the room and laughs))
g    R_____Te___R_____
```

The excerpt shows how the robot's attraction is materialized in the children's subtle body movements, facial expressions, and their not-yet-linguistic vocalizations. Perry's reaction to the robot's movement is emblematic (see Fig. 6). The robot starts to move in line 14: Its head tilts down to then pan across the room, while its computer starts to display a real-time video of its surrounding. Just after the onset of the robot's movement, Perry swiftly but cautiously retracts. Aiming for the door, she keeps a close eye on the robot as she moves backwards. She then leaves the room in a hurry. Tansy's conduct in lines 14–15 is similar: When the robot turns its head toward her, Tansy looks down (line 15). Once she drops her gaze and the robot moves its head away, Tansy looks back at it. Somewhat analogous is Joy's reaction in line 17: When the robot starts to emanate laughing sounds, Joy laughs back.

The toddlers' conduct indicates the importance of the robot's physical features to its status [17]. The excerpt shows how the timing and morphology of the robot's movements impact the reaction to its presence. As soon as the robot moves, Perry leaves the room while Tansy initiates a "looking" action. But is this all? In other words, can we account for aliveness and the social character of the machine entirely in terms of its physical body and the response to its movements?

In her study of children's relationships with computers, Sherry Turkle [18] refers to the work of Jean Piaget to discuss the emotional experience of very young children regarding their tendency to attribute life to physical objects on the basis of behaviors such as autonomous motion or reactivity:

> Young children see almost everything in the world as alive in one way or another. This "animism" pervades the child's thinking until the development of concepts that help draw the line between the alive and the not-alive. Childhood animism has two faces: it makes the world friendly and understandable, but it can make it frightening as well. Emerging from animism is more than a chapter in the intellectual development of the child—it is a struggle against the insecurities that come from not knowing what objects can act independently and potentially antagonistically. Children spend a great deal of energy trying to get such matters under control, and thus it is not surprising that they are disturbed when a computer behaves halfway between a person and a thing ([18], pp. 33–34)

While Turkle studies children's responses to animate objects through interviews, here we see a group of toddlers enacting such responses through their prelinguistic, embodied conduct as they accomplish close coordination with the robot at the level of spatial positioning and gaze management. In this setup, the robot's movements—and their achievement of the agential effects—are inseparable from their situational grounding. The doings of the research team, the arrangements of the multiple pieces of technology, the spatial organization of Area B, the subtle semiotic acts of the preschool's inhabitants, and their coordination participate in the articulation of the robot's agency.

The way the robot moves has to do with a chain of careful decisions and ad hoc design solutions that are not only about technological specifications, but also concern cultural modes of present-day significations [19]. For example, when building the robot, the researchers' goal was to construct an inexpensive, easily assembled robot that would quickly appeal to children. To do so, they built the robot's body themselves: They assembled, mended, took apart and put together the pieces of the robot's future body as they spent hours of work in what they call the "tinkering room." The PI's children and the preschool educators also got involved: While the PI's six-year-old daughter considered the robot a girl, the educators denominated the robot "Mama RUBI." In building this version of the robot, the goal for the practitioners was to design a sociable machine that can move around the room. Because this made the robot much larger and heavier the researchers were worried that the bulkier size made the robot appear threatening. In designing the following model, they opted for a smaller design and gave up on locomotion for the time being. These expectations placed into the construction and imagining of the technology clearly put it apart from other objects, such as children's toys (dolls, for example) and functional equipment (door closers and hammers).

Yet, in addition to the robot's movements, humanoid morphology, the inscriptions of its gender, social role and age-markers [20, 21], the effects of the robot's agency are also relative to the situated actions and interactional moves through which the robot becomes legible as a living actor. In this paper, we show how, to be considered "alive," or even "social," the robot's movements have to be made publicly available as relevant —the robot's potential interlocutors need to be able to recognize the robot's movements as something that they already know, can attribute meaning to, and should act upon. As the excerpt indicates, one of the crucial components of this process is accomplished through the PI's interactional work. We first see how the PI selects the robot's features as particularly significant. He then categorizes the robot's behaviors or lack thereof in terms of human action. Finally, the PI formulates the toddlers' actions as enactments of the robot's social character.

Perry's retracting from Area B (line 14, Fig. 6) is preceded by the PI's high pitched "Oh," uttered as the robot starts to move and its screen switches to video. The PI's utterance *highlights* [8] that the change in the robot's state is supposed to be noticed, drawing the group's attention to it. In contrast, when the PI notices the change in the appearance of the Unix shells on the robot's computer screen, he frequently touches the screen, but he does not produce any semiotic action that would render such changes more evident. In other words, he treats these kinds of changes as events that are only pertinent to the practical actions of the team, while being semiotically irrelevant, and thus not of interest to the toddlers.

The PI's semiotic moves also *code* [8] the robot as an alive, human being. In lines 7 and 10, for example, when commenting that the robot is asleep, the PI attributes aliveness to the machine. In a similar vein, in lines 1, 4, 5, 6, and 10, the researcher refers to the technological object by calling it "RUBI," while in lines 5 and 6 he indicates that the object needs to be greeted, as he touches its face and hand. Aug Nishizaka has studied the involvement of touch in interaction to describe how a tactile reference to specific locations is shaped by the action sequences in which it is contingently embedded [21]. In line 5, the PI directs Tansy's attention toward the robot's face by accomplishing the reference through touch. The touched face is interactionally framed not only as a place to be visually oriented to, but—in combination with the linguistically expressed greeting— as a focal point of interaction. In fact, in line 6 the PI follows by enacting the greeting procedure (see Fig. 3), whereby uttering "Hi RUBI" and shaking the robot's hand. This enactment of the greeting ritual, that models the appropriate and expected way of acting and interacting with the machine, constitutes the addressee as a particular kind of entity—a social actor. By greeting the robot, the PI projects an expected action, namely a greeting, performed by a social actor. In this sense, the robot is an actor not only because of the features intrinsic to its body, but also because of its involvement in the PI's semiotic actions performed during the research session.

The achievement of the robot's social character also involves the participation of the toddlers [2]. The toddlers participate through their own actions, but we can also read their involvement in the PI's conduct. When the PI talks about and treats the robot as a living, social actor, he responds to the toddlers' presence in Area B and the specific actions that they initiate. When the PI says that the robot is asleep, this is not because the robot's eyes are closed, or because it is lying in bed. One cannot even say that the robot is asleep because its body fails to exhibit any activity. During the interaction, the robot's computer screen is in fact turned on, displaying various Unix shells with flickering commands on it. Rather, the sleeping has to do with the robot's inappropriate functioning in respect to the presence of the toddlers in the room. The robot is not yet set up to sing, run educational games and track faces, as expected when approached by the toddlers. The practitioners, however, cannot just treat it as any other piece of technology. Their "sleeping" explanation is an answer to the complexity of the situation to which they are responding. With children too young to understand what death is, "sleeping" is often used as a comforting euphemism when they express curiosity toward a dead creature. It seems that the PI relies on this culturally available organization of experience, and uses the sleeping explanation to sustain the animated *frame* [23] around the robot. In presenting the robot as sleeping (rather than non-functioning), the PI also engages in "face-work" [24]. He "gives face" to the robot as he arranges for it "to take a better line that he might otherwise have been able to take" ([24], p. 9). In doing so the PI not only preserves the face of the robot, but also maintains his own face ([24], pp. 11–14) and the face of his team. With an alive piece of technology, the practitioners position themselves as roboticists.

Immersed in the scene, the toddlers initiate acts of shared attention toward the robot. They cheerfully articulate proto-words and direct deictic gestures toward the robot while they monitor how the PI's reacts to their moves. In doing so, they shape the PI's actions. In line 3, Tansy points while saying "Ah." In line 10, both Tansy and Perry point while Perry utters "Ta ta." In line 11, Perry points again while uttering "Ah-uh." In lines 15 and 16, Tansy enacts a series of indexical gestures while saying "Uh uh uh" (line 16). As the toddlers manifest their excitement toward the robot, the PI readily responds to their actions. Mardi Kidwell and Don Zimmerman [25] emphasized that the acts of shared attention do not only involve drawing and sustaining other's attention toward an object, as suggested by psychologists (e.g., Tomasello [26]).[4] When immersed in social settings, those acts indicate what another should do in response. A child that shows an object to an adult projects a social action that s/he expects the recipient of the show to accomplish. When Perry and Tansy look at the PI while vocalizing and pointing, the PI treats their actions as "shows" to which he diligently responds by identifying and appreciating what they are showing. When Tansy utters "Ah" in line 3, the PI responds

---

[4] Developmental psychologists point out that engagements with attention-organizing behavior start to rapidly evolve from the end of the child's first year and early into the second year. They link these attention-organizing behaviors with the capacity of intention-attribution, and consider them to be a prerequisite for the development of human language.

(line 4) with "Yes RUBI." When she says "Eh" in line 7, the PI readily answers with "RUBI is sleeping yes" (line 7) (Fig. 4). Similarly, in lines 4 and 7, Tansy's "Uh uh uh" (in line 16) is immediately followed by PI's "RUBI is there." And when Perry utters "To-toui:!" in line 8, and "Toui:! Toui!" in line 9, the PI repeats "RUBI is- RUBI is sleeping sleeping (.) RUBI is sleeping" (Line 10). When Perry says "Ah-uh" (line 11), the PI cheerfully follows with "Can we have music?" (line 11).

By responding to what the toddlers are drawing attention to, the PI configures the toddlers' actions in terms of the robot's agency. He does not only orient to the robot as he would to another object, but makes the toddlers' actions intelligible by framing them in terms of the robot's social character. Through the PI's utterances, the toddlers actively participate in the scene. Their preverbal expressions assume verbal forms so that their actions can be read as addressing the robot as "RUBI," greeting it, asking for music, and talking about the robot's sleeping state. In other words, through the coordination with the PI's actions, the toddlers engage the robot, inscribing it with the traits of an animate social actor. Notice also that the PI's expressions of appreciation often have a form of affirmation. By saying "Yes RUBI" or "RUBI is sleeping yes," the PI assigns the content of "RUBI" or "RUBI is sleeping" to the toddlers' "Ah" and "Eh," respectively.

A further example of how the PI elaborates toddlers' utterances—and thus insures that they participate in configuring the robot as an actor—is his participation in line 13. After Perry utters "Bo" in line 12, the PI follows with another "Bo:" in the consecutive line. Instead of attributing an already existing linguistic form to the toddlers' prelinguistic utterances, the PI adopts the toddler's idiosyncratic expression to talk about the robot. Through this uptake, the PI talks through Perry's words. The example, however, also illustrates how the toddler's actions are shaped by the PI's multimodal intervention. As soon as he repeats Perry's "Bo:," the PI follows his utterance with a gesture that looks like an act of taking off of the head (see Fig. 5). Because the PI's "Bo:" voices Perry's expression, the PI's semiotic action further elaborates Perry's utterance, allowing it to take up the content of the gesture. Through the PI's gesture, Perry's "Bo:" is performed as an expression that indicates the robot's inappropriate functioning.

This framing can be understood as the PI's manipulation of whatever the toddlers do to make it appear as part of a coherent series of expressions that implicate an animate "other" (see, for example, Melvin Pollner and Lynn McDonald-Wiker's discussion of how a family attributes competence to a severely retarded child [27]). This kind of interpretation, however, would miss some of the essential features of the interaction. First, it would privilege the linguistic performance over the multimodal interaction that characterizes the toddlers' conduct. By erasing the toddlers' positioning in the space, their gestures, and their vocalizations, it would make them out to be the passive objects of a puppeteer. What is more, this interpretation would not take into account the trajectory of interaction, where the PI's account of the robot's sleeping is shaped by the toddler's presence in the room. As the PI indicated, when the toddlers entered the space, he had no choice but to say the robot was sleeping. Rather, his actions are constantly sensitive to the intersubjective life-world that he and the toddlers inhabit. Finally, the interpretation of the toddlers as passive objects of a puppeteer would overlook the members' understanding of the RUBI project. As the PI points out, the goal of the preschool visits is not to show that the toddlers treat the robot as an interactant (if this was the goal, the PI would probably perform a controlled laboratory experiment), but to observe how the toddlers respond to the robot in their everyday setting so that he and his team can improve on the robot's design. In this sense, instead of understanding the roboticists' moves as intentional manipulation of the activity, we see them as part of the ongoing scene in Area B. In the situation where the practitioners are trying to explain the robot's lack of proper functioning in terms of its being asleep, and the toddlers are constantly pointing toward what they recognize as important (based on their previous encounters with the team), the PI translates the toddlers' proto-words and shapes his utterances to toddlers', so that his semiotic acts fit the common course of action. Perry's "Bo" and the PI's further elaboration of the utterance are performed as legitimate moves in the *language game* [28] that configures the robot as an alive, social actor.

The functioning of this language game is organized around the robot's movements. By following the interaction reported in the excerpt, we saw how the absence of those movements is framed in terms of a sleeping state, while their occurrence is marked by the participants' visible orientation toward (or away from) the robot, and by treating it as *somebody* to be greeted. We can, thus, say that, similarly to what Turkle reports, the robot's movement appears to generate effects (e.g., the movement is immediately followed by Perry's leaving the room), but, as Suchman suggests, the effects of a robot's agency also concern human actions and technologies that go beyond the robot's body. Here, we dealt with this "extended network" by describing spatial arrangements and how the preschool inhabitants employ touch and embodied multimodal communicative means to engage each other and the robot on the occasion of the research visit to the preschool. This allowed us to indicate how the effects of the robot's movement are achieved as a part of the larger, continuously updated articulation of a historically shaped interactional situation.

Acknowledgment. We would like to thank Mayumi Bono, Yelena Gluzman, Charles Goodwin, John Haviland, Shimako Iwasaki, Michael Lynch, Maurizio Marchetti, Susanna Messier, Aug Nishizaka, Paul Ruvolo, Masaki Suwa, Cynthia Taylor, Paul Ruvolo and the participants in the ethnographic study for their contribution to this paper.

References

1. Breazeal, C.: Designing Sociable Robots. MIT Press, Cambridge (2002)
2. Alač, M., Movellan, J., Tanaka, F.: When a robot is social: enacting a social robot through spatial arrangements and multimodal semiotic engagement in robotics practice. Soc. Stud. Sci. **41**(6), 126–159 (2011)
3. Garfinkel, H.: Studies in Ethomethodology. Polity Press; Englewood Cliffs (1984[1967])
4. Jefferson, G.: Glossary of transcript symbols with an introduction. In: Lerner, G.H. (ed.) Conversation Analysis: Studies from the First Generation, pp. 13–31. John Benjamins, New York (2004)
5. Sacks, H., Schegloff, E.A., Jefferson, G.: A simplest systematics for the organization of turn-taking for conversation. Language **50**, 696–735 (1974)

6. Suchman, L.: Human-Machine Reconfigurations: Plans and Situated Actions, 2nd expanded edn. Cambridge University Press, New York (2007)
7. Turkle, S.: Alone Together. Basic Books, New York (2011)
8. Goodwin, C.: Professional vision. Am. Anthropol. 96(3), 606–633 (1994)
9. Heath, C., Hindmarsh, J.: In: May, T. (ed.) Analyzing Interaction Video Ethnography and Situated Conduct. Qualitative Research in Action, pp. 99–121. Sage, London (2002)
10. Lynch, M.: Art and Artifact in Laboratory Science: A Study of Shop Work and Shop Talk in a Research Laboratory. Routledge & Kegan Paul, London (1985)
11. Movellan, J., Tanaka, F., Taylor, C., Ruvolo, P., Eckhardt, M.: The RUBI project: a progress report. In: Proceedings of the 2nd ACM/IEEE International Conference of Human-Robot Interaction, Washington, D.C., 9–11 March 2007. ACM, New York (2007)
12. Tanaka, F., Cicourel, A., Movellan, J.: Socialization between toddlers and robots at an early childhood education center. Proc. Nat. Acad. Sci. 104(46), 17954–17958 (2007)
13. Lave, J., Murtaugh, M., De La Rocha, O.: The dialectic of arithmetic in grocery shopping. In: Rogoff, B., Lave, J. (eds.) Everyday Cognition. Harvard UP, Cambridge (1984)
14. Goffman, E.: The Presentation of Self in Everyday Life. Anchor Books, New York (1959)
15. Hindmarsh, J., Heath, C.: Embodied reference: a study of deixis in workplace interaction. J. Pragmat. 32, 1855–1878 (2000)
16. Schegloff, E.A.: On some gestures relation to talk. In: Atkinson, J.M., Heritage, J. (eds.) Structures of Social Action: Studies in Conversation Analysis, pp. 266–296. Cambridge University Press, Cambridge (1984)
17. Suwa, M., Kato, F.: Pattern language and storytelling: a methodology for describing embodied experience and encouraging others to learn. In: Proceedings of AAAI Spring Symposium on Shikakeology, Stanford, March 2013
18. Turkle, S.: The Second Self: Computers and the Human Spirit. Simon and Schuster, New York (1984)
19. Barthes, R.: Mythologies, pp. 88–90. Vintage, London (1957, 1972)
20. Robertson, J.: Gendering humanoid robots: robo-sexism in Japan. Body Soc. 16(2), 1–36 (2010)
21. Castaneda, C., Suchman, L.: Robot Visions. Social Studies of Science (2013)
22. Nishizaka, A.: Hand touching hand: referential practice at a Japanese midwife house. Hum. Stud. 30(3), 199–217 (2007)
23. Goffman, E.: Frame Analysis: An Essay on the Organization of Experience. Northeastern UP, Boston (1974)
24. Goffman, E.: Interaction Ritual: Essays on Face-to-Face Behavior. Pantheon Books, New York (1967)
25. Kidwell, M., Zimmerman, D.: Joint attention in action. Pragmatics 39(3), 592–611 (2007)
26. Tomasello, M.: Constructing a Language. Harvard University Press, Cambridge (2003)
27. Pollner, M., McDonald-Wikler, L.: The social construction of unreality: a case of a family's attribution of competence to a severely retarded child. Fam. Process 24(1985), 241–254 (1985)
28. Wittgenstein, L.: Philosophical Investigations. Basil Blackwell, Oxford (1953)

AAA

Abduction in Argumentation Frameworks and Its Use in Debate Games

Chiaki Sakama(✉)

Department of Computer and Communication Sciences,
Wakayama University, Sakaedani, Wakayama 640-8510, Japan
sakama@sys.wakayama-u.ac.jp

Abstract. This paper studies an *abduction* problem in formal argumentation frameworks. Given an argument, an agent verifies whether the argument is justified or not in its argumentation framework. If the argument is not justified, the agent seeks conditions to explain the argument in its argumentation framework. We formulate such abductive reasoning in argumentation semantics and provide its computation in logic programming. Next we apply abduction in argumentation frameworks to reasoning by players in *debate games*. In debate games, two players have their own argumentation frameworks and each player builds claims to refute the opponent. A player may provide false or inaccurate arguments as a tactic to win the game. We show that abduction is used not only for seeking counter-claims but also for building dishonest claims in debate games.

1 Introduction

Arguments and *explanations* play different roles in human reasoning and have been distinguished in philosophy of science. According to [17], "the purpose of an explanation is to show *why and how* some phenomenon occurred or some event happened; the purpose of an argument is to show *that* some view or statement is correct or true." In other words, "argument is the mechanism by which we produce knowledge" and "explanation is the mechanism by which we produce understanding" [22]. On the other hand, an argument is used for knowing whether an explanation is appropriate and an explanation is used for understanding how an evidence occurs in an argument. In this sense, arguments and explanations are mutually supportive, so "arguments and explanations have a complementary relationship and reasoning is normally perceived as incomplete when one occurs in the absence of the other" [22]. In the field of artificial intelligence, argumentation and *abduction* are implicitly related in [13] where Dung provides an argumentation-theoretic semantics of abductive logic programs. The framework has been later extended to *assumption-based argumentation* [9]. Dung also introduces *formal argumentation* [14] as an abstract framework for argumentative reasoning, and the framework has been extended in various ways to incorporate explanatory reasoning [5,20,28,29].

© Springer International Publishing Switzerland 2014
Y. Nakano et al. (Eds.): JSAI-isAI 2013, LNAI 8417, pp. 285–303, 2014.
DOI: 10.1007/978-3-319-10061-6_19

This paper studies an abductive framework based on Dung's abstract argumentation. Different from previous studies, we combine an argumentation framework and *extended abduction* proposed by Inoue and Sakama [18]. In extended abduction, hypotheses can not only be added to background knowledge but also be removed from it to explain (or unexplain) an observation. In the context of argumentation, extended abduction is used for verifying whether a particular argument is justified or not, and seeking conditions to explain a particular argument in an argumentation framework. We next apply the abductive framework to reasoning by players in *debate games* [26]. A debate game provides an abstract model of dialogue between two players based on a formal argumentation framework. A unique feature of debate games is that a player may claim false or inaccurate arguments as a tactic to win the game. The proposed framework combines abduction and argumentation in a way different from existing studies, and exploits a new application of abduction in a formal dialogue system based on argumentation frameworks.

The rest of this paper is organized as follows. Section 2 reviews abstract argumentation frameworks. Section 3 introduces abduction to argumentation frameworks, and Sect. 4 applies the framework to debate games. Section 5 discusses related issues and Sect. 6 concludes the paper.

2   Argumentation Framework

Definition 2.1 (argumentation framework). [10,14] Let U be the universe of all possible arguments. An *argumentation framework* (AF) is a pair (Ar, att) where Ar is a finite subset of U and $att \subseteq Ar \times Ar$. An argument A *attacks* an argument B iff $(A, B) \in att$. A set $S \subseteq Ar$ is *conflict-free* if there is no $A, B \in S$ such that $(A, B) \in att$. A set $S \subseteq Ar$ is *admissible* iff it is conflict-free and for any $A \in S$ such that $(B, A) \in att$ for some $B \in Ar$, there is $C \in S$ such that $(C, B) \in att$.

An argumentation framework (Ar, att) is associated with a directed graph (called an *argumentation graph*) in which vertices are arguments in Ar and directed arcs from A to B exist whenever $(A, B) \in att$. An argumentation framework is identified with its argumentation graph.

Definition 2.2 (labelling). [10] Let $AF = (Ar, att)$ be an argumentation framework. A *labelling* of AF is a (total) function $\mathcal{L} : Ar \to \{\texttt{in}, \texttt{out}, \texttt{undec}\}$.

When $\mathcal{L}(A) = \texttt{in}$ (resp. $\mathcal{L}(A) = \texttt{out}$ or $\mathcal{L}(A) = \texttt{undec}$) for $A \in Ar$, it is written as $\texttt{in}(A)$ (resp. $\texttt{out}(A)$ or $\texttt{undec}(A)$). In this case, the argument A is *accepted* (resp. *rejected* or *undecided*). We call $\texttt{in}(A)$, $\texttt{out}(A)$ and $\texttt{undec}(A)$ *labelled arguments*.

Definition 2.3 (complete labelling). [10] Let $AF = (Ar, att)$ be an argumentation framework. A labelling $\mathcal{L}$ of AF is a *complete labelling* if for each argument $A \in Ar$, it holds that:

- $\mathcal{L}(A) = \text{in}$ iff $\mathcal{L}(B) = \text{out}$ for every $B \in Ar$ such that $(B, A) \in att$.
- $\mathcal{L}(A) = \text{out}$ iff $\mathcal{L}(B) = \text{in}$ for some $B \in Ar$ such that $(B, A) \in att$.
- $\mathcal{L}(A) = \text{undec}$ iff $\mathcal{L}(A) \neq \text{in}$ and $\mathcal{L}(A) \neq \text{out}$.

Let $\text{in}(\mathcal{L}) = \{A \mid \mathcal{L}(A) = \text{in}\}$, $\text{out}(\mathcal{L}) = \{A \mid \mathcal{L}(A) = \text{out}\}$ and $\text{undec}(\mathcal{L}) = \{A \mid \mathcal{L}(A) = \text{undec}\}$.

Definition 2.4 (stable, semi-stable, grounded, preferred labelling). [10] Let AF be an argumentation framework and $\mathcal{L}$ a complete labelling of AF. Then, (1) $\mathcal{L}$ is a *stable labelling* iff $\text{undec}(\mathcal{L}) = \emptyset$. (2) $\mathcal{L}$ is a *semi-stable labelling* iff $\text{undec}(\mathcal{L})$ is minimal wrt set inclusion among all complete labellings of AF. (3) $\mathcal{L}$ is a *grounded labelling* iff $\text{in}(\mathcal{L})$ is minimal wrt set inclusion among all complete labellings of AF. (4) $\mathcal{L}$ is a *preferred labelling* iff $\text{in}(\mathcal{L})$ is maximal wrt set inclusion among all complete labellings of AF.

There is a one-to-one correspondence between the set $\text{in}(\mathcal{L})$ with a complete (resp. stable, semi-stable, grounded, preferred) labelling $\mathcal{L}$ of an argumentation framework AF and a *complete* (resp. *stable, semi-stable, grounded, preferred*) *extension* of AF [10,14]. In this paper, the distinction between different labellings is often unimportant and *S-labelling* means one of the five labellings introduced above.

Definition 2.5 (justify). [2] Let AF be an argumentation framework. Then, a labelled argument L is *skeptically* (resp. *credulously*) *justified* by AF under the S-labelling if L is included in every (resp. some) S-labelling $\mathcal{L}$ of AF.

3 Abduction in Argumentation Framework

3.1 Explanations

Suppose the following dialogue between Alice and Bob:

Alice: "I think Mary can speak Japanese because she has stayed in Japan."
Bob: "I don't think so because her staying in Japan was too short to learn Japanese."

The situation is represented by the argumentation framework $AF = (\{A, B\}, \{(B, A)\})$ where A represents the argument "Mary speaks Japanese" by Alice and B represents the argument "Mary does not speak Japanese" by Bob. The AF has the complete labelling $\{\text{out}(A), \text{in}(B)\}$ which means that the argument A is rejected and the argument B is accepted. In another day, Bob observes that Mary speaks Japanese. To explain this, he assumes an argument C that Mary studied Japanese hard to be able to speak it well. The revised argumentation becomes $AF' = (\{A, B, C\}, \{(C, B), (B, A)\})$ and is represented by the argumentation graph below.

$$A \qquad B \qquad C$$

After introducing the new argument C, the situation changes: the revised AF' has the complete labelling $\{\text{in}(A), \text{out}(B), \text{in}(C)\}$, where A and C are now accepted and B is rejected. It illustrates the situation in which a new argument is introduced to explain a new observation. Suppose another dialogue such that

Alice: "I think the new iPhone will be selling well."
Bob: "I don't think so because few people will get interested in this new model."

The situation is represented by $AF = (\{A, B\}, \{(B, A)\})$ where A is rejected and B is accepted. Later it is observed that the new iPhone breaks the sales record. Bob then withdraws his argument B and the revised AF becomes $AF' = (\{A\}, \emptyset)$. Then, the argument A is now accepted in AF'. It illustrates the situation in which a previously believed argument is removed in face of a new observation.

To realize such explanatory reasoning in argumentation frameworks, it is necessary to introduce assumptions to an argumentation framework. In Definition 2.1, the set Ar of arguments is a subset of the universe U of all possible arguments. We then consider the notion of the universal argumentation framework which consists of the set of all possible arguments and attack relations over them.

Definition 3.1 (universal AF). The *universal argumentation framework* (UAF) is an argumentation framework (U, att_U) in which U is the set of all possible arguments and $att_U \subseteq U \times U$ is the set of fixed attack relations over U.

The UAF specifies a world which consists of arguments and attack relations over them. An *agent* has (partial) knowledge about the world as an argumentation framework $AF = (Ar, att)$ where $Ar \subseteq U$ is *finite* and $att = att_U \cap (Ar \times Ar)$. In this sense, AF is often called a *subargumentation framework* (*sub-AF* for short) of the UAF. The agent has a belief on the labelling of every argument in Ar based on the attack relations in att under the designated semantics S. On the other hand, an agent can recognize the possibility of arguments in $U \backslash Ar$, but does not know whether those arguments are valid or not. The agent has no information on labelling of any argument in $U \backslash Ar$ and each argument in $U \backslash Ar$ is called a *hypothesis*. In what follows, an agent is identified with its AF.

Definition 3.2 (observation). Let $UAF = (U, att_U)$ and $AF = (Ar, att)$ a sub-AF. An *observation* O by AF is either $\text{in}(A)$ or $\text{out}(A)$ for some $A \in U$ such that $(A, A) \notin att_U$. When $O = \text{in}(A)$ or $O = \text{out}(A)$, define $arg(O) = A$.

When $O = \text{in}(A)$ is observed, it means that there is an evidence for A. When $O = \text{out}(A)$ is observed, on the other hand, it means that there is an evidence against A. In each case, an agent tries to skeptically or credulously justify O in his/her AF under a designated labelling. We consider that any meaningful observation contains no self-attacking argument, which is represented by the condition $(A, A) \notin att_U$.[1] If an agent fails to justify O in his/her AF, it implies

---

[1] A reviewer comments that "an argument A attacking itself is a very natural explanation for the observation that there is evidence against A, i.e. that A is out." However,

that AF believed by the agent is inaccurate or incomplete. In this case, the agent performs *abduction* to explain O.

Definition 3.3 (explanation). Let $UAF = (U, att_U)$ and $AF = (Ar, att)$ a sub-AF. An observation O (by AF) is *skeptically* (resp. *credulously*) *explained* by $E = (I, J)$ under the S-labelling of AF_E if O is included in every (resp. some) S-labelling $\mathcal{L}_E$ of the argumentation framework $AF_E = (Ar_E, att_E)$ where $Ar_E = (Ar \backslash J) \cup I$, $I \subseteq U \backslash Ar$, $J \subseteq Ar$, and $att_E = att_U \cap (Ar_E \times Ar_E)$. In this case, E is called a *skeptical* (resp. *credulous*) *explanation* of O (under the S-labelling of AF_E), and we say that O has a skeptical (resp. credulous) explanation E in AF.

An explanation (I, J) of an observation O is *minimal* if $I' \subseteq I$ and $J' \subseteq J$ imply $I' = I$ and $J' = J$ for any explanation (I', J') of O. An explanation (I, J) is *empty* if $I = J = \emptyset$; otherwise, (I, J) is *non-empty*.

If E is a skeptical explanation of an observation O, then E is also a credulous explanation of O, but not vice versa. The notions of skeptical and credulous explanations coincide when AF_E has the unique S-labelling. A skeptical/credulous explanation is simply called an *explanation* if the distinction between the two is unimportant in the context. In Definition 3.3, if $O = \text{in}(A)$ (resp. $O = \text{out}(A)$) for some argument A, the goal of abduction is to produce a labelling of AF in which A is labelled **in** (resp. **out**). To this end, arguments in J are removed from Ar and hypotheses in I are introduced to Ar to explain O. Removal of J means that an agent does not believe arguments in J anymore, or an agent has some reason to withdraw J. Introduction of I means that an agent learns new arguments in I. When O is observed by AF, it is either $arg(O) \in Ar$ or $arg(O) \in U \backslash Ar$. In case of $arg(O) \in Ar$, the argument $arg(O)$ is known by AF but its labelling in O may be different from the labelling of the argument in AF. In case of $arg(O) \in U \backslash Ar$, on the other hand, the argument $arg(O)$ is a hypothesis for AF and AF has no labelling of the argument.

Example 3.1. Let $UAF = (\{A, B, C, D, F\},$ $\{(B, A), (B, C), (C, B), (D, C), (C, F)\})$ and $AF = (\{A, B, C\}, \{(B, A), (B, C), (C, B)\})$ where AF has three complete labellings: $\mathcal{L}_1 = \{\text{in}(A), \text{out}(B), \text{in}(C)\}$, $\mathcal{L}_2 = \{\text{out}(A), \text{in}(B), \text{out}(C)\}$, and $\mathcal{L}_3 = \{\text{undec}(A), \text{undec}(B), \text{undec}(C)\}$.

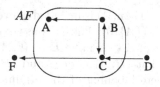

Then the following facts hold.

– Two observations $O_1 = \text{in}(A)$ and $O_2 = \text{out}(A)$ have the single minimal credulous explanation $E_0 = (\emptyset, \emptyset)$ under the complete labelling of $AF_{E_0} = AF$.

A's attacking itself does not explain that "A is out" but explains that "A is *not* in." In fact, A is labelled **undec** in $AF = (\{A\}, \{(A, A)\})$ under the complete, semi-stable, grounded and preferred semantics. We exclude such "undecided" observations. (AF has no stable labelling.)

– $O_2 = \text{out}(A)$ has two minimal skeptical explanations $E_1 = (\emptyset, \{C\})$ under the complete labelling of $AF_{E_1} = (\{A, B\}, \{(B, A)\})$, and $E_2 = (\{D\}, \emptyset)$ under the complete labelling of $AF_{E_2} = (\{A, B, C, D\}, \{(B, A), (B, C), (C, B), (D, C)\})$.

– $O_3 = \text{in}(F)$ has two minimal skeptical explanations: $E_3 = (\{F\}, \{C\})$ under the complete labelling of $AF_{E_3} = (\{A, B, F\}, \{(B, A)\})$, and $E_4 = (\{D, F\}, \emptyset)$ under the complete labelling of $AF_{E_4} = UAF$.

– $O_4 = \text{out}(D)$ has no credulous/skeptical explanation.

In Example 3.1, the observation $O_1 = \text{in}(A)$ has the credulous empty explanation in AF. This means that the labelled argument $\text{in}(A)$ is credulously justified in the argumentation framework AF under the complete labelling. On the other hand, $O_2 = \text{out}(A)$ has two minimal skeptical explanations in AF and both of them are non-empty explanations. This means that the labelled argument $\text{out}(A)$ is not skeptically justified in AF under the complete labelling. To skeptically justify $\text{out}(A)$, it is necessary to remove the argument C from Ar or to introduce the hypothesis D to Ar in AF.

By Definitions 2.5 and 3.3, an observation O is skeptically (resp. credulously) justified by AF_E under the S-labelling iff O has a skeptical (resp. credulous) explanation E under the S-labelling of AF_E. In particular, O is skeptically (resp. credulously) justified by AF under the S-labelling iff O has the skeptical (resp. credulous) empty explanation under the S-labelling of AF. An observation may have none, one, or multiple explanations in general. In particular, the next proposition holds.

Proposition 3.1. *Let $UAF = (U, att_U)$ and AF a sub-AF. For any $A \in U$,*

1. *an observation $O = \text{in}(A)$ has a skeptical/credulous explanation in AF.*
2. *an observation $O = \text{out}(A)$ has a credulous explanation in AF under the complete, (semi-)stable, preferred labelling iff there is an argument $B \in U$ such that $(B, A) \in att_U$ and $(B, B) \notin att_U$. Moreover, O has a skeptical explanation in AF under S-labelling iff the additional condition $(A, B) \notin att_U$ is satisfied.*

Proof. (1) $\text{in}(A)$ is included in every S-labelling of the argumentation framework $AF_E = (\{A\}, \emptyset)$. Thus, O has the skeptical/credulous explanation $E = (\{A\}, Ar)$ in case of $A \notin Ar$; and $E' = (\emptyset, Ar \setminus \{A\})$ in case of $A \in Ar$. (2) If there is $B \in U$ such that $(B, A) \in att_U$ and $(B, B) \notin att_U$, then $\text{out}(A)$ is included in some complete labelling of the argumentation framework $AF_E = (\{A, B\}, \{(B, A)\})$ in case of $(A, B) \notin att_U$; or $AF_E = (\{A, B\}, \{(B, A), (A, B)\})$ in case of $(A, B) \in att_U$. Thus, O has a credulous explanation $E = (\{A, B\}, Ar)$ under the complete, (semi-)stable, preferred labelling. In case of $(A, B) \notin att_U$, E is also a skeptical explanation under S-labelling. The only-if part follows by definition. □

When an observation O does not have the empty skeptical/credulous explanation, O is not skeptically/credulously justified by AF under the S-labelling. In this case, a non-empty explanation E is likely to change not only the labelling of the argument $arg(O)$ but the labellings of arguments other than $arg(O)$ in AF_E. In Example 3.1, for instance, the complete labelling $\mathcal{L}_1 = \{\text{in}(A), \text{out}(B), \text{in}(C)\}$

of AF changes into $\{\texttt{out}(A), \texttt{in}(B)\}$ of AF_{E_1}. Thus, the explanation E_1 changes not only the labelling of A but also the labellings of B and C. The change of labellings between two argumentation frameworks is defined as follows.

Definition 3.4 (minimal change). Let $AF = (Ar, att)$ be an argumentation framework and $\mathcal{L}$ any S-labelling of it. For any S-labelling $\mathcal{L}_E$ of AF_E, define

$$\Delta(\mathcal{L}, \mathcal{L}_E) = \{A \mid \mathcal{L}(A) \neq \mathcal{L}_E(A) \text{ for } A \in Ar\} \cup \{A \mid A \in (Ar \backslash Ar_E) \cup (Ar_E \backslash Ar)\}.$$

A skeptical (resp. credulous) explanation E of an observation O *minimally changes* AF if for any skeptical (resp. credulous) explanation F of O in AF, the following condition is satisfied: for any S-labelling $\mathcal{L}_F$ of AF_F which includes O, there is an S-labelling $\mathcal{L}_E$ of AF_E which includes O such that $\Delta(\mathcal{L}, \mathcal{L}_F) \subseteq \Delta(\mathcal{L}, \mathcal{L}_E)$ implies $\Delta(\mathcal{L}, \mathcal{L}_E) \subseteq \Delta(\mathcal{L}, \mathcal{L}_F)$ for some S-labelling $\mathcal{L}$ of AF.

If O has the empty explanation E in AF, then E minimally changes AF.

Example 3.2. In Example 3.1, the skeptical explanation $E_1 = (\emptyset, \{C\})$ of O_2 produces the complete labelling $\mathcal{L}_{E_1} = \{\texttt{out}(A), \texttt{in}(B)\}$, and the skeptical explanation $E_2 = (\{D\}, \emptyset)$ of O_2 produces the complete labelling $\mathcal{L}_{E_2} = \{\texttt{out}(A), \texttt{in}(B), \texttt{out}(C), \texttt{in}(D)\}$. Then, $\Delta(\mathcal{L}_1, \mathcal{L}_{E_1}) = \{A, B, C\}$ and $\Delta(\mathcal{L}_1, \mathcal{L}_{E_2}) = \{A, B, C, D\}$, so that E_1 minimally changes AF.

When an observation has more than one explanations, explanations that minimally change the labellings of arguments in AF are preferred.

Definition 3.5 (preferred explanation). Given an argumentation framework AF and an observation O, an explanation E is a *preferred explanation* of O if E minimally changes AF. A preferred explanation E is *most preferred* if it is also minimal (in the sense of Definition 3.3) among all of the preferred explanations of O.

Definition 3.5 says that there are two conditions for selecting the best explanations. The first condition requests that such explanations minimally change the labellings of the original AF. The second condition requests that the minimality of explanations. The first condition precedes the second one, that is, non-minimal preferred explanations are considered better than minimal non-preferred explanations. In particular, the empty explanation is always most preferred. By definition, we have the next result.

Proposition 3.2. *If an observation O has an explanation in an argumentation framework AF, then there is a most preferred explanation of O in AF.*

3.2   Computation

Next we provide a method of computing abduction in AF using *logic programming*. A *normal logic program* (or simply a *program*) is a set of rules of the form

$$A \leftarrow B_1, \ldots, B_m, \, not \, B_{m+1}, \ldots, not \, B_n$$

where A and B_i's are ground atoms ($n \geq m \geq 0$), and *not* represents the *negation as failure* operator. Let $\mathcal{B}_P$ be the Herbrand base of a program P. Then, a *3-valued interpretation* of a program P is defined as a pair $I = \langle T, F \rangle$ where T contains all ground atoms *true* in I, F contains all ground atoms *false* in I, and the remaining atoms in $W = \mathcal{B}_P \backslash (T \cup F)$ are *unknown*. Let $I(A) = 1$ (resp. $I(A) = \frac{1}{2}$, $I(A) = 0$) if $A \in T$ (resp. $A \in W$, $A \in F$), and $I(not\,A) = 1 - I(A)$. Then, a 3-valued interpretation I is a *model* of a program P if $I(A) \geq min\{I(L_i) \mid 1 \leq i \leq n\}$ holds for every rule $A \leftarrow L_1, \ldots, L_n$ in P where L_i is either B_i or $not\,B_i$. Among models of a program, the following models are important: *partial stable models, stable models, L-stable models, regular models,* and *well-founded models.*[2]

An argumentation framework $AF = (Ar, att)$ is transformed into the logic program P_{AF} by identifying each argument with a ground atom as follows [30]: $P_{AF} = \{A \leftarrow not\,B_1, \ldots, not\,B_n \mid A, B_1, \ldots, B_n \in Ar\,(n \geq 0) \text{ and } (B_i, A) \in att\,(1 \leq i \leq n)\}$. Then, there is a one-to-one correspondence between complete (resp. stable, semi-stable, grounded, preferred) labellings of AF and partial stable (resp. stable, L-stable, well-founded, regular) models of P_{AF} [11,30]. We modify the transformation to characterize abduction in argumentation frameworks.

Definition 3.6 (transformation). Given $UAF = (U, att_U)$, the associated logic program P_{UAF} is defined as follows.

$$P_{UAF} = \{A \leftarrow not\,B_1, \ldots, not\,B_n, N_A \mid A, B_1, \ldots, B_n \in U\,(n \geq 0) \text{ and }$$
$$(B_i, A) \in att_U\,(1 \leq i \leq n)\} \cup \{N_A \leftarrow not\,N'_A, \quad N'_A \leftarrow not\,N_A \mid A \in U\}$$

where N_A and N'_A are new ground atoms uniquely associated with each atom A.

Each atom N_A or N'_A has one of the truth values *true, false* or *unknown*. If N_A is *true* (resp. *false*) in a partial stable model M of P_{UAF}, N'_A is *false* (resp. *true*) in M. Otherwise, both N_A and N'_A are *unknown* in M. If N_A is *true*, the rule $A \leftarrow not\,B_1, \ldots, not\,B_n, N_A$ is identified with $A \leftarrow not\,B_1, \ldots, not\,B_n$. In other words, by switching the truth values of N_A and N'_A, we can simulate introduction/removal of arguments $A, B_1, \ldots, B_n$ and attack relations (B_i, A) to/from a sub-AF of the UAF. For convenience, define $choice(U) = \{N_A \leftarrow not\,N'_A, \quad N'_A \leftarrow not\,N_A \mid A \in U\}$.

Example 3.3. Consider $UAF = (\{A, B, C\}, \{(C, B), (B, A)\})$ and $AF = (\{A, B\}, \{(B, A)\})$. Then, $P_{UAF} = \{A \leftarrow not\,B, N_A, \quad B \leftarrow not\,C, N_B, \quad C \leftarrow N_C\} \cup choice(\{A, B, C\})$ where the partial stable model $\langle \{B, N_A, N_B, N'_C\}, \{A, C, N'_A, N'_B, N_C\} \rangle$ corresponds to the complete labelling $\{out(A), in(B)\}$ of AF. On the other hand, the partial stable model $\langle \{A, C, N_A, N_B, N_C\}, \{B, N'_A, N'_B, N'_C\} \rangle$ corresponds to the complete labelling $\{in(A), out(B), in(C)\}$ of $AF_E = UAF$ with $E = (\{C\}, \emptyset)$, and the partial stable model $\langle \{A, N_A, N'_B, N'_C\}, \{B, C, N'_A,$

---

[2] We refer the readers to the references in [11] for the precise definition of each semantics.

$N_B, N_C\}\rangle$ corresponds to the complete labelling $\{\mathtt{in}(A)\}$ of $AF_{E'} = (\{A\}, \emptyset)$ with $E' = (\emptyset, \{B\})$.

Lemma 3.3. [30] *Let* $AF = (Ar, att)$ *and* P_{AF} *its transformed logic program. If* AF *has a complete labelling* $\mathcal{L}$*, then* $\langle T, F\rangle = \langle \mathtt{in}(\mathcal{L}), \mathtt{out}(\mathcal{L})\rangle$ *where* $\mathcal{B}_{P_{AF}} \setminus (T \cup F) = \mathtt{undec}(\mathcal{L})$ *is a partial stable model of* P_{AF}*. Conversely, if* $\langle T, F\rangle$ *is a partial stable model of* P_{AF}*, then a labelling* $\mathcal{L}$ *such that* $\mathtt{in}(\mathcal{L}) = T$*,* $\mathtt{out}(\mathcal{L}) = F$ *and* $\mathtt{undec}(\mathcal{L}) = \mathcal{B}_{P_{AF}} \setminus (T \cup F)$ *is a complete labelling of* AF*.*

For a set S of atoms, let $\mathcal{N}_S = \{N_A \mid A \in S\}$; in particular, $\mathcal{N}_S = \emptyset$ if $S = \emptyset$.

Theorem 3.4. *Let* $UAF = (U, att_U)$ *and* $AF = (Ar, att)$ *a sub-AF. Also let* $IN = \{N_A \mid A \in U \setminus Ar\}$ *and* $OUT = \{N_A \mid A \in Ar\}$*. Then, an observation* $O = \mathtt{in}(A)$ *(resp.* $O = \mathtt{out}(A)$*) has a credulous explanation* $E = (I, J)$ *under a complete (or stable, semi-stable, grounded, preferred) labelling of* AF_E *iff* P_{UAF} *has a partial stable (or stable, L-stable, well-founded, regular) model* $\langle T, F\rangle$ *such that* $A \in T$ *(resp.* $A \in F$*),* $\mathcal{N}_I = T \cap IN$ *and* $\mathcal{N}_J = F \cap OUT$*. In particular,* E *is also a skeptical explanation of* O *iff* $A \in T$ *(resp.* $A \in F$*) for any* $\langle T, F\rangle$ *such that* $\mathcal{N}_I = T \cap IN$ *and* $\mathcal{N}_J = F \cap OUT$*.*

Proof. We show the result for complete labelling. If $O = \mathtt{in}(A)$ has a credulous explanation $E = (I, J)$ under a complete labelling of AF_E, then O is included in some complete labelling $\mathcal{L}_E$ of $AF_E = (Ar_E, att_E)$ where $Ar_E = (Ar \setminus J) \cup I$ with $I \subseteq U \setminus Ar$ and $J \subseteq Ar$. By Lemma 3.3, $\langle T, F\rangle$ with $T = \mathtt{in}(\mathcal{L}_E) \cup \{N_B \mid B \in Ar_E\} \cup \{N'_C \mid C \in U \setminus Ar_E\}$ and $F = \mathtt{out}(\mathcal{L}_E) \cup \{N'_B \mid B \in Ar_E\} \cup \{N_C \mid C \in U \setminus Ar_E\}$ becomes a partial stable model of P_{UAF}, and $A \in \mathtt{in}(\mathcal{L}_E)$. In this case, $\mathcal{N}_I = T \cap IN$ and $\mathcal{N}_J = F \cap OUT$ hold. In particular, if O is included in every complete labelling $\mathcal{L}_E$ of $AF_E = (Ar_E, att_E)$ with $E = (I, J)$, then $A \in T$ for any $\langle T, F\rangle$ such that $\mathcal{N}_I = T \cap IN$ and $\mathcal{N}_J = F \cap OUT$. The converse also holds by the fact that a partial stable model $\langle T, F\rangle$ of P_{UAF} is translated into a complete labelling $\mathtt{in}(\mathcal{L}_E) = \{A \mid A \in T \text{ and } N_A \in T\}$ and $\mathtt{out}(\mathcal{L}_E) = \{B \mid B \in F \text{ and } N_B \in T\}$ of AF_E. The results hold for (semi-) stable, grounded and preferred labelling using their equivalence to respective logic programming semantics [11]. The result of $O = \mathtt{out}(A)$ is shown in a similar way. □

Finally, we remark some complexity results on abduction in AF. By Proposition 3.1, an observation $O = \mathtt{in}(A)$ always has a skeptical/credulous explanation, and $O = \mathtt{out}(A)$ has a skeptical/credulous explanation if A is attacked by some argument B which satisfies simple conditions. Thus, deciding the existence of an explanation given an observation is trivial or done in polynomial time. On the other hand, given a pair of arguments $E = (I, J)$, the problem of deciding whether E is a credulous (or skeptical) explanation of an observation O under $\mathcal{S}$-labelling has different complexities under different semantics. In case of $O = \mathtt{in}(A)$, E is a credulous (resp. skeptical) explanation of O under $\mathcal{S}$-labelling of AF_E iff A is included in some (resp. every) $\mathcal{S}$-extension of AF_E. In case of $O = \mathtt{out}(A)$, put $UAF' = (U \cup \{X\}, att_U \cup \{(A, X)\})$ where X is a new argument such that $X \notin U$. For $AF = (Ar, att)$, put $AF' = (Ar \cup \{A, X\}, att \cup \{(A, X)\})$. Then, for any $A \in U$, E is a credulous (resp. skeptical) explanation of $O = \text{out}(A)$ under S-labelling of AF_E iff E is a credulous (resp. skeptical) explanation of $O' = \text{in}(X)$ under S-labelling of AF'_E. The next results hold by the complexity results in [16].

Theorem 3.5. *Let* $UAF = (U, att_U)$ *and* $AF = (Ar, att)$ *a sub-AF. Given* $E = (I, J)$, *deciding whether* E *is a credulous (resp. skeptical) explanation of an observation* O *under* S-*labelling of* AF_E *is NP-complete (resp. polynomial) for complete labelling, NP-complete (resp. coNP-complete) for stable labelling, NP-complete (resp.* Π_2^P-*complete) for preferred labelling, and* Σ_2^P-*complete (resp.* Π_2^P-*complete) for semi-stable labelling. In case of grounded labelling, it is decided in polynomial time.*

4 Debate Games

Suppose a debate between a prosecutor (P) and a defense (D) in court.

P_1: The suspect is guilty because he had a grudge against the murder victim.
D_1: There is no evidence that the suspect killed the victim. No one is guilty until proven guilty.
P_2: There is an eyewitness who saw the suspect leaving the victim's apartment on the night of the crime.
D_2: The testimony is incredible because it was dark at night.

Given the argument P_1 by a prosecutor, the defense seeks an argument against P_1. Once the defense successfully refutes P_1 by the argument D_1, the prosecutor tries to refute D_1. A debate continues until one cannot refute the other. An appropriate modelling of debate should allow for the following three properties: (i) players have different beliefs and opinions in general; (ii) during a debate, each player may revise its own beliefs by new information provided by the opponent; (iii) a player may use inaccurate or even false arguments to win a debate [27].

Sakama [26] introduced a *debate game* based on an argumentation framework, which provides an abstract model of debates between two players and satisfies all three of the above requirements. We first review definitions of debate games. A *player* is an agent who has its own AF as a sub-AF of the given UAF.

Definition 4.1 (claim). [26] A *claim* is a pair of the form: $(\text{in}(A), \_)$ or $(\text{out}(B), \text{in}(A))$ where A and B are different arguments. $(\text{in}(A), \_)$ is read "A is labelled in", while $(\text{out}(B), \text{in}(A))$ is read "B is labelled out because A is labelled in". A claim $(\text{in}(A), \_)$ or $(\text{out}(B), \text{in}(A))$ by a player is *refuted* by the claim $(\text{out}(A), \text{in}(C))$ with some argument C by another player.

Definition 4.2 (revision). [26] Let $UAF = (U, att_U)$ and $AF = (Ar, att)$ a sub-AF of the UAF. Then, a *revision* of AF with an argument $X \in U$ is defined as

$$AF \circ X = \begin{cases} (Ar \cup \{X\}, att \cup att_X) & \text{if } X \notin Ar \\ AF & \text{otherwise} \end{cases}$$

where $att_X = \{(X, Y), (Z, X) \mid Y, Z \in Ar \text{ and } (X, Y), (Z, X) \in att_U \setminus att\}$.

Definition 4.3 (debate game). [26] Let $UAF = (U, att_U)$, and $AF_1 = (Ar_1, att_1)$ and $AF_2 = (Ar_2, att_2)$ argumentation frameworks of two players P_1 and P_2, respectively. Then, an *admissible debate* is a sequence of claims $[(\text{in}(X_0), \_),$ $(\text{out}(X_0), \text{in}(Y_1)), (\text{out}(Y_1), \text{in}(X_1)), \ldots, (\text{out}(X_i), \text{in}(Y_{i+1})), (\text{out}(Y_{i+1}), \text{in}(X_{i+1})), \ldots]$ such that

- $X_0 \in Ar_1$ and $X_k \in Ar_1^k$ where $AF_1^k = (Ar_1^k, att_1^k) = AF_1^{k-1} \circ Y_k$ ($k \geq 1$) and $AF_1^0 = AF_1$.
- $Y_k \in Ar_2^k$ where $AF_2^k = (Ar_2^k, att_2^k) = AF_2^{k-1} \circ X_{k-1}$ ($k \geq 1$) and $AF_2^0 = AF_2$.
- for each $\text{out}(Z_j)$ in a claim by P_1 (resp. P_2), there is $\text{in}(Z_i)$ ($i \leq j$) in a claim by P_2 (resp. P_1) such that $Z_j = Z_i$.
- $(V_j, U_i) \in att_U$ for each $(\text{out}(U_i), \text{in}(V_j))$.

For a player P_1 (resp. P_2), the player P_2 (resp. P_1) is called the *opponent*.

Let Γ_n ($n \geq 0$) be any claim. A *debate game* Δ (*for an argument* X_0) is an admissible debate between two players $[\Gamma_0, \Gamma_1, \ldots]$ where the initial claim is $\Gamma_0 = (\text{in}(X_0), \_)$. A debate game Δ for an argument X_0 *terminates* with Γ_n if $\Delta = [\Gamma_0, \Gamma_1, \ldots, \Gamma_n]$ is an admissible debate and there is no claim Γ_{n+1} such that $[\Gamma_0, \Gamma_1, \ldots, \Gamma_n, \Gamma_{n+1}]$ is an admissible debate. In this case, the player P_i who makes the claim Γ_n *wins* the game.

The player P_1 starts a debate with the claim $\Gamma_0 = (\text{in}(X_0), \_)$ based on its argumentation framework AF_1. The player P_2 then revises its argumentation framework AF_2 by X_0, and responds to the player P_1 with a counter-claim $\Gamma_1 = (\text{out}(X_0), \text{in}(Y_1))$ based on the revised argumentation framework AF_2^1. A debate continues by iterating revisions and claims. A debate game Δ terminates if each player does not repeat the same claim in the game ($\Gamma_i \neq \Gamma_{i+2k}$ ($k = 1, 2, \ldots$) for any Γ_i ($i \geq 1$) in Δ). AF_i^k means an AF of a player P_i after k-th revision. We often omit k of AF_i^k and just call an argumentation framework AF_i of a player P_i when no confusion arises.

Example 4.1. Let $UAF = (\{A, B, C, D\}, \{(D, C),$ $(C, B), (B, A)\})$, $AF_1 = (\{A, B, C\}, \{(C, B),$ $(B, A)\})$ and $AF_2 = (\{A, B, D\}, \{(B, A)\})$. AF_1 and AF_2 have the complete labellings: $\{\text{in}(A), \text{out}(B), \text{in}(C)\}$ and $\{\text{out}(A), \text{in}(B),$ $\text{in}(D)\}$, respectively. The argumentation graph of two players is on the right.

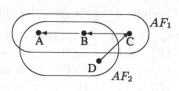

A debate game for the argument A between two players proceeds as follows:

AF_1: $(\text{in}(A), \_)$ "I claim that A is in."
AF_2^1: $(\text{out}(A), \text{in}(B))$ "A is out because B is in."
AF_1^1: $(\text{out}(B), \text{in}(C))$ "B is out because C is in."
AF_2^2: $(\text{out}(C), \text{in}(D))$ "C is out because D is in."

Here, "AF_i^k: $(\text{out}(X), \text{in}(Y))$" means that a player P_i makes a claim $(\text{out}(X),$ $\text{in}(Y))$ based on the argumentation framework AF_i^k. At first, the player P_1 has no information on the argument D, while the player P_2 has no information on the argument C. During the debate, the player P_2 learns the argument C by AF_1^1, then introduces it to AF_2^2 together with the attack relations (C, B) and (D, C). The player P_1 learns the argument D by AF_2^2 but cannot refute it. As a result, the player P_2 wins the game.

During a game, a player may make false or inaccurate claims to win the game.

Example 4.2. (1) Let $UAF = (\{A, B, C, D,$ $E, F\}, \{(F, E), (E, D), (D, C), (C, B),$ $(B, A)\})$, $AF_1 = (\{A, B, C, E, F\}, \{(F, E),$ $(C, B), (B, A)\})$ and $AF_2 = (\{A, B, D\},$ $\{(B, A)\})$. AF_1 and AF_2 have the complete labellings: $\{\texttt{in}(A), \texttt{out}(B), \texttt{in}(C), \texttt{out}(E),$ $\texttt{in}(F)\}$ and $\{\texttt{out}(A), \texttt{in}(B), \texttt{in}(D)\}$, respectively.

Consider a debate game for the argument A between two players as follows:

$AF_1: (\texttt{in}(A), \_)$ "I claim that A is $\texttt{in}$."
$AF_2^1: (\texttt{out}(A), \texttt{in}(B))$ "A is $\texttt{out}$ because B is $\texttt{in}$."
$AF_1^1: (\texttt{out}(B), \texttt{in}(C))$ "B is $\texttt{out}$ because C is $\texttt{in}$."
$AF_2^2: (\texttt{out}(C), \texttt{in}(D))$ "C is $\texttt{out}$ because D is $\texttt{in}$."
$AF_1^2: (\texttt{out}(D), \texttt{in}(E))$ "D is $\texttt{out}$ because E is $\texttt{in}$."

The player P_2 cannot refute AF_1^2, then the player P_1 wins the game. In AF_1^2, however, P_1 provides a *false* claim on E because E is $\texttt{out}$ in his/her labelling.

(2) Let $UAF = (\{A, B, C, D, G\}, \{(G, D),$ $(D, C), (C, B), (B, A)\})$, $AF_1 = (\{A, B, C\},$ $\{(C, B), (B, A)\})$ and $AF_2 = (\{A, B, D\},$ $\{(B, A)\})$. AF_1 and AF_2 have the complete labellings: $\{\texttt{in}(A), \texttt{out}(B), \texttt{in}(C)\}$ and $\{\texttt{out}(A), \texttt{in}(B), \texttt{in}(D)\}$, respectively.

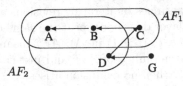

Consider a debate game for the argument A between two players as follows:

$AF_1: (\texttt{in}(A), \_)$ "I claim that A is $\texttt{in}$."
$AF_2^1: (\texttt{out}(A), \texttt{in}(B))$ "A is $\texttt{out}$ because B is $\texttt{in}$."
$AF_1^1: (\texttt{out}(B), \texttt{in}(C))$ "B is $\texttt{out}$ because C is $\texttt{in}$."
$AF_2^2: (\texttt{out}(C), \texttt{in}(D))$ "C is $\texttt{out}$ because D is $\texttt{in}$."
$AF_1^2: (\texttt{out}(D), \texttt{in}(G))$ "D is $\texttt{out}$ because G is $\texttt{in}$."

The player P_2 cannot refute AF_1^2, then the player P_1 wins the game. In AF_1^2, however, P_1 provides an *inaccurate* claim on G because G is not included in his/her labelling. In this sense, P_1 breaks the rule of admissibility of claims but P_2 cannot know it.

Definition 4.4 (honest/dishonest claim). [26] Let $UAF = (U, att_U)$ and $AF_i = (Ar, att)$ an argumentation framework of a player P_i in a debate game. Then,

- a claim $(\mathtt{in}(A), \_)$ or $(\mathtt{out}(B), \mathtt{in}(A))$ is *honest* wrt AF_i if $A \in Ar$ and $\mathcal{L}(A) = \mathtt{in}$ for some complete labelling $\mathcal{L}$ of AF_i.
- a claim $(\mathtt{in}(A), \_)$ or $(\mathtt{out}(B), \mathtt{in}(A))$ is a *lie* wrt AF_i if $A \in Ar$ and $\mathcal{L}(A) \neq \mathtt{in}$ for any complete labelling $\mathcal{L}$ of AF_i.
- a claim $(\mathtt{in}(A), \_)$ or $(\mathtt{out}(B), \mathtt{in}(A))$ is *bullshit* wrt AF_i if $A \in U \backslash Ar$.

A claim is called *dishonest* if it is either a lie or bullshit. A player is *honest* if every claim by the player is honest. Otherwise, a player is *dishonest*.[3]

A player P_i makes a claim under the complete labelling of his/her argumentation framework AF_i. A claim is honest if arguments included in the claim are credulously justified by AF_i. On the other hand, a player lies if he/she brings $\mathtt{in}(A)$ while believing $\mathtt{out}(A)$ or $\mathtt{undec}(A)$ in his/her labelling (AF_1^2 of Example 4.2(1)). A player bullshits if he/she brings $\mathtt{in}(A)$ while none of $\mathtt{in}(A)$, $\mathtt{out}(A)$ nor $\mathtt{undec}(A)$ is in his/her labelling (AF_1^2 of Example 4.2(2)). To allow the existence of dishonest players who may bullshit, Definition 4.3 of debate games is slightly modified in a way that each player may claim an argument which is not in his/her AF [26].

In a debate game, a player seeks a counter-claim which refutes a claim given by the opponent player. Viewing an argument given by the opponent player as an observation, computation of a counter-claim by a player is characterized by abduction as follows.

Theorem 4.1. *Let $UAF = (U, att_U)$ and $(\mathtt{out}(B), \mathtt{in}(A))$ (or $(\mathtt{in}(A), \_)$) be a claim made by a player P_1 under AF_1^k in a debate game.*

1. *If $O = \mathtt{out}(A)$ has the empty credulous explanation in AF_2^{k+1}, then a player P_2 can make an honest claim $(\mathtt{out}(A), \mathtt{in}(C))$ that refutes the claim by P_1.*
2. *Else if $O = \mathtt{out}(A)$ has no empty credulous explanation but has a non-empty credulous explanation E in AF_2^{k+1}, then a player P_2 cannot make an honest claim but can make a dishonest claim $(\mathtt{out}(A), \mathtt{in}(C))$ that refutes the claim by P_1.*
3. *Otherwise, if $O = \mathtt{out}(A)$ has no explanation, then P_2 cannot refute the claim by P_1 and loses the game.*

A similar result holds for a player P_1 against a claim made by a player P_2.

Proof. (1) If O has the empty credulous explanation in AF_2^{k+1}, then $\mathtt{out}(A)$ is credulously justified by AF_2^{k+1} under the complete labelling. In this case, P_2 can make an honest claim $(\mathtt{out}(A), \mathtt{in}(C))$ with an argument $C \in Ar_2^{k+1}$ such that $(C, A) \in att_2^{k+1}$. (2) Else if O has a non-empty credulous explanation $E = (I, J)$ in AF_2^{k+1}, then $\mathtt{out}(A)$ is credulously justified by $(AF_2^{k+1})_E = ((Ar_2^{k+1})_E, (att_2^{k+1})_E)$ under the complete labelling where $(Ar_2^{k+1})_E = (Ar_2^{k+1} \backslash J) \cup I$ and $(att_2^{k+1})_E = att_U \cap (Ar_2^{k+1})_E \times (Ar_2^{k+1})_E$. In this case, P_2 can make a dishonest claim $(\mathtt{out}(A),$

---

[3] We use the notion of (dis)honest claims based on credulous justification under the complete labelling in [26], while alternative definitions are considered based on skeptical justification or different labellings.

$\text{in}(C)$) with an argument $C \in (Ar_2^{k+1})_E$ such that $(C, A) \in (att_2^{k+1})_E$. (3) Otherwise, if O has no explanation in AF_2^{k+1}, P_2 cannot make a counter-claim ($\text{out}(A)$, $\text{in}(C)$). □

In this characterization, an observation is always labelled out. This is because the goal of a player is to justify $O = \text{out}(A)$ or to explain it. When O has the empty credulous explanation, it is a most preferred explanation and a player makes an honest counter-claim. When O has multiple non-empty explanations, most preferred explanations are selected as best strategies. This is because a dishonest claim makes labellings of arguments deviate from those believed by the player. In Example 4.2(1), P_1 makes the dishonest claim ($\text{out}(D)$, $\text{in}(E)$) but P_1 believes $\text{in}(D)$ and $\text{out}(E)$. A dishonest claim which increases such deviation is undesirable for a player because it would make difficult for the player to keep consistency during a debate and also increases the chance of dishonest claims being detected. However, selection of most preferred explanations as dishonest claims is not always successful. For instance, if the only explanation given for an observation needs to remove an argument that has already been used in the previous exchanges, then the player cannot hope to refute the opponent by hiding that argument. Comparing the lie ($\text{out}(D)$, $\text{in}(E)$) by AF_1^2 in Example 4.2(1) with the bullshit ($\text{out}(D)$, $\text{in}(G)$) by AF_1^2 in Example 4.2(2), lies are considered worse than bullshit. This is because the player P_1 knows the falsehood of ($\text{out}(D)$, $\text{in}(E)$), while he/she does not know the truthfulness of ($\text{out}(D)$, $\text{in}(G)$). There is no possibility of $\text{in}(E)$ as far as F is in, while there is a possibility of $\text{in}(G)$ as far as there is no attacker of it. These behavioral rules are summarized as strategies of a player P_i as follows:

- If $O = \text{out}(A)$ has the empty explanation in AF_i^k ($i = 1, 2$; $k \geq 1$), then make an honest claim ($\text{out}(A)$, $\text{in}(B)$) based on AF_i^k. Else if O has a preferred explanation E in AF_i^k then make a dishonest claim ($\text{out}(A)$, $\text{in}(B)$) based on $(AF_i^k)_E$.
- If $O = \text{out}(A)$ has non-empty multiple preferred explanations in AF_i^k, then select one $E = (I, J)$ such that for any $B \in J$, $\text{in}(B)$ does not appear in any claim made by AF_i^j ($j < k$).
- If $O = \text{out}(A)$ has non-empty multiple preferred explanations in AF_i^k, then select one $E = (I, J)$ such that there is $B \in I \cap (U \backslash Ar_i^k)$ and $(B, A) \in att_U$ if any.

The first item says selecting honest claims at first. The other two items provide criteria for selecting dishonest claims. The second one is used for avoiding lie detection, while the third one presents preference of bullshit to lies.

5   Related Work

Abduction and argumentation have been combined in different ways in the literature. Dung [13] introduces the preferred extension semantics of abductive logic programs, which is defined as a maximally consistent set of hypotheses that contains its own defense against all attacks. The semantics is analyzed from the argumentation-theoretic viewpoint [19] and extended to *assumption-based argumentation* (ABA) [9]. In ABA an argument is a deduction of a conclusion (claim) c from a set of assumptions S represented as a tree, with c at the root and S at the leaves [15]. The goal of ABA is to construct an argument (tree) such that c is deduced from S using inference rules $(S \vdash c)$. In ABA both a claim and assumptions are parts of an argument, which is different from our problem setting where arguments play the role of assumptions to explain another observed (labelled) argument.

Wakaki *et al.* [29] introduce hypothetical arguments to Dung's argumentation framework. They introduce *abductive argumentation framework* (AAF) which computes explanations to skeptically justify or not to credulously justify the argument supporting a claim. They consider concrete argumentation frameworks associated with *abductive logic programs* [19] under the answer set semantics. This is in contrast to our approach for abduction in abstract argumentation frameworks that have no restriction to any particular representation for arguments nor argumentation semantics. Moreover, in the AAF arguments are introduced to explain observations, while they cannot be removed from the knowledge base of an agent. In this sense, the AAF is based on the normal setting of abduction [19], while our current proposal is based on *extended abduction* of [18]. Extended abduction is particularly useful when a knowledge base is *nonmonotonic*. In nonmonotonic theories, deletion of formulas may introduce new formulas. Thus, addition and deletion of hypotheses play a complementary role in accounting for an observation in nonmonotonic theories. Since an argumentation framework is inherently nonmonotonic (i.e., introduction/removal of arguments changes labelling in general), the use of extended abduction is more natural and appropriate. Deletion of arguments would happen when one notices that his/her previous argument was incorrect (see the example at the beginning of Sect. 3). For another case, one would withdraw his/her argument and make a concession (to reach an agreement), even if he/she has a counter-argument against the opponent.

Kakas and Moraitis [20] use abduction to seek conditions to support arguments. An *argumentation theory* is defined as a pair (T, P) where T is a set of argument rules and P represents priorities over T. Then, a *supported argument* is defined as a tuple (Δ, S) where Δ is a set of argument rules from T and S is a set of hypothetical explanations. In their framework, an argument is a set of rules of the form $l_0 \leftarrow l_1, \ldots, l_n$ where l_i is a positive or negative literal. Each literal l_i $(1 \le i \le n)$ in the conditional part can be a hypothetical explanation but it is not an argument. This is different from our setting where explanations are also arguments. For another difference, abduction considered in their framework is normal setting of abduction, which is different from our setting of extended abduction. A supported argument is also used for building a proposal or responding to a proposal in argumentation-based negotiation [21]. Argumentation-based negotiation is studied by other researchers as well (for instance, [1]). A debate game is similar to argumentation-based negotiation in the sense that they use argumentation frameworks for formulating dialogues between competitive agents. However, the goal of negotiation is slightly different from debate—the goal of negotiation is to reach an agreement among players, while the goal of debate is to defeat the opponent player.

Šešelja and Straßer [28] integrate abduction and argumentation in their *explanatory argumentation framework* (EAF). An EAF is defined as a tuple $\langle \mathcal{A}, \chi, \rightarrow, \dashrightarrow, \sim \rangle$ where $\langle \mathcal{A}, \rightarrow \rangle$ is an AF, χ is a set of *explananda*, $\dashrightarrow$ is the *explanatory relation* over $\mathcal{A} \times (\mathcal{A} \cup \chi)$, and $\sim$ is the *incompatible relation* over $\mathcal{A} \times \mathcal{A}$. Thus, they distinguish attack relations and explanatory relations, and explananda and arguments. On the other hand, they do not distinguish arguments and hypotheses. Bex *et al.* [5] combine abduction and argumentation in the context of evidential reasoning. An argumentation framework is given as a pair (G, E) where G is a set of *evidential generalisations* and E is a set of *evidences*. The set O of observations is produced by applying evidential generalisations to evidences, and explanations (causal rules plus hypotheses) which account for the set of explananda $F \subseteq O$ are selected. In this study, argumentation and abduction are combined in a way different from ours: arguments are used for generating observations supported by evidences and justifying explanations against observations. Bex and Prakken [6] apply the framework to a formal dialogue game. In the game, players try to find a plausible and evidentially well-supported explanation for the explananda. None of the players wants to win, since they have the joint goal to find the best explanation of the explananda. This is in contrast with debate games where each player seeks explanations to justify its own individual argument to win a game.

Rotstein *et al.* [24] study argumentation theory change in abstract argumentation framework. A *dynamic argumentation framework* (DAF) has the universe U of arguments and the set $A \subseteq U$ of active arguments. Given an argument X to be warranted, a *dialectical tree* rooted in X is modified by activating nodes in $U \backslash A$ and by deactivating nodes in A to make X justified. They introduce argument change operators which expand the set A of arguments and contract some arguments from A. The goal of their study differs from ours in that their framework is dedicated to characterize dynamics of AF while abduction in AF is intended to reason explanations for/against a particular argument. Technically, their revision operators do not distinguish skeptical and credulous justifications. Baumann and Brewka [3] consider the problem of modifying an argumentation framework in a way that a desired set of arguments becomes an extension. To this end, they add new arguments and attack relations to an AF, while they do not delete arguments because one could delete everything and add the wanted arguments without any attacks. We consider deleting arguments (and corresponding attack relations) as well as introducing ones, while preferring explanations that minimally change the original AF. Baumann [4] enforces a desired set of arguments by adding/removing a minimal number of attack relations to an AF. He then introduces value functions to compute different types of modification. In this study, the distance between two argumentation frameworks is measured by counting added/removed attacks. On the other hand, we measure the distance by comparing labelling of arguments in two AFs. In this sense, minimal change considered in [4] is syntax-based, while minimal change considered in this paper is semantic-based. Boella *et al.* [7, 8] consider the effect of adding/removing arguments or attack relations under the grounded semantics. Cayrol *et al.* [12] study the effect of an addition of an argument on the outcome of the argumentation semantics. The goal of these studies [7, 8, 12] is identifying possible changes of extensions after revising an argumentation framework, which is in contrast with our goal of identifying possible changes of an AF to have a particular outcome. Rahwan *et al.* [23] introduce a formal argumentation theory in which an agent may hide arguments or make up new arguments to accept a particular argument. The purpose of their study is to develop a game-theoretic argumentation mechanism design and to characterize strategy-proofness under graph-theoretic conditions. However, they do not provide any computational mechanism of dishonest arguments. We show the use of abduction in debate games based on formal argumentation frameworks, especially computing dishonest arguments. Extended abduction is also used for dishonest reasoning in logic programming [25]. In [25] an agent reasons dishonestly to have a particular goal at the individual level. The current study shows that extended abduction in AFs is used for computing dishonest arguments in debate games between two players.

6 Conclusion

We introduced extended abduction to abstract argumentation frameworks and provided its computational method in logic programming. Next we showed its application to computing (dis)honest claims in debate games. The result of this paper realizes extended abduction in argumentation frameworks, and provides a strong link between abduction, argumentative reasoning, and dishonest reasoning in a formal dialogue system based on AF. The abduction mechanism proposed in this paper will also be applied to revision of AF and will be realized in argumentation systems associated with logic programming. These issues are left for future work.

References

1. Amgoud, L., Dimopoulos, Y., Moraitis, P.: A unified and general framework for argumentation-based negotiation. In: Proceedings of the AAMAS-07, pp. 1018–1025 (2007)
2. Baroni, P., Giacomin, M.: Semantics of abstract argument systems. In: Rahwan, I., Simari, G.R., et al. (eds.) Argumentation in Artificial Intelligence, pp. 25–44. Springer, Heidelberg (2009)
3. Baumann, R., Brewka, G.: Expanding argumentation frameworks: enforcing and monotonicity results. In: Proceedings of the 3rd COMMA. Frontiers in AI, vol. 216, pp. 75–86. IOS Press, Amsterdam (2010)

4. Baumann, R.: What does it take to enforce an argument? Minimal change in abstract argumentation. In: Proceedings of the 20th European Conference on Artificial Intelligence, pp. 127–132. IOS Press, Amsterdam (2012)
5. Bex, F.J., Prakken, H., Verheij, B.: Formalising argumentation story-based analysis of evidence. In: Proceedings of the 11th International Conference on Artificial Intelligence and Law, pp. 1–10 (2007)
6. Bex, F.J., Prakken, H.: Investigating stories in a formal dialogue game. In: Proceedings of the 2nd International Conference on Computational Models of Argument, pp. 73–84. IOS Press, Amsterdam (2008)
7. Boella, G., Kaci, S., van der Torre, L.: Dynamics in argumentation with single extensions: attack refinement and the grounded extension. In: Proceedings of the AAMAS-09, pp. 1213–1214 (2009)
8. Boella, G., Kaci, S., van der Torre, L.: Dynamics in argumentation with single extensions: abstraction principles and the grounded extension. In: Sossai, C., Chemello, G. (eds.) ECSQARU 2009. LNCS (LNAI), vol. 5590, pp. 107–118. Springer, Heidelberg (2009)
9. Bondarenko, A., Dung, P.M., Kowalski, R.A., Toni, F.: An abstract, argumentation-theoretic approach to default reasoning. Artif. Intell. **93**, 63–101 (1997)
10. Caminada, M., Gabbay, D.M.: A logical account of formal argumentation. Stud. Log. **93**, 109–145 (2009)
11. Caminada, M., Sá, S., Alcântara, J.: On the equivalence between logic programming semantics and argumentation semantics. Technical report ABDN-CS-13-01, University of Aberdeen, 2013. A shorter version in: van der Gaag, L.C. (ed.) ECSQARU 2013. LNCS (LNAI), vol. 7958, pp. 97–108. Springer, Heidelberg (2013)
12. Cayrol, C., Dupin de Saint-Cyr, F., Lagasquie-Schiex, M.-C.: Change in abstract argumentation frameworks: adding an argument. J. Artif. Intell. Res. **38**, 49–84 (2010)
13. Dung, P.M.: Negation as hypothesis: an abductive foundation for logic programming. In: Proceedings of the ICLP, pp. 3–17. MIT Press, Cambridge (1991)
14. Dung, P.M.: On the acceptability of arguments and its fundamental role in nonmonotonic reasoning, logic programming and n-person games. Artif. Intell. **77**, 321–357 (1995)
15. Dung, P.M., Kowalski, R.A., Toni, F.: Assumption-based argumentation. In: Rahwan, I., Simari, G.R. (eds.) Argumentation in Artificial Intelligence, pp. 199–218. Springer, Heidelberg (2009)
16. Dvořák, W., Woltran, S.: On the intertranslatability of argumentation semantics. J. Artif. Intell. Res. **41**, 445–475 (2011)
17. Hughes, W.: Critical Thinking: An Introduction to the Basic Skills. Broadview Press, Peterborough (1992)
18. Inoue, K., Sakama, C.: Abductive framework for nonmonotonic theory change. In: Proceedings of the IJCAI-95, pp. 204–210 (1995)
19. Kakas, A.C., Kowalski, R.A., Toni, F.: Abductive logic programming. J. Log. Comput. **2**(6), 719–770 (1992)
20. Kakas, A.C., Moraitis, P.: Argumentative agent deliberation, roles and context. Electr. Notes Theor. Comput. Sci. **70**, 39–53 (2002)
21. Kakas, A.C., Moraitis, P.: Adaptive agent negotiation via argumentation. In: Proceedings of the AAMAS-06, pp. 384–391 (2006)
22. Mayes, G.R.: Argument-explanation complementarity and the structure of informal reasoning. Informal Log. **30**, 92–111 (2010)

23. Rahwan, I., Larson, K., Tohmé, F.: A characterisation of strategy-proofness for grounded argumentation semantics. In: Proceedings of the IJCAI-09, pp. 251–256 (2009)
24. Rotstein, N.D., Moguillansky, M.O., Falappa, M.A., García, A.J., Simari, G.R.: Argument theory change: revision upon warrant. In: Proceedings of the 2nd COMMA, pp. 336–347. IOS Press, Amsterdam (2008)
25. Sakama, C.: Dishonest reasoning by abduction. In: Proceedings of the IJCAI-11, pp. 1063–1068 (2011)
26. Sakama, C.: Dishonest arguments in debate games. In: Proceedings of the 4th International Conference on Computational Models of Argument. Frontiers in AI and Applications, vol. 245, pp. 177–184. IOS Press, Amsterdam (2012)
27. Schopenhauer, A.: The Art of Controversy. Originally published in 1896 and is translated by T. Bailey Saunders. Cosimo Classics, New York (2007)
28. Šešelja, D., Straßer, C.: Abstract argumentation and explanation applied to scientific debates. Synthese **190**(12), 2195–2217 (2013)
29. Wakaki, T., Nitta, K., Sawamura, H.: Computing abductive argumentation in answer set programming. In: McBurney, P., Rahwan, I., Parsons, S., Maudet, N. (eds.) ArgMAS 2009. LNCS (LNAI), vol. 6057, pp. 195–215. Springer, Heidelberg (2010)
30. Wu, Y., Caminada, M., Gabbay, D.M.: Complete extensions in argumentation coincides with 3-valued stable models in logic programming. Stud. Logica. **93**(2–3), 383–403 (2009)

Mechanized Support
for Assurance Case Argumentation

John Rushby[✉]

Computer Science Laboratory, SRI International, 333 Ravenswood Avenue,
Menlo Park, CA 94025, USA
Rushby@csl.sri.com

Abstract. An assurance case provides an argument that certain claims
(usually concerning safety or other critical properties) are justified, based
on given evidence concerning the context, design, and implementation of
a system. An assurance case serves two purposes: reasoning and communication. For the first, the argument in the case should approach
the standards of mathematical proof (though it may be grounded on
premises—i.e., evidence—that are equivocal); for the second it must
assist human stakeholders to grasp the essence of the case, to explore
its details, and to challenge it. Because of the scale and complexity of
assurance cases, both purposes benefit from mechanized assistance. We
propose simple ways in which an assurance case, formalized in a mechanized verification system to support the first purpose, can be adapted
to serve the second.

1 Introduction

An assurance case provides an *argument* that certain *claims* about a system
(usually concerning safety or other critical properties) are justified, based on
evidence concerning its context, design, and implementation [1,2].

The assurance case for a real system is a massive artifact: typically thousands
of pages of documentation, diagrams, analyses, and tests. It is surely difficult to
evaluate the argument that binds such a large amount of evidence together and
connects it to the claims. Greenwell and colleagues examined three industrial
safety cases and discovered logical fallacies in all of them [3]. Furthermore, each
case was examined by two reviewers and there were considerable differences in
the flaws detected by each reviewer.

Thus, it seems that human review is not particularly reliable for assurance
case arguments and that mechanized support could add precision to their construction and analysis. Modern formal verification systems (such as Acl2, Agda,
Coq, Isabelle, or PVS) provide notations adequate to the formalization and specification of complex systems and the automation (based on theorem proving and
model checking) to analyze them (see, e.g., [4–6]). As we will explain, verification
is a narrower problem than system assurance, but it seems plausible that the
application of formal verification systems might be extended from verification

© Springer International Publishing Switzerland 2014
Y. Nakano et al. (Eds.): JSAI-isAI 2013, LNAI 8417, pp. 304–318, 2014.
DOI: 10.1007/978-3-319-10061-6_20

to the analysis of assurance case arguments, and there are proposals for doing this [7–9].

These proposals presuppose that the argument of an assurance case should be deductively sound, but there are differing views on this. Some believe such arguments are quintessentially inductive: they provide strong evidence that the conclusion is highly probable, not proof that it is certain [10]. My view is that we may have doubts about some of the premises (i.e., evidence) used in the argument, but that the reasoning, *given these premises*, should be logically or deductively sound. I call this the *reasoning aspect* of the assurance case and argue that formal verification methods can eliminate *logic doubt* concerning this aspect of the argument, allowing attention to be focused on *epistemic doubt* about the accuracy and completeness of our knowledge of the system, as represented in the premises to the argument [11].

But, whereas logic doubt can be eliminated by mechanized verification, epistemic doubt requires human review. There is much evidence that human individuals and groups are prone to confirmation bias, so reviews should actively challenge and explore the assumptions and claimed knowledge underlying a case. Prior to, and in addition to, review, a case also serves as a vehicle for communication and shared understanding among its stakeholders and these purposes, too, are likely to be best served by active exploration and "what-if" inquiry, rather than passive appraisal. Thus, the reasoning aspect of a case is complemented by a *communication aspect* that focuses on exploration of its epistemic foundation.

In support of its communication aspect, the epistemic foundation of an assurance case will be explored, modified, and revised—possibly many times—and its reasoning aspect will be adjusted correspondingly. Thus, seen in the large, the reasoning in an assurance case, although it can be supported locally by the tools of formal verification, is not a proof, but an *argument*: the distinction being that an argument is *defeasible*—i.e., it is subject to revision in the light of objections or new information, or for the purposes of exploration. Just as the reasoning or deductive aspect of an assurance case can benefit from mechanized support, so can its communication or defeasible aspect—and, of course, the mechanized support for each aspect must somehow coexist with the other.

In the following section, I review some topics in applying mechanized support to the reasoning and the communication aspects of an assurance case, respectively. Then, in Sect. 3, I propose a simple way in which a formalized assurance case, whose mechanization is primarily intended to support reasoning, can be augmented to allow defeasibility and used to support the communication aspect also. The proposal is illustrated with a simple example. Section 4 provides brief conclusions and suggestions for further research.

2   Mechanized Support for Assurance Cases

Assurance cases are large and complex artifacts and so it is necessary to have automated support for managing the overall structure of a case, and for providing representations in graphical notations such as GSN [12] to aid comprehension.

Several such tools exist [13,14], and there are emerging standards to assist their interoperation [15,16].

However, my focus here is on mechanized support for the logical aspects of an assurance case. In the next subsection, I review some topics in applying mechanized support to the reasoning or deductive aspect of an assurance case, and in the subsection that follows I review topics in providing support for the communication or defeasible aspect of a case.

2.1   Mechanized Support for Reasoning in Assurance Cases

Modern formal methods tools such as verification systems, model checkers, and SMT solvers have sufficient expressiveness and automation to undertake the task of providing mechanized support to the reasoning or deductive aspect of an assurance case. But just as its deductive aspect is not the whole purpose of an assurance case, so classical formal verification is not quite the same as the deductive part of an assurance case, so some care is needed in the way in which an assurance case is represented as a formal verification.

A formal verification differs from the deductive aspect of an assurance case in that verification takes the specification (i.e., premises) as given and verifies correctness of the conclusions that are derived from it, whereas an assurance case must also justify (often by citing the evidence of the case) the premises and rules of inference that it uses, and it must also justify that the verified conclusion bears an interpretation that is relevant to the claim of the case. For example, in applying formal verification to an assurance case we might use a proof rule that says "a system is safe if it is shown to be safe for each of its hazards"; in applying this rule we would show safety for each hazard that has been explicitly identified and would have a premise that says these are *all* the hazards. Justification of this premise would be a major part of the safety case, yet is outside the formal verification. In [7], I proposed a way in which an assurance case can be represented in a formal verification system so that these aspects are at least recorded. In essence, I conjoin to each premise a predicate that is set *true* only when a reviewer accepts its supporting evidence, which can be attached to the predicate as a comment. A formal verification system such as PVS requires small enhancements to fully support this proposal (basically, support for referencing documents as comments), but recent tool-integration frameworks such as the Evidential Tool Bus (ETB) [17] allow nonformal justifications to be attached to claims as a basic capability.

The premises to an assurance case represented within a formal verification system record our knowledge about the system or, to use a fancier term, its *epistemology*. The soundness of the deductive aspect of an assurance case rests on two pillars: how complete and accurate is our knowledge about the system, and how accurate is our reasoning about the case, given our knowledge. Concern about the second of these (*logic doubt*) is largely eliminated by the soundness guarantee of formal verification, so concern should mainly focus on the first item (*epistemic doubt*), especially its completeness.

As suggested above, justification for many of the premises in a formalized assurance case will be references to the evidence of the case. In developing a formalized assurance case we can choose how abstractly to represent the system and, in consequence, the granularity of the evidence that is explicitly represented. Since evidence is opaque to the formal analysis, there is much to be said for refining the level of abstraction and breaking large items of evidence into more tightly focused pieces connected by explicit reasoning. In essence, this means we should represent our knowledge in logic. Software *is* logic, so there is, in principle, no obstacle to representing its epistemology (requirements, specification, code, semantics) in logic: that is why formal verification is feasible—and increasingly practical and cost-effective—for software.

The world with which the software interacts—the world of devices, machines, people and institutions—has not traditionally been represented in logic, but indirectly it is becoming so, for it is increasingly common that system developers build models of the world using simulation environments such as Simulink/ Stateflow. These models represent their epistemology, which they refine and validate by conducting simulation experiments.

It is feasible to import models from simulation environments such as Stateflow/Simulink into verification environments (see, e.g., [18]). However, simulation models are not the best representation of the epistemology for an assurance case. Simulation models are designed for that purpose and simultaneously say too much and too little for the purposes of assurance and minimization of epistemic doubt. For example, the Simulink model for a car braking system will provide equations that allow calculation of the exact rate of deceleration in given circumstances (which is more information than we need), but will not provide (other than indirectly) the maximum stopping distance—which is an example of a property that may be needed in an assurance case. The crucial point is that it should be easier to resolve epistemic doubts about a simple constraint, such as maximum stopping distance, than the detailed equations that underlie a full simulation model.

A proposal, developed in [11], is that for the purpose of recording the epistemology of a safety case, models should be expressed as systems of constraints rather than as simulation models. Until fairly recently, it would have been difficult to validate systems of constraints: unlike simulation models, it was not feasible to run experimental calculations to check the predictions of the model against intuition and reality. But now we have technology such as "infinite bounded model checkers," based on highly effective constraint solvers for "satisfiability modulo theories" (SMT) that allow effective exploration of constraint-based models [19].

2.2 Mechanized Support for Communication in Assurance Cases

As noted earlier, a formal verification is not the same as an assurance case and, even with the adjustments proposed above, there are purposes served by an assurance case that are not supported by its embedding in a formal verification system, as currently envisaged.

In particular, an assurance case is not purely about deductive reasoning: it is also about communication and, ultimately, persuasion. That is, an assurance case is constructed by humans and embodies their understanding and beliefs about the system and these need to be communicated to other stakeholders, including regulators and certifiers. Effective communication is unlikely to be one-way; it is more likely to be a dialog and the process of developing a common understanding may lead to revisions in the assurance case. The revisions may adjust some of the premises at the bottom of the argument, and they may adjust some of the reasoning expressed in its rules or axioms.[1]

In addition to revisions that represent adjustments to the argument, reviewers of the case may wish to temporarily change elements of the argument (i.e., conduct "what if" experiments) to assist their comprehension of the case. These permanent and temporary revisions to an assurance case suggest that its argument should be viewed as provisional, or contingent, and should therefore be developed in a framework that supports such "defeasible" reasoning.

Defeasible reasoning is well-studied in philosophy and in AI (where it is generally referred to as nonmonotonic reasoning), and there are rich bodies of work on belief revision, commonsense reasoning, truth maintenance, and so on. Closely related are the fields of reasoning under uncertainty, where we find fuzzy logic, Dempster-Shafer belief functions and so on, and probabilistic methods, where we find probabilistic and Markov logics, Bayesian Belief Networks (BBNs) and so on.[2] The field of Argumentation frames similar issues in a (generally) more abstract setting [21] where different agents may employ different sets of premises so that a premise of one may "defeat" that of another, and entire arguments may "attack" one another.

Most of these methods for defeasible reasoning, and the tools that support them, are framed as augmentations to propositional logic, whereas we earlier made the case that mechanized support for deductive reasoning in assurance cases should build on the much more powerful logics and theories of modern verification systems, model checkers, and SMT solvers.

One could imagine a two-pronged approach to mechanized support for development and evaluation of assurance cases: a powerful deductive system for analyzing the reasoning in detail, and a defeasible or argumentation-based system to support exploration and experiment on the overall argument at an abstract level for the purposes of understanding and communication. Such an approach could be viable, and it might even be possible to automate abstraction from the deductive to the defeasible levels of detail (though the reverse might be more difficult), but I believe there could be benefits in augmenting the representation

---

[1] There are persuasive claims that human consciousness evolved to enable communication and cooperative behavior, and that reasoning evolved to evaluate the epistemic claims of others [20]. Thus, argument is a fundamental human capability, constructive reasoning is an epiphenomenon, and confirmation bias is intrinsic.

[2] I prefer not to cite specific works from the vast repertoire of articles and books on these topics; an Internet search will provide many good references.

and tools proposed for deductive analysis of assurance cases so that they can also support the defeasible level. This is the topic of the following section.

3   Supporting Defeasible Reasoning in Mechanized Verification

In defeasible reasoning, we may draw a conclusion based on a state of knowledge that is subsequently revised, invalidating the previous conclusion. A standard example is

(1) Tweety is a bird,
(2) Birds can fly,
(3) Therefore Tweety can fly.

Subsequently, we learn that Tweety is a penguin and penguins cannot fly.

This new information contradicts our prior knowledge and there are many proposals how to adjust our logic and our reasoning to accommodate such revisions. Often, we will have both a general rule "birds can fly" and a revision "penguins cannot fly" that each apply to Tweety, who is both a bird and a penguin, and we need some method (such as "circumscription" [22]) for resolving the apparent inconsistency and preferring one conclusion over another. In other cases, a revision may flatly deny some prior rule (e.g., Tweety is not a bird but a bat) and defeasible reasoning provides ways to handle these inconsistencies, too.

While this kind of sophistication is valuable when representing commonsense reasoning, or when resolving arguments where different parties advance different premises, I do not believe it is necessary or desirable in the evaluation of assurance cases. In evaluating an assurance case about Tweety, we would wish to be alerted to the potential inconsistency in our epistemology concerning his ability to fly, but would surely then seek to reach consensus on the point and then reason classically from there. That is to say, rather than rely on logics for default or defeasible reasoning to cope with inconsistencies resulting from different opinions or conflicting evidence, we would revise our assurance case to resolve or eliminate the inconsistencies so that classical deductive reasoning provides a single conclusion (this is similar to Pollock's notion of a "warrant" [23]).

I propose that one simple way to allow exploration and challenge while still using classical reasoning is to introduce explicit "defeater" predicates into the premises of an assurance case.[3] Then, our premises concerning Tweety become

(1) Absent a defeater about Tweety, Tweety is a bird, and
(2) Absent a defeater about flying, birds can fly.

More formally, any premise p becomes $\neg d_p \supset p$ where d_p denotes the defeater for p (and $\neg$ and $\supset$ are logical negation and implication, respectively). Initially, all defeaters are absent (i.e., false) and we conclude that Tweety can fly.

---

[3] Some treatments of defeasible reasoning distinguish "undercutting," "undermining," and "rebutting" defeaters, but the distinctions are not sharp and are not used here.

A reviewer who has doubts about the universality of the premise "birds can fly" may turn on its defeater, observe the consequences, and revise the argument by adding additional premises and constraints about penguins. I provide an example below where the consequences of a defeater are a little less obvious, and the benefits more significant.

In addition to turning defeaters on and off and then reasoning "forwards" to deduce the consequences, we could instead assert that Tweety is a bird but cannot fly, and then reason "backwards" to seek an explanation. Observe that this is exactly the basis for model-based diagnosis [24], where our "defeater" predicates take the role of the "abnormal" predicates used in diagnosis (and the related "reconfig" predicates used in model-based repair [25]). Some of the tools that underlie modern model checkers and formal verification systems provide capabilities that directly support this kind of examination. For example, our Yices SMT solver [26] not only can generate counterexamples as well as verify large formulas in a rich combination of theories (i.e., it does SAT as well as UNSAT for SMT), but it can also generate UNSAT Cores, and perform Weighted MAXSAT for SMT.

In the following section, I illustrate the use of explicit "defeater" predicates, and also some of the other points made earlier, in a simple example.

3.1 Example

I illustrate the proposal above using a small example from [27]. Below, I reproduce the "structured prose" rendition of the assurance case from that example, to which I have added paragraph numbers.

(1) This argument establishes the following **claim**: the control system is acceptably safe, within the context of a definition of acceptably safe. To establish the top-level claim, two **sub-claims** are established: (a) all identified hazards have been eliminated or sufficiently mitigated and (b) the software has been developed to the integrity levels appropriate to the hazards involved.

(2) Within the **context** of the tolerability targets for hazards (from reference Z) and the list of hazards identified from the functional hazard analysis (from reference Y), we follow the **strategy** of arguing over all three of the identified hazards (H1, H2, and H3) to establish sub-claim 1, yielding three additional **claims**: H1 has been eliminated; H2 has been sufficiently mitigated; and H3 has been sufficiently mitigated.

(3) The **evidence** that H1 has been eliminated is formal verification.

(4) The **evidence** that catastrophic hazard H2 has been sufficiently mitigated is a fault tree analysis showing that its probability of occurrence is less than 1×10^{-6} per annum. The **justification** for using this evidence is that the acceptable probability in our environment for a catastrophic hazard is 1×10^{-6} per annum.

(5) The **evidence** that the major hazard H3 has been sufficiently mitigated is a fault tree analysis showing that its probability of occurrence is less than 1×10^{-3} per annum. The **justification** for using this evidence is that the

acceptable probability in our environment for a major hazard is 1×10^{-3} per annum.

(6) We establish sub-claim (b) within the **context** of the list of hazards identified from the functional hazard analysis in reference Y, and the integrity level (IL) process guidelines defined in reference X. The process **evidence** shows that the primary protection system was developed to the required IL 4. The process **evidence** also shows that the secondary protection system was developed to the required IL 2.

I present a few highlights from a formalization of this argument in PVS [28, 29].

As soon as we start to formalize the argument, we recognize that paragraph (6) is not well connected to the rest of the case. This illustrates one of the benefits in applying mechanized checking to an assurance case: we are forced to ensure that the argument "connects up" and is deductively sound. Presumably, a more fully developed version of the argument would say that part of the fault tree analysis cited in (4) is an assumption that the software of the primary protection system has a failure rate below some threshold, and development to Integrity Level 4 (IL4) is considered to ensure that. Similarly for paragraph (5) and development of the secondary protection system to IL2.

To formalize this in PVS, we introduce the enumerated types `hazlevels` and `intlevels` to represent hazard and integrity levels respectively, and we provide axioms asserting that hazard H2 is `catastrophic` and that process evidence attests that the system was developed to integrity level IL4 with respect to this hazard, and similarly for hazard H3 (we omit specifications for the signatures of the functions `hazlev` and `process`).

```
hazlevels: TYPE = { minor, major, catastrophic }
intlevels: TYPE = { IL2, IL4 }

H2hlev: POSTULATE hazlev(system, H2) = catastrophic
H3hlev: POSTULATE hazlev(system, H3) = major

H2ilev: POSTULATE process(system, H2) = IL4
H3ilev: POSTULATE process(system, H3) = IL2
```

We use the keyword `POSTULATE` to indicate premises justified by evidence; in contrast the keyword `AXIOM` indicates premises that represent the reasoning or "proof rules" employed. PVS treats these keywords as synonyms, but the distinction is useful for communication with human readers.

Next, we provide the "proof rule" axiom `pr` that relates hazard and integrity levels to the claim that a given hazard is adequately "handled." Here `sy` and `hz` are variables ranging over systems and hazards, respectively.

```
pr: AXIOM
   (hazlev(sy, hz) = catastrophic AND process(sy, hz) = IL4
OR hazlev(sy, hz) = major AND process(sy, hz) = IL2)
   => handles(sy, hz)
```

From these we can prove the lemmas that hazards H2 and H3 are adequately handled. There is a similar (omitted) treatment for H1 on the basis that it has been formally verified.

```
H1OK: LEMMA handles(system, H1)
H2OK: LEMMA handles(system, H2)
H3OK: LEMMA handles(system, H3)
```

We then assert that H1, H2, and H3 are *all* the hazards for this system and claim in this context

```
H1, H2, H3: hazards

hazard_ax: POSTULATE
  allhazards(claim, system, context) = {: H1, H2, H3 :}
```

We employ the argument strategy that a system is safe if each of its hazards is adequately handled. Here cl and co are variables ranging over claims and contexts, respectively.

```
strategy: AXIOM
  LET hset = allhazards(cl, sy, co) IN
    (FORALL (h: (hset)): handles(sy, h))
        IMPLIES safe(cl, sy, co)
```

With these specifications, we can easily prove that the system is safe.

```
sysOK: THEOREM safe(claim, system, context)
```

Skeptical reviewers who examine this formalized assurance case might suggest that the level of abstraction is too high: they might be concerned about independence of H2 and H3 and be disappointed that the fault tree analyses are opaque items of evidence.[4] This illustrates the point made in Sect. 2.1 concerning epistemic doubt and the granularity of evidence.

There are two plausible approaches at this juncture: one is to elaborate the formalized case to include the top levels of the fault tree analyses so that the crucial topic of independence is exposed in the formal representation of the case; the other is to introduce a new hazard H23 that represents joint occurrence of H2 and H3. We will pursue the latter course here.

The developers of the assurance case might then introduce a premise that states that the joint hazard H23 is catastrophic and must be mitigated to a probability of occurrence less than 1×10^{-6} per annum, and claim that this is ensured by the combination of process evidence of IL4 for the primary system and IL2 for the secondary. The relevant changes are shown below and the formal verification of the case succeeds as before.

---

[4] The prose description in [27] suggests that the system under consideration has a primary and a secondary protection system; a standard concern in these kinds of system is that both protection systems fail on the same demand [30].

```
hazard_ax: POSTULATE
  allhazards(theclaim, system, context) = {: H1, H2, H3, H23 :}

H23hlev: POSTULATE hazlev(system, H23) = catastrophic

H23pr: AXIOM
  handles(system, H2) AND handles(system, H3)
    => handles(system, H23)

H23OK: LEMMA handles(system, H23)
```

Reviewers might be skeptical that the conjunction of process evidence for the primary and secondary systems, each considered in isolation, is sufficient to ensure mitigation of the *joint* occurrence represented by H23. To explore this they could turn on the defeater dfH23pr for premise H23pr. (To keep the presentation simple, I have not included the defeaters until now.)

```
dfH23pr: boolean = TRUE

H23pr: AXIOM NOT dfH23pr =>
  handles(system, H2) and handles(system, H3)
    => handles(system, H23)
```

The formal verification now fails to guarantee safety.

The developers of the case could then introduce new evidence that the combined primary and secondary system has been used previously in a different, but similar context (with a system called otherS and hazard otherH).

```
previous: POSTULATE
  similar((otherS, otherH),(system, H23))
    AND handles(otherS, otherH)
```

They assert this is sufficient to claim that the present system handles H23.

```
dfprior23: boolean = FALSE

prior23: AXIOM NOT dfprior23 =>
  similar((otherS, otherH), (system, H23)) AND handles(otherS, otherH)
    => handles(system, H23)
```

They turn off the defeater for this new premise and are once again able to verify safety.

We now have two ways to justify safety of the system: one citing evidence of integrity levels and fault tree analyses, and another citing prior experience. In a conventional formal verification there is little purpose in such redundancy of argument, but in an assurance case it can be useful. Here, the reviewers might be skeptical of the evidence by prior experience because they are uncertain that the context of the previous system is sufficiently similar to the present one, and so they turn on the defeater for this argument.

```
dfprior23: boolean = TRUE
```

Once again, the formal verification fails to guarantee safety. But now the developers might argue that although both lines of safety justification have their flaws, in combination they constitute a "multi-legged" case (with independent legs) that is surely sufficient. The reviewers might accept this and can adjust their intervention in the formalized assurance case to state that either defeater may be true, but not both together.

```
dfH23pr, dfprior23: boolean

notboth: AXIOM NOT (dfH23pr AND dfprior23)
```

Now the formal verification succeeds once again, and the reviewers are satisfied.

Here, "inspection" was sufficient to see that our epistemic foundation remained consistent as we introduced new premises and toggled defeaters, but in larger examples it will be important to use mechanized assistance to ensure this.

That concludes our small example. Its purpose was to illustrate the idea, but its small size means that it cannot illustrate what I believe is the main attraction in this approach: namely, that it can exploit the full power of modern formal methods tools and should therefore scale to large examples that use rich logics and theories.

4   Discussion and Conclusion

This paper has reviewed some topics in providing mechanized support for the analysis and exploration of arguments in assurance cases. We saw that powerful modern tools for formal methods can provide useful support for the deductive or reasoning aspect of an assurance case and we explored some of the issues in representing cases so that epistemic doubts are minimized. We then considered support for the communication aspect of assurance cases and concluded that this requires some element of defeasible reasoning. However, we suggested that the purposes served by assurance cases are such that special logics for defeasible reasoning or abstract argumentation are unnecessary—in fact, undesirable—and that adequate support can be obtained by simply adding explicit "defeater" predicates to the premises of the formalized case. We illustrated this with a simple example. Related work includes similar proposals by Kinoshita and Takeyama [31].

Notice that our defeater predicates are not the same as the defeaters of Pollock [23,32], where defeaters are premises that contradict other premises and some mechanism is required to derive a preferred conclusion in the face of these inconsistencies. Our defeater predicates are used to turn premises on or off so that classical reasoning can be used. It is therefore important to check, for any given assignment of values to defeater predicates, that the enabled premises are not contradictory; unlike in our example, this check should be automated.

Notice also that our defeater predicates are either given explicit truth assignments or the conclusions to the verification are true under all interpretations (possibly subject to constraints, such as notboth in the example) so, contrary to [33, section 6.1], philosophical objections to "logically-uninterpreted conditions" do not apply.

Future research could include comparison with proposals that do employ more sophisticated treatments of defeasible argumentation, such as [32, 34]. Some treatments of argumentation, beginning with Dung [35], relate this to logic programming, and it would be interesting to explore the extent to which this can be supported in the Evidential Tool Bus, where the underlying framework is Datalog [17].

The comparison with argumentation frameworks should consider philosophy as well as technology. Argumentation generally presupposes a context where participants have different points of view and there may be no single "correct" conclusion (for example, arguments about ethics or aesthetics), or where participants have limited access to ground truth (e.g., drawing conclusions on the basis of imperfect sensors). Argumentation methods will evaluate proffered arguments and their defeaters or their attack relations and will derive conclusions, but these may not be deductively sound. In contrast, I believe that while argumentation may be an appropriate framework during development of an assurance case, the finished case should be one in which every credible objection has been anticipated and incorporated into the argument in such a way that the conclusion is deductively sound. Exploration and examination of the case then focuses on epistemic doubt about the premises, aided by the presence of defeater predicates that enable what-if experimentation.

Related to the philosophy and the purpose of an assurance case, Steele and Knight [36] provide a very illuminating account of certification, which I formulate as follows. The system under consideration is a designed artifact and may have flaws that could lead to accidents. The task of safety-critical design is to identify and either eliminate or mitigate all hazards to its safe deployment. The task of an assurance case is to provide confidence that this has been done, correctly and completely. But the assurance case itself is a designed artifact and may have flaws that could lead to a "certification accident": that is, the decision to approve and allow deployment of a potentially unsafe system. So the principles of safety-critical design should be applied "recursively" to the assurance case itself. That is, we should use systematic methods, inspired by those used for systems (e.g., fault tree analysis), actively to seek hazards (i.e., defeaters) to the assurance case, and should then seek to eliminate or mitigate them. Mitigation could take the form of a multi-legged case, as used in the example of the previous section, where an attractive method of justification could be that the defeaters of each leg are independent [37].

An excellent topic for future research is to explore the application and consequences of Steele and Knight's insight and its representation within the framework proposed here. A related topic is to explore the novel structure for assurance cases proposed by Hawkins and colleagues [10], who divide the overall case into a safety argument and a confidence argument.

A final topic for future research is to explore whether it may be feasible to derive some measure for the confidence in a case from the number and the nature of the defeaters that are accommodated. One way to do this would be to attach subjective probabilities to defeaters and then use calculation in some suitable probabilistic framework such as Bayesian Belief Networks (BBNs); another might be to employ "Baconian probabilities" as proposed in [38].

Acknowledgements. I am grateful for helpful comments by the reviewers that caused me to rethink some of the presentation, and to stimulating discussions with Michael Holloway and John Knight.

This work was supported by NASA under contracts NNA13AB02C with Drexel University and NNL13AA00B with the Boeing Company, and by SRI International. The content is solely the responsibility of the author and does not necessarily represent the official views of NASA.

References

1. Bishop, P., Bloomfield, R.: A methodology for safety case development. In: Safety-Critical Systems Symposium, Birmingham, UK (1998)
2. Kelly, T.: arguing safety–a systematic approach to safety case management. Ph.D. thesis, Department of Computer Science, University of York, UK (1998)
3. Greenwell, W.S., Knight, J.C., Holloway, C.M., Pease, J.J.: A taxonomy of fallacies in system safety arguments. In: Proceedings of the 24th International System Safety Conference, Albuquerque, NM (2006)
4. Klein, G., Elphinstone, K., Heiser, G., Andronick, J., Cock, D., Derrin, P., Elkaduwe, D., Engelhardt, K., Kolanski, R., Norrish, M., et al.: seL4: formal verification of an OS kernel. In: Proceedings of the ACM SIGOPS 22nd Symposium on Operating Systems Principles, pp. 207–220. ACM (2009)
5. Miner, P., Geser, A., Pike, L., Maddalon, J.: A unified fault-tolerance protocol. In: Lakhnech, Y., Yovine, S. (eds.) FORMATS 2004 and FTRTFT 2004. LNCS, vol. 3253, pp. 167–182. Springer, Heidelberg (2004)
6. Narkawicz, A., Muñoz, C.: Formal verification of conflict detection algorithms for arbitrary trajectories. Reliable Comput. **17**, 209–237 (2012)
7. Rushby, J.: Formalism in safety cases. In: Dale, C., Anderson, T. (eds.) Making Systems Safer: Proceedings of the Eighteenth Safety-Critical Systems Symposium, Bristol, UK, pp. 3–17. Springer (2010)
8. Basir, N., Denney, E., Fischer, B.: Deriving safety cases from automatically constructed proofs. In: 4th IET International Conference on System Safety, London, UK. The Institutions of Engineering and Technology (2009)
9. Takeyama, M., Kido, H., Kinoshita, Y.: Using a proof assistant to construct assurance cases: correctness by construction (fast abstract). In: The International Conference on Dependable Systems and Networks, Boston, MA. IEEE Computer Society (2012)
10. Hawkins, R., Kelly, T., Knight, J., Graydon, P.: A new approach to creating clear safety arguments. In: Dale, C., Anderson, T. (eds.) Advances in System Safety: Proceedings of the Nineteenth Safety-Critical Systems Symposium, Southampton, UK. Springer (2011)

11. Rushby, J.: Logic and epistemology in safety cases. In: Bitsch, F., Guiochet, J., Kaâniche, M. (eds.) SAFECOMP 2013. LNCS, vol. 8153, pp. 1–7. Springer, Heidelberg (2013)
12. Spriggs, J.: GSN–The Goal Structuring Notation. Springer, London (2012)
13. Denney, E., Pai, G., Pohl, J.: AdvoCATE: an assurance case automation toolset. In: Proceedings of the Workshop on Next Generation of System Assurance Approaches for Safety Critical Systems (SASSUR), Magdeburg, Germany (2012)
14. ASCE: ASCE. http://www.adelard.com/web/hnav/ASCE/index.html
15. SACM: OMG Structured Assurance Case Metamodel (SACM). http://www.omg.org/spec/SACM/
16. MACL: OMG Machine-Checkable Assurance Case Language (MACL). http://www.omg.org/cgi-bin/doc?sysa/2012-9-4/
17. Cruanes, S., Hamon, G., Owre, S., Shankar, N.: Tool integration with the evidential tool bus. In: Giacobazzi, R., Berdine, J., Mastroeni, I. (eds.) VMCAI 2013. LNCS, vol. 7737, pp. 275–294. Springer, Heidelberg (2013)
18. Miller, S.P., Whalen, M.W., Cofer, D.D.: Software model checking takes off. Commun. ACM **53**, 58–64 (2010)
19. Rushby, J.: Harnessing disruptive innovation in formal verification. In: Hung, D.V., Pandya, P. (eds.) Fourth International Conference on Software Engineering and Formal Methods (SEFM), India, Pune, pp. 21–28. IEEE Computer Society (2006)
20. Mercier, H., Sperber, D.: Why do humans reason? Arguments for an argumentative theory. Behav. Brain Sci. **34**, 57–111 (2011); See also the commentary on page 74 by Baumeister, R.F., Masicampo, E.J., Nathan DeWall, C.: Arguing, Reasoning, and the Interpersonal (Cultural) Functions of Human Consciousness
21. Chesñevar, C.I., Maguitman, A.G., Loui, R.P.: Logical models of argument. ACM Comput. Surv. **32**, 337–383 (2000)
22. McCarthy, J.: Circumscription-a form of non-monotonic reasoning. Artif. Intell. **13**, 27–39 (1980)
23. Pollock, J.L.: Defeasible reasoning. Cogn. Sci. **11**, 481–518 (1987)
24. Reiter, R.: A theory of diagnosis from first principles. Artif. Intell. **32**, 57–95 (1987)
25. Crow, J., Rushby, J.: Model-based reconfiguration: toward an integration with diagnosis. In: Proceedings of AAAI-91, Anaheim, CA, vol. 2, pp. 836–841 (1991)
26. Yices: Yices. http://yices.csl.sri.com/
27. Holloway, C.M.: Safety case notations: alternatives for the non-graphically inclined? In: 3rd IET International Conference on System Safety, Birmingham, UK. The Institutions of Engineering and Technology (2008)
28. Owre, S., Rushby, J., Shankar, N., von Henke, F.: Formal verification for fault-tolerant architectures: prolegomena to the design of PVS. IEEE Trans. Softw. Eng. **21**, 107–125 (1995)
29. PVS: PVS. http://pvs.csl.sri.com/
30. Littlewood, B., Rushby, J.: Reasoning about the reliability of diverse two-channel systems in which one channel is "possibly perfect". IEEE Trans. Softw. Eng. **38**, 1178–1194 (2012)
31. Kinoshita, Y., Takeyama, M.: Assurance case as a proof in a theory: towards formulation of rebuttals. In: Dale, C., Anderson, T. (eds.) Assuring the Safety of Systems: Proceedings of the 21st Safety-Critical Systems Symposium, SCSC, pp. 205–230 (2013)
32. Pollock, J.L.: Defeasible reasoning with variable degrees of justification. Artif. Intell. **133**, 233–282 (2001)
33. Staples, M.: Critical rationalism and engineering: Ontology. Synthese (to appear, 2014)

34. Caminada, M.W.A.: A formal account of Socratic-style argumentation. J. Appl. Logic **6**, 109–132 (2008)
35. Dung, P.M.: On the acceptability of arguments and its fundamental role in nonmonotonic reasoning, logic programming and n-person games. Artif. Intell. **77**, 321–357 (1995)
36. Steele, P., Knight, J.: Analysis of critical system certication. In: 15th IEEE International Symposium on High Assurance Systems Engineering, Miami, FL (2014)
37. Goodenough, J.B., Weinstock, C.B., Klein, A.Z., Ernst, N.: Analyzing a multi-legged argument using eliminative argumentation. In: Layered Assurance Workshop, New Orleans, LA (2013)
38. Weinstock, C.B., Goodenough, J.B., Klein, A.Z.: Measuring assurance case confidence using Baconian probabilities. In: 1st International Workshop on Assurance Cases for Software-Intensive Systems (ASSURE), San Francisco, CA (2013)

DDS13

Agreement Subtree Mapping Kernel for Phylogenetic Trees

Issei Hamada[1], Takaharu Shimada[1,4], Daiki Nakata[1], Kouichi Hirata[2($\boxtimes$)], and Tetsuji Kuboyama[3]

[1] Graduate School of Computer Science and Systems Engineering,
Kyushu Institute of Technology, Kawazu 680-4, Iizuka 820-8502, Japan
{hamada,shimada,nakata}@dumbo.ai.kyutech.ac.jp
[2] Department of Artificial Intelligence,
Kyushu Institute of Technology, Kawazu 680-4, Iizuka 820-8502, Japan
hirata@dumbo.ai.kyutech.ac.jp
[3] Computer Center, Gakushuin University,
Mejiro 1-5-1, Toshima, Tokyo 171-8588, Japan
kuboyama@gakushuin.ac.jp
[4] Mazda Motor Corporation, Hiroshima, Japan

Abstract. In this paper, we introduce an *agreement subtree mapping kernel* counting all of the agreement subtree mappings and design the algorithm to compute it for *phylogenetic trees*, which are unordered leaf-labeled full binary trees, in quadratic time. Then, by applying the agreement subtree mapping kernel to trimmed phylogenetic trees obtained from all the positions in nucleotide sequences for A (H1N1) influenza viruses, we classify pandemic viruses from non-pandemic viruses and viruses in one region from viruses in the other regions. On the other hand, for leaf-labeled trees, we show that the problem of counting all of the agreement subtree mappings is #P-complete.

1 Introduction

A *tree kernel* is one of the fundamental method to classify *rooted labeled trees* (*trees*, for short) through support vector machines (SVMs). Many researches to design the tree kernel for *ordered* trees, in which the order of sibling nodes is given, have been developed (*cf.*, [4,15]). In particular, a *mapping kernel* [15] is a powerful and general tree kernel counting all of the *mappings* [17] (and their variations) as the set of one-to-one node correspondences.

On the other hand, few researches to design the tree kernel for *unordered* trees, in which the order of sibling nodes is arbitrary, have been developed, where we call it an *unordered tree kernel*. One of the reason is that the problem of counting all of the subtrees for unordered trees is #P-complete [6].

This work is partially supported by Grant-in-Aid for Scientific Research 22240010, 24240021, 24300060 and 25540137 from the Ministry of Education, Culture, Sports, Science and Technology, Japan.

In order to avoid such difficulty, the unordered tree kernel have been developed as counting all of the specific substructures, instead of subtrees. For example, Kuboyama *et al.* [8] and Kimura *et al.* [7] have designed the unordered tree kernel counting all of the *bifoliate q-grams* and all of the *subpaths*, respectively. Note that the problem of counting their substructures is tractable and the labels of internal nodes are essential for these researches.

A *leaf-labeled* tree is a labeled tree such that all of the labels are assigned to just leaves. Also a *full binary tree* is a tree that the number of children of internal nodes is just two. A *phylogenetic tree* is an unordered leaf-labeled full binary tree to represent the phylogeny of taxa assigned to leaves [3,5,16]. Then, it is natural for the related works to the phylogenetic trees such as phylogeny reconstruction and comparison (*cf.*, [3,5,16]) to assume that every label in leaves is different. In particular, the purpose of phylogeny reconstruction is to reconstruct just one phylogenetic tree. Hence, for bioinformatics or machine learning, no researches to classify many phylogenetic trees have developed, because no problem setting with many phylogenetic trees have arisen yet.

Recently, as the method of analyzing positions in nucleotide sequences to influence the phylogeny, our previous works [11–13] have dealt with many phylogenetic trees as follows. First, we reconstruct a phylogenetic tree from nucleotide sequences, of which labels of leaves are indices of nucleotide sequences. Then, for every position in nucleotide sequences, we replace labels of leaves with nucleotides at the position. As a result, we obtain the phylogenetic trees whose number is same as the length of nucleotide sequences and whose labels are duplicated, and call them *relabeled phylogenetic trees*. Next, after applying the *label-based closest-neighbor trimming method* [11] to every relabeled phylogenetic tree, we obtain the *trimmed phylogenetic trees*. Hence, in our previous works, we have compared the trimmed phylogenetic trees to analyze the positions [11,12] and applied the method of clustering to them [13].

As the research related above, in this paper, we design a tree kernel for phylogenetic trees to classify the trimmed phylogenetic trees. Note that we cannot apply the same reduction to show the above #P-completeness [6] to phylogenetic trees, because they are full binary. Also it is ineffective to apply the unordered tree kernels counting all of the specific substructures [7,8] to phylogenetic trees, because they are leaf-labeled and no internal node in them is assigned to a label.

Hence, in this paper, we introduce a new tree kernel for phylogenetic trees counting *agreement subtree mappings*, as a kind of mapping kernels [15] and an extension of [14] from ordered trees to unordered trees, which we call an *agreement subtree mapping kernel*. Then, motivated by the algorithm to compute the number of all leaves in the *maximum agreement subtree* [5,16], we design the algorithm to compute the agreement subtree mapping kernel in quadratic time. Furthermore, since the value of agreement subtree mapping kernel is very large with respect to the number of nodes in phylogenetic trees, we introduce the *normalized* agreement subtree mapping kernel to classify the trimmed phylogenetic trees obtained from nucleotide sequence of influenza A (H1N1) viruses.

On the other hand, when relaxing the conditions of phylogenetic trees, we show that the problem of counting all of the agreement subtree mappings for leaf-labeled trees is #P-complete. Here, this proof is different from one in [6].

This paper is organized as follows. In Sect. 2, we prepare the notions including an *agreement subtree mapping* necessary to later discussion. In Sect. 3, we introduce an *agreement subtree mapping kernel* and design the algorithm to compute it. In Sect. 4, we give experimental results of the agreement subtree mapping kernel to classify pandemic viruses from non-pandemic viruses and viruses in one region from viruses in the other regions obtained from nucleotide sequences of influenza A (H1N1) viruses provided from NCBI database [1]. In Sect. 5, we show that the problem of counting all of the agreement subtree mappings for leaf-labeled trees is #P-complete. Section 6 concludes this paper.

2   Agreement Subtree Mapping

A *tree* is a connected graph without cycles. For a tree $T = (V, E)$, we denote V and E by $V(T)$ and $E(T)$, respectively. We sometime denote $v \in V(T)$ by $v \in T$. A *rooted tree* is a tree with one node r chosen as its *root*. We denote the root of a rooted tree T by $r(T)$.

For each node v in a rooted tree with the root r, let $UP_r(v)$ be the unique path from r to v. The *parent* of $v(\neq r)$, which we denote by $par(v)$, is its adjacent node on $UP_r(v)$ and the *ancestors* of $v(\neq r)$ are the nodes on $UP_r(v) - \{v\}$. We say that u is a *child* of v if v is the parent of u, and u is a *descendant* of v if v is an ancestor of u.

In this paper, we use the ancestor orders $<$ and $\leq$, that is, $u < v$ if v is an ancestor of u and $u \leq v$ if $u < v$ or $u = v$. We say that w is the *least common ancestor* of u and v, denoted by $u \sqcup v$, if $u \leq w, v \leq w$, and there exists no w' such that $w' < w, u \leq w'$ and $v \leq w'$. A *(complete) subtree of T rooted by v*, denoted by T^v, is a tree $T' = (V', E')$ such that $r(T') = v$, $V' = \{w \in V \mid w \leq v\}$ and $E' = \{(u, w) \in E \mid u, w \in V'\}$.

Two nodes with the common parent are called *siblings*. A *leaf* is a node having no children and an *internal node* otherwise. We denote the set of leaves of a rooted tree T by $lv(T)$.

A rooted tree is *unordered* if an order between siblings is not given. A rooted unordered tree is *leaf-labeled* if just leaves are labeled by some symbols drawn from an alphabet Σ, and *full binary* if every internal node has just 2 children. We denote the label of a leaf v in Σ by $l(v)$. We call a rooted unordered leaf-labeled full binary tree a *phylogenetic tree* [5].

We say that a phylogenetic tree is the *restricted subtree* of T with respect to $L \subseteq lv(T)$, which we denote by $T|L$ if:

1. $lv(T|L) = L$ and the internal nodes in $T|L$ are the least common ancestors of such leaves, and
2. the edges of $T|L$ preserve the ancestor order $\leq$ of T.

Note that this definition of $T|L$ is different from the standard definition [16] which uses a set $L \subseteq \Sigma$, instead of $L \subseteq lv(T)$, under the assumption that every label of leaves is different. On the other hand, in this paper, since we formulate a phylogenetic tree as a leaf-labeled full binary tree possible to have the same label in leaves, we adopt the set $L \subseteq lv(T)$.

We can obtain the restricted subtree $T|L$ of T by first removing leaves not in L and then contracting all internal nodes with at most one child. Hence, the restricted subtree is also a phylogenetic tree.

Definition 1 (Agreement subtree [16]). Let T_1 and T_2 be phylogenetic trees. Then, T' is an *agreement subtree* of T_1 and T_2 if there exist two sets $L_1 \subseteq lv(T_1)$ and $L_2 \subseteq lv(T_2)$ having a one-to-one and label-preserving correspondence between L_1 and L_2 such that $T' = T_1|L_1 = T_2|L_2$. A *maximum agreement subtree* of T_1 and T_2 is an agreement subtree of T_1 and T_2 with maximum number of leaves. We denote the number of leaves in the maximum agreement subtree of T_1 and T_2 by $\mathrm{MAST}(T_1, T_2)$.

Example 1. Consider phylogenetic trees T_1 and T_2 illustrated in Fig. 1 (left). Let $\{v_1, v_2, v_3, v_4\}$ and $\{w_1, w_2, w_3, w_4\}$ be the sets of leaves in T_1 and T_2 from left to right. Also let $L_1^1 = \{v_1, v_3\}$, $L_2^1 = \{w_1, w_3\}$, $L_1^2 = \{v_3, v_4\}$, $L_2^2 = \{w_3, w_4\}$, $L_1^3 = \{v_1, v_3, v_4\}$ and $L_2^3 = \{w_1, w_3, w_4\}$.

For T_1', T_2' and T_3' illustrated in Fig. 1 (right), it holds that $T_1' = T_1|L_1^1 = T_2|L_2^1$, $T_2' = T_1|L_1^2 = T_2|L_2^2$ and $T_3' = T_1|L_1^3 = T_2|L_2^3$. In particular, T_3' is the maximum agreement subtree of T_1 and T_2.

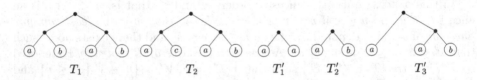

Fig. 1. Phylogenetic trees T_1 and T_2 (left) and agreement subtrees T_1', T_2' and T_3' of T_1 and T_2 (right).

Let T_1 and T_2 be phylogenetic trees such that $|T_1| = n > m = |T_2|$. In Fig. 2, we show the recurrence equations from the literature [16], which gives the value of $\mathrm{MAST}(T_1, T_2)$ in a bottom-up way in $O(n^2)$ time. Here, $v \in T_1$, $w \in T_2$, v_1 and v_2 are the children of v and w_1 and w_2 are the children of w. The fastest algorithm to compute $\mathrm{MAST}(T_1, T_2)$ is known as [5], which runs in $O(n^{1.5} \log n)$ time.

Next, we introduce an agreement subtree mapping to characterize agreement subtrees.

Definition 2 (Mapping [17]). Let T_1 and T_2 be phylogenetic trees. Then, we say that $M \subseteq T_1 \times T_2$ is a *(Tai) mapping* between T_1 and T_2 if M satisfies the following conditions.

$$\mathrm{mast}(v, w) = \begin{cases} 1, & l(v) = l(w), \\ 0, & l(v) \neq l(w), \end{cases}$$

$$\mathrm{mast}(T_1^v, T_2^w) = \max \left\{ \begin{array}{ll} \mathrm{mast}(T_1^v, T_2^{w_1}), & \mathrm{mast}(T_1^v, T_2^{w_2}), \\ \mathrm{mast}(T_1^{v_1}, T_2^w), & \mathrm{mast}(T_1^{v_2}, T_2^w), \\ \mathrm{mast}(T_1^{v_1}, T_2^{w_1}) + \mathrm{mast}(T_1^{v_2}, T_2^{w_2}), \\ \mathrm{mast}(T_1^{v_1}, T_2^{w_2}) + \mathrm{mast}(T_1^{v_2}, T_2^{w_1}) \end{array} \right\}.$$

Fig. 2. The equations to compute $\mathrm{MAST}(T_1, T_2)$.

1. $\forall (v_1, w_1), (v_2, w_2) \in M \Big(v_1 = v_2 \iff w_1 = w_2 \Big).$
2. $\forall (v_1, w_1), (v_2, w_2) \in M \Big(v_1 \leq v_2 \iff w_1 \leq w_2 \Big).$

For a mapping M phylogenetic trees T_1 and T_2, let M^{lv} be $M \cap (lv(T_1) \times lv(T_2))$.

Definition 3 (Agreement subtree mapping). Let T_1 and T_2 be phylogenetic trees and M a mapping between T_1 and T_2. Then, we say that M is an *agreement subtree mapping* if M satisfies the following conditions.

1. $\forall (v, w) \in M \Big(v \in lv(T_1) \iff w \in lv(T_2) \Big).$
2. $\forall (v, w) \in M^{lv} \Big(l(v) = l(w) \Big).$
3. $\forall (v_1, w_1), (v_2, w_2) \in M^{lv} \Big((v_1 \sqcup v_2, w_1 \sqcup w_2) \in M \Big).$
4. $\forall (v, w) \in M - M^{lv} \ \exists (v_1, w_1), (v_2, w_2) \in M^{lv} \Big((v = v_1 \sqcup v_2) \wedge (w = w_1 \sqcup w_2) \Big).$

Example 2. Consider the same phylogenetic trees T_1 and T_2 in Example 1 (Fig. 1). Also consider four mappings M_1, M_2, M_3 and M_4 between T_1 and T_2 illustrated in Fig. 3.

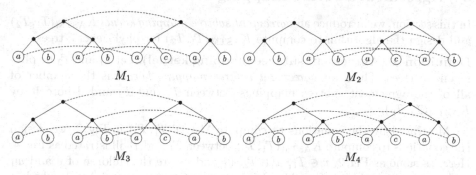

Fig. 3. Mappings M_1, M_2, M_3 and M_4 between T_1 and T_2.

Then, since M_1, M_2 and M_3 satisfies all the conditions in Definition 3, they are agreement subtree mappings between T_1 and T_2. They are corresponding to the agreement subtrees T_1', T_2' and T_3' in Fig. 1, respectively. On the other hand, M_4 satisfies the conditions 1, 2 and 3 but it does not satisfy the condition 4. Then, M_4 is not an agreement subtree mapping[1].

The following theorem shows that the agreement subtree of T_1 and T_2 is corresponding to the agreement subtree mapping between T_1 and T_2 and vice versa.

Theorem 1. *Let T_1 and T_2 be phylogenetic trees. Then, the following statements hold.*

1. *If T is an agreement subtree of T_1 and T_2 such that $T = T_1|L_1 = T_2|L_2$, then there exists an agreement subtree mapping M between T_1 and T_2 such that $M^{lv} = L_1 \times L_2$.*
2. *If M is an agreement subtree mapping between T_1 and T_2, then there exist two sets $L_1 \subseteq lv(T_1)$ and $L_2 \subseteq lv(T_2)$ such that $T_1|L_1(= T_2|L_2)$ is an agreement subtree of T_1 and T_2.*

Proof. 1. Let φ be the one-to-one and label-preserving correspondence between L_1 and L_2. Also let M be the mapping $\{(v, \varphi(v)) \mid v \in L_1\} \cup \{(v_1 \sqcup v_2, \varphi(v_1) \sqcup \varphi(v_2)) \mid v_1, v_2 \in L_1\}$. Then, since M satisfies all of the conditions in Definition 3, M is an agreement subtree mapping between T_1 and T_2. Also it holds that $M^{lv} = L_1 \times L_2$.

2. For an agreement subtree mapping M, let $L_1 = \{v \in T_1 \mid (v, w) \in M^{lv}\}$ and $L_2 = \{w \in T_2 \mid (v, w) \in M^{lv}\}$. From the conditions 1 and 2 in Definition 3, every element in M^{lv} represents the one-to-one and label-preserving correspondence between L_1 and L_2. Also the condition 3 guarantees that $(v_1 \sqcup v_2, w_1 \sqcup w_2) \in M$ for every $(v_1, w_1), (v_2, w_2) \in M^{lv}$, and the condition 4 guarantees that $M - M^{lv}$ contain no pairs of internal nodes at least one of which is not the least common ancestors of two leaves. Hence, it holds that $T_1|L_1 = T_2|L_2$ and $T_1|L_1$ is an agreement subtree of T_1 and T_2. □

3   Agreement Subtree Mapping Kernel

In this section, we introduce an *agreement subtree mapping kernel* $K_{\text{ASTM}}(T_1, T_2)$ and design the algorithm to compute $K_{\text{ASTM}}(T_1, T_2)$ for phylogenetic trees.

Definition 4 (Agreement subtree mapping kernel). Let T_1 and T_2 be phylogenetic trees. Then, an *agreement subtree mapping kernel* is the number of all of the agreement subtree mappings between T_1 and T_2 and denote it by $K_{\text{ASTM}}(T_1, T_2)$.

Motivated by the equations to compute $\text{MAST}(T_1, T_2)$ in Fig. 2, we design the equations to compute $K_{\text{ASTM}}(T_1, T_2)$ between T_1 and T_2 illustrated as Fig. 4. Here, as same as Fig. 2, $v \in T_1$, $w \in T_2$, v_1 and v_2 are the children of v and w_1 and w_2 are the children of w. Also λ is a constant such that $0 \leq \lambda \leq 1$, which is called a *decay factor* (*cf.*, [10]).

---

[1] The mapping satisfying the conditions 1 and 2 and the condition "$\forall (v_1, w_1), (v_2, w_2) \in M((v_1 \sqcup v_2, w_1 \sqcup w_2) \in M)$" is called an *LCA-preserving mapping* (*cf.*, [5,13]). Then, every M_i is an LCA-preserving mapping.

$$K_{\mathrm{ASTM}}(T_1, T_2) = \sum_{(v,w) \in T_1 \times T_2} K'(T_1^v, T_2^w),$$

$$K'(v, w) = K(v, w),$$

$$K'(v, T_2^w) = K'(T_1^v, w) = 0,$$

$$K'(T_1^v, T_2^w) = K(T_1^{v_1}, T_2^{w_1}) \cdot K(T_1^{v_2}, T_2^{w_2}) + K(T_1^{v_1}, T_2^{w_2}) \cdot K(T_1^{v_2}, T_2^{w_1}),$$

$$K(v, w) = \begin{cases} 1, & l(v) = l(w), \\ 0, & l(v) \neq l(w), \end{cases}$$

$$K(v, T_2^w) = K(v, T_2^{w_1}) + K(v, T_2^{w_2}),$$

$$K(T_1^v, w) = K(T_1^{v_1}, w) + K(T_1^{v_2}, w),$$

$$K(T_1^v, T_2^w) = \left(K(T_1^{v_1}, T_2^{w_1}) \cdot K(T_1^{v_2}, T_2^{w_2}) + K(T_1^{v_1}, T_2^{w_2}) \cdot K(T_1^{v_2}, T_2^{w_1}) \right) \lambda$$
$$+ K(T_1^v, T_2^{w_1}) + K(T_1^v, T_2^{w_2}) + K(T_1^{v_1}, T_2^w) + K(T_1^{v_2}, T_2^w).$$

Fig. 4. The equations to compute $K_{\mathrm{ASTM}}(T_1, T_2)$.

Theorem 2. *Let T_1 and T_2 be phylogenetic trees. When $\lambda = 1$, the equations of $K_{\mathrm{ASTM}}(T_1, T_2)$ in Fig. 4 count all of the agreement subtree mappings between T_1 and T_2 correctly in $O(|T_1||T_2|)$ time.*

Proof. The equation of K_{ASTM} in Fig. 4 counts the number of agreement subtree mappings between T_1^v and T_2^2 for every $(v, w) \in T_1 \times T_2$ by using the equations of K'. When counting the number of agreement subtree mappings between T_1^v and T_2^w in $K'(T_1^v, T_2^w)$, they contain no pairs consisting of ancestors of v in T_1 and w in T_2. Furthermore, $K'(T_1^v, T_2^w)$ counts the number of agreement subtree mappings when both v in T_1^v and w in T_1^w are either internal nodes or leaves by using the equation of K. Hence, it is sufficient to show that the equations of K counts the number of agreement subtree mappings uniquely.

Suppose that both T_1 and T_2 are leaves v and w. Then, $K(T_1, T_2)$ is 1 if both labels are same; 0 otherwise, which is represented by the first equation of K.

Suppose that T_1 is a leaf v and T_2 is not a leaf such that the root of T_2 is w, one descendant of w is $T_2^{w_1}$ and the other descendant of w is $T_2^{w_2}$. In this case, since no agreement subtree mapping contains a pair (v, w), $K(T_1, T_2)$ is the sum of $K(T_1, T_2^{w_1})$ and $K(T_1, T_2^{w_2})$, which is represented by the second equation of K, where $T_1 = v$. Under the same discussion, the case that T_2 is a leaf and T_1 is not a leaf is represented by the third equation of K.

Suppose that both T_1 and T_2 are phylogenetic trees that are not leaves such that the root of T_1 is v, the root of T_2 is w, one descendant of v is $T_1^{v_1}$, the other descendant of v is $T_1^{v_2}$, one descendant of w is $T_2^{w_1}$ and the other descendant of w is $T_2^{w_2}$. Then, since the number of agreement subtree mappings between $T_1^{v_i}$ and $T_2^{w_j}$ is $K(T_1^{v_i}, T_2^{w_j})$ for $i, j = 1, 2$, the number of agreement subtree mappings containing a pair (v, w) is $K(T_1^{v_1}, T_2^{w_1}) \cdot K(T_1^{v_2}, T_2^{w_2}) + K(T_1^{v_1}, T_2^{w_2}) \cdot K(T_1^{v_2}, T_2^{w_1})$. Also, since the number of agreement subtree mappings between T_1 and $T_2^{w_i}$ (*resp.*, $T_1^{v_i}$ and T_2) is $K(T_1, T_2^{w_i})$ (*resp.*, $K(T_1^{v_i}, T_2)$) for $i = 1, 2$, the number of agreement subtree mappings not containing a pair (v, w) is $K(T_1, T_2^{w_1}) + K(T_1, T_2^{w_2}) + K(T_1^{v_1}, T_2) + K(T_1^{v_2}, T_2)$. Hence, this case is represented by the fourth equation of K.

Since every pair $(v, w) \in T_1 \times T_2$ is called just once in computing $K_{\mathrm{ASTM}}(T_1, T_2)$ and the number of subproblems for (v, w) is bounded by eight, we can compute $K(T_1, T_2)$ in $O(|T_1||T_2|)$ time by using dynamic programming. □

Example 3. Consider the phylogenetic trees $T(a, n)$ in Fig. 5 (left). Then, Fig. 5 (right) illustrates the value of $K_{\mathrm{ASTM}}(T(a, n), T(a, n))$ for $1 \leq n \leq 12$ when $\lambda = 1$. In this case, the running time is about 0.1 seconds, while the value of K_{ASTM} is rapidly increasing when n is increasing.

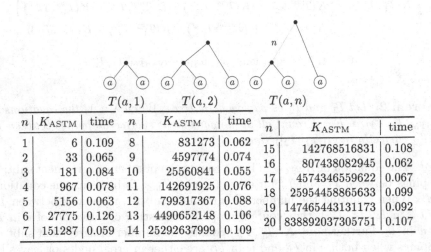

n	K_{ASTM}	time	n	K_{ASTM}	time
1	6	0.109	8	831273	0.062
2	33	0.065	9	4597774	0.074
3	181	0.084	10	25560841	0.055
4	967	0.078	11	142691925	0.076
5	5156	0.063	12	799317367	0.088
6	27775	0.126	13	4490652148	0.106
7	151287	0.059	14	25292637999	0.109

n	K_{ASTM}	time
15	142768516831	0.108
16	807438082945	0.062
17	4574346559622	0.067
18	25954458865633	0.099
19	147465443131173	0.092
20	838892037305751	0.107

Fig. 5. The phylogenetic trees $T(a, 1)$, $T(a, 2)$ and $T(a, n)$ (upper) and the value of $K_{\mathrm{ASTM}}(T(a, n), T(a, n))$ and its running time (sec.) for $1 \leq n \leq 20$ (lower) in Example 3.

As shown in Example 3, the value of $K_{\mathrm{ASTM}}(T_1, T_2)$ is very large. In order to apply $K_{\mathrm{ASTM}}(T_1, T_2)$ to SVMs, we normalize it as follows.

Definition 5 (Normalized agreement subtree mapping kernel). Let T_1 and T_2 be phylogenetic trees. Then, the *normalized agreement subtree mapping kernel* $N_{\mathrm{ASTM}}(T_1, T_2)$ is defined as follows. Here, $N_{\mathrm{ASTM}}(T_1, T_2)$ varies from 0 to 1.

$$N_{\mathrm{ASTM}}(T_1, T_2) = \frac{K_{\mathrm{ASTM}}(T_1, T_2)}{\sqrt{K_{\mathrm{ASTM}}(T_1, T_1)}\sqrt{K_{\mathrm{ASTM}}(T_2, T_2)}}.$$

4 Experimental Results

In this section, we give experimental results for the agreement subtree mapping kernel of phylogenetic trees.

First, in order to obtain many phylogenetic trees from nucleotide sequences, we explain trimmed phylogenetic trees according to [11,12].

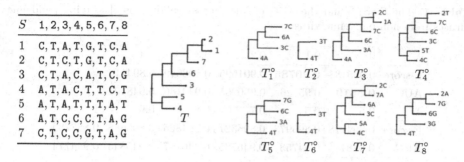

S	1, 2, 3, 4, 5, 6, 7, 8
1	C, T, A, T, G, T, C, A
2	C, T, C, T, G, T, C, A
3	C, T, A, C, A, T, C, G
4	A, T, A, C, T, T, C, T
5	A, T, A, T, T, T, A, T
6	A, T, C, C, C, T, A, G
7	C, T, C, C, G, T, A, G

Fig. 6. Nucleotide sequences S (left), the phylogenetic tree T of S (center) and the trimmed phylogenetic tree T_i° for $1 \le i \le 8$ (right) in Example 4.

As a reconstruction of a phylogenetic tree from nucleotide sequences, we adopt a *neighbor joining method* (*cf.*, [3, 16]) based on the Hamming distance between nucleotide sequences. The neighbor joining method first draws leaves labeled by an index for every nucleotide sequence. Then, it finds two nodes such that the sum of the length of the paths is minimum, draws a new node and connects the new node to the two nodes. Finally, it halts if the number of connected component is just two, and then connects the remained two components.

Next, we introduce a *label-based closest-neighbor trimming method* [11, 12]. A *branch* in a phylogenetic tree is a subtree represented by (v_1, v_2, w, d_1, d_2) such that v_1 and v_2 are leaves, w is the parent of v_1 and v_2 and d_i is the distance from v_i to w. We call a branch (v_1, v_2, w, d_1, d_2) such that $l(v_1) = l(v_2)$ and $d_1 + d_2$ is minimum in all the branches the *minimum label branch*. For a branch $b = (v_1, v_2, w, d_1, d_2)$ such that $d_i < d_j$ $(i, j = 1, 2, i \ne j)$, we call a new branch obtained by deleting the leaf v_j and the edge between w and v_j from b the *minimum leaf branch* of b. Then, a *label-based closest-neighbor trimming method* is a procedure to replace the minimum label branch b with the minimum leaf branch of b until no minimum label branch exists or the number of leaves is 2.

Let S be a set of nucleotide sequences with length n and T a phylogenetic tree reconstructed from S. Then, we can obtain n phylogenetic trees by relabeling an index of S as the leaves in T with the i-th nucleotide in S $(1 \le i \le n)$, which we call a *relabeled phylogenetic tree* at the position i and denote it by T_i. Furthermore, we call the phylogenetic tree obtained by applying the label-based closest-neighbor trimming method to T_i the *trimmed phylogenetic tree* at the position i and denote it by T_i°.

Example 4. Consider the set $S = \{1, 2, 3, 4, 5, 6, 7\}$ of nucleotide sequences illustrated in Fig. 6 (left) and suppose that T in Fig. 6 (center) is a phylogenetic tree reconstructed from S. Then, Fig. 6 (right) illustrates the trimmed phylogenetic trees T_i° for $1 \le i \le 8$ by applying label-based closest-neighbor trimming method to T. Here, the label of a leaf denotes the index of S and its nucleotide at the position i. In this case, it holds that $T_2^\circ = T_6^\circ$.

Table 1. The F-score and the AUC of 5-fold cross validations classifying pandemic viruses from non-pandemic viruses.

λ	0	0.1	0.2	0.3	0.4	
F-score	0.905188	**0.90785**	0.901695	0.898649	0.894118	
AUC	**0.9519**	0.95126	0.950261	0.949186	0.948401	
λ	0.5	0.6	0.7	0.8	0.9	1
F-score	0.894118	0.895973	0.896321	0.894825	0.893688	0.887789
AUC	0.9481	0.94753	0.946735	0.946197	0.945843	0.944714

In the remainder of this section, we give experimental results to classify influenza A (H1N1) viruses, by applying the normalized agreement subtree kernel to trimmed phylogenetic trees.

In our experiment, we use the positions in nucleotide sequences for HA segments of influenza A (H1N1) viruses at 2008 as non-pandemic viruses and at 2009 as pandemic viruses, provided from NCBI [1]. Here, the number of nucleotide sequences at 2008 is 326 and that at 2009 is 3344, respectively. After deleting the positions with the same nucleotide at 2008 or 2009, the number of positions commonly occurring in both 2008 and 2009 is 305. Then, we can obtain 305 relabeled and trimmed phylogenetic trees for 2008 and 2009, respectively.

In the first experiment, we classify pandemic viruses from non-pandemic viruses by using LIBSVM [2] through a gram matrix of N_{ASTM} between all of the pairs of the 305 trimmed phylogenetic trees. Table 1 illustrates the F-score and the AUC of 5-fold cross validations classifying pandemic viruses from non-pandemic viruses for every λ from 0 to 1 within 0.1 span. Here, the bold face denotes the maximum values.

Table 1 shows that the F-score is more than 0.90 and the AUC is more than 0.95 if $\lambda < 0.2$. In this case, the F-score is maximum for $\lambda = 0.1$ and the AUC is maximum for $\lambda = 0$.

Figure 7 illustrates the graphs of the F-score and the AUC by varying λ from 0 to 1. Figure 7 shows that, while the AUC decreases monotonically when λ increases, the F-score have two peaks near to 0.1 and 0.7.

In the second experiment, we compare the normalized agreement subtree mapping kernel N_{ASTM} with non-structured kernels as array kernels K_A and K_{A^w}, multiset kernels $K_\times$ and $K_\cap$ [4] and a string kernel K_S^k [9].

Let Σ be $\{\text{A}, \text{C}, \text{G}, \text{T}\}$. Then, for $x, y \in \Sigma$, we define $\delta_1(x, y) = 1$ if $x = y$; 0 otherwise. Also we define $\delta_2(x, y) = 1$ if $x = y$; $1/2$ if $(x, y) = (\text{A}, \text{T}), (\text{T}, \text{A}), (\text{C}, \text{G}), (\text{G}, \text{C})$ (that is, base pairs are weighted); 0 otherwise. Then, by regarding a nucleotide sequence as an array on Σ, we define two *array kernels* K_A and K_{A^w} for two arrays $X = (x_1, \ldots, x_n)$ and $Y = (y_1, \ldots, y_n)$ ($x_i, y_i \in \Sigma$) on Σ as follows.

$$K_A(X, Y) = \frac{1}{n} \sum_{i=1}^{n} \delta_1(x_i, y_i), \quad K_{A^w}(X, Y) = \frac{1}{n} \sum_{i=1}^{n} \delta_2(x_i, y_i)$$

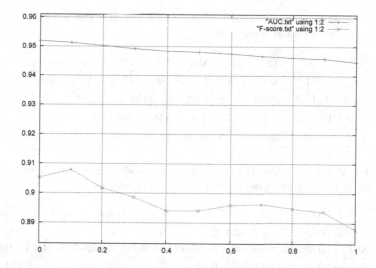

Fig. 7. The graphs of the F-score and the AUC by varying λ from 0 to 1.

We call $X \subseteq \Sigma \times \mathbf{N}$ a *multiset* on Σ. For a multiset X, we call an n in $(x, n) \in X$ the number of occurrences of x in X and denote it by $\Gamma_X(x)$. Then, by regarding a nucleotide sequence as a multiset on Σ, we define a *multiset product kernel* $K_\times$ and a *multiset intersection kernel* $K_\cap$ [4] for two multisets X and Y on Σ as follows.

$$K_\times(X, Y) = \sum_{a \in \Sigma} \Gamma_X(a) \cdot \Gamma_Y(a), \quad K_\cap(X, Y) = \sum_{a \in \Sigma} \min\{\Gamma_X(a), \Gamma_Y(a)\}.$$

For a string $X \in \Sigma^*$ and a substring $s \in \Sigma^*$ of X, let $\Gamma_X(s)$ be the number of occurrences of s in X. Also, for $k \in \mathbf{N}$, let Σ^k be $\{s \in \Sigma^* \mid |s| = k\}$. Then, by regarding a nucleotide sequence as a string on Σ, we define a *spectrum string kernel* K_S^k [9] for two strings X and Y on Σ as follows.

$$K_S^k(X, Y) = \sum_{s \in \Sigma^k} \Gamma_X(s) \cdot \Gamma_X(s).$$

In the second experiment, we divide 3670 nucleotide sequences at 2008 and 2009 provided from NCBI [1] into seven regions as Africa (AF), Asia (AS), Europe (EU), Middle East (ME), North America (NA), Oceania (OC) and South America (SA). Here, Table 2 shows the number of nucleotide sequences (#NS) and the number of trimmed phylogenetic trees (#PT) obtained by deleting the positions with the same nucleotide in seven regions.

When applying the non-structured kernels K_A, K_{A^w}, $K_\times$, $K_\cap$ and K_S^k, we construct a gram matrix with size 3670 direct from nucleotide sequences, and set one region as positive examples and the other regions as negative examples. For example, when classifying the region AF, the number of positive examples is 61 and the number of negative examples is $3670 - 61 = 3609$.

Table 2. The number of nucleotide sequences (#NS) and the number of trimmed phylogenetic trees (#PT) in seven regions.

	AF	AS	EU	ME	NA	OC	SA	Total
#NS	61	949	965	71	1403	47	174	3670
%	1.66	25.86	26.29	1.93	38.23	1.28	4.74	
#PT	289	593	487	311	538	290	344	2852
%	10.13	20.79	17.08	10.90	18.86	10.17	12.06	

Table 3. The F-value and the AUC of 5-fold cross validation classifying viruses in one region given at the first line from viruses in the other regions.

		AF	AS	EU	ME	NA	OC	SA
K_A	F-value	0	0.0290456	0	0	0	0	0
	AUC	0.622401	0.690439	0.657017	0.636456	0.662675	0.743741	0.64512
K_{A^w}	F-value	0	0.0125654	0	0	0	0	0
	AUC	0.628624	0.689144	0.650094	0.55904	0.662412	0.745045	0.646902
$K_\times$	F-value	0	0	0	0	0	0	0
	AUC	0.437949	0.50171	0.541192	0.437949	0.544779	0.549163	0.470115
$K_\cap$	F-value	0	0.127741	0.094518	0	0.257143	0	0
	AUC	0.445103	0.550467	0.616156	0.499689	0.593366	0.637869	0.562636
K_S^1	F-value	0	0	0	0	0	0	0
	AUC	0.516112	0.519856	0.498595	0.537559	0.542415	0.51617	0.463544
K_S^2	F-value	0	0.022798	0	0	0.478452	0	0
	AUC	0.495716	0.612197	0.596828	0.452125	0.666733	0.531046	0.660334
K_S^3	F-value	0	0.38897	0.351121	0	0.480042	0	0.12766
	AUC	0.713779	0.708284	0.708518	0.550685	0.720158	0.624082	0.825156
K_S^4	F-value	0.382353	0.53468	0.507246	0	0.546072	0.15534	0.375
	AUC	0.713779	0.708284	0.708518	0.550685	0.720158	0.624082	0.825156
K_S^5	F-value	0.361446	0.600724	0.544811	0.152381	0.593873	0.282132	0.361446
	AUC	0.793359	0.786081	0.759964	0.653691	0.763135	0.763571	0.934137
N_{ASTM}	F-value	0.911864	0.766284	0.929425	0.031348	0.830266	0.300493	0.753783
	AUC	0.947608	0.898753	0.978779	0.814773	0.955031	0.9333	0.919454

When applying the normalized agreement subtree mapping kernel N_{ASTM}, we construct a gram matrix with size 2852 from trimmed phylogenetic trees, and set one region as positive examples and the other regions as negative examples. For example, when classifying the region AF, the number of positive examples is 289 and the number of negative examples is $2852 - 289 = 2563$.

Table 3 illustrates the F-value and the AUC of 5-fold cross validation classifying viruses in one region given at the first line from viruses in the other regions by using LIBSVM through the kernels of K_A, K_{A^w}, $K_\times$, $K_\cap$, K_S^k (where $1 \le k \le 5$) and N_{ASTM} (under $\lambda = 0.1$).

Table 3 shows that, except the regions of ME and OC, the F-value and the AUC of N_{ASTM} are near to 1, while those of other non-structured kernels are near to 0 and 0.5, respectively. Hence, for the regions of AF, AS, EU, NA and SA, the normalized agreement subtree mapping kernel N_{ASTM} succeeds to classify viruses in one region from viruses in the other regions, while other non-structured kernels fail to classify. On the other hand, for the regions of ME and OC, every kernel fails to classify.

5   #P-Completeness of Computing Agreement Subtree Mapping Kernel for Leaf-Labeled Trees

One of the reason that the problem of computing the agreement subtree mapping kernel is tractable (Theorem 2) is that a phylogenetic tree is full binary. In this section, we discuss the same problem for leaf-labeled trees with unbounded degrees. Note that we can apply Definitions 1, 2, 3 and 4 to leaf-labeled trees. Also refer to [6, 18] to the notion of #P-completeness.

Theorem 3. *The problem of counting all of the agreement subtree mappings between two leaf-labeled trees is #P-complete.*

Proof. Valiant [18] has shown that the problem of counting all of the matchings in a bipartite graph, which we denote #BipartiteMatching, is #P-complete. Then, we give two leaf-labeled trees such that the number of all of the agreement subtree mappings between them is equal to the output of #Bipartite Matching. Here, for a set $S = \{T_1, \ldots, T_n\}$ of leaf-labeled trees, we denote a leaf-labeled tree such that the root is a node $\bullet$ and the children of the root is $T_1, \ldots, T_n$ by $\bullet(S)$.

Let $G = (X \cup Y, E)$ be a bipartite graph. For $v \in X \cup Y$, we denote a neighbor of v by $N(v)$. It is obvious that $N(v) \subseteq Y$ if $v \in X$ and $N(v) \subseteq X$ if $v \in Y$.

Then, we construct $T_x = \bullet(\{xy \mid y \in N(x)\})$ for every $x \in X$ and $T_1 = \bullet(\{T_x \mid x \in X\})$. Similarly, we construct $T_y = \bullet(\{xy \mid x \in N(y)\})$ for every $y \in Y$ and $T_2 = \bullet(\{T_y \mid y \in Y\})$. Here, we regard xy as a leaf in T_x and T_y and its label. Figure 8 illustrates an example of the above construction of T_1 and T_2 from a bipartite graph G.

For a matching $B \subseteq E$ in G, we construct the agreement subtree mapping M between T_1 and T_2 such that $M_B = \{(xy, xy) \mid (x, y) \in B\}$ and $M = M_B \cup \{(r(T_1), r(T_2))\}$ if $|M_B| \geq 2$; $M = M_B$ if $|M_B| \leq 1$. For example, let B be a matching $\{(1, 2), (2, 1), (3, 3)\}$ in G illustrated in Fig. 8 as think lines. Then, the agreement subtree mapping M between T_1 and T_2 is $\{(12, 12), (21, 21), (33, 33), (r(T_1), r(T_2))\}$, illustrated by dashed lines.

Note that all of the labels in leaves in T_1 and T_2 are mutually distinct, because the label is corresponding to an edge in G. Then, by the definition of a matching, xy such that $(xy, xy) \in M_B$ is selected from T_x in T_1 and T_y in T_2 uniquely, and M_B also selects at most one mapped node from T_x in T_1 and T_y in T_2. Furthermore, when $|M_B| \geq 2$, it holds that $v_1 \sqcup v_2 = r(T_1)$ for every

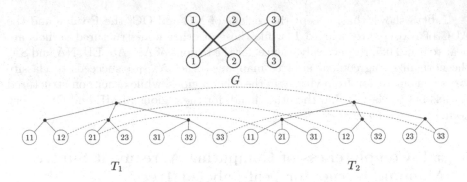

Fig. 8. A bipartite graph G and the leaf-labeled trees T_1 and T_2.

distinct $v_1, v_2 \in \{v \in T_1 \mid (v, w) \in M\}$ and $w_1 \sqcup w_2 = r(T_2)$ for every distinct $w_1, w_2 \in \{w \in T_2 \mid (v, w) \in M\}$.

Hence, a matching B in G determines the agreement subtree mapping M between T_1 and T_2 uniquely and vice versa. Then, the number of all of the matchings in G which is the output of #BipartiteMatching is equal to the number of all of the agreement subtree mappings between T_1 and T_2.   □

In the proof of Theorem 3, it is essential that all of the labels in leaves are mutually distinct and the degrees are unbounded. Also the proof of Theorem 3 is simple and direct rather than the proof of #P-completeness in [6].

Note that the leaf-labeled tree is a restricted unordered tree and the agreement subtree mapping is a restricted variation of mappings, that is, the restriction of an LCA-preserving mapping [19]. Hence, Theorem 3 also suggests the impossibility of designing a tractable mapping kernel for unordered trees, in particular, a mapping kernel containing the problem of counting pairs of leaves.

6   Conclusion

In this paper, in order to classify many phylogenetic trees reconstructed from nucleotide sequences, we have introduced an agreement subtree mapping kernel counting all of the agreement subtree mappings and designed the algorithm to compute it for phylogenetic trees in quadratic time. Then, we have given experimental results by using nucleotide sequences of influenza A (H1N1) viruses, and classified pandemic viruses from non-pandemic viruses and viruses in one region from viruses in the other regions by using LIBSVM. In particular, in the second experiment, for the five regions, we have succeeded to classify viruses in one region from viruses in the other regions by using the agreement subtree mapping kernel, while failed to classify by using non-structured kernels. On the other hand, we have shown that the problem of counting all of the agreement subtree mappings for leaf-labeled trees is #P-complete.

It is a future work to apply another nucleotide sequences to classify phylogenetic trees, for example, positions in packaging signals (*cf.*, [12]). Also it is a future work to design another kernel for phylogenetic trees based on counting all of the paths between leaves similar as [8] and compare the performance with one of the agreement subtree mapping kernel. Furthermore, it is an important theoretical future work to investigate whether the problem of counting all of the agreement subtree mappings for leaf-labeled trees either with duplicated labels or with bounded degrees greater than 2 is tractable or #P-complete.

References

1. Bao, Y., Bolotov, P., Dernovoy, D., Kiryutin, B., Zaslavsky, L., Tatusova, T., Ostell, J., Lipman, D.: The influenza virus resource at the national center for biotechnology information. J. Virol. **82**, 596–601 (2008). http://www.ncbi.nlm. gov/genomes/FLU/
2. Chang, C.-C., Lin, C.-J.: LIBSVM - A library for support vector machine (version 3.17) (2013). http://www.csie.ntu.edu.tw/~cjlin/libsvm
3. Durbin, R., Eddy, S., Krogh, A., Mitchison, G.: Biological Sequence Analysis: Probablistic Models of Proteins and Nucleic Acids. Cambridge University Press, Cambridge (1998)
4. Gärtner, T.: Kernels for Structured Data. World Scientific, Norwich (2008)
5. Kao, M.-Y., Lam, T.-W., Sung, W.-K., Ting, H.-F.: An even faster and more unifying algorithm for comparing trees via unlabeled bipartite matchings. J. Algo. **40**, 212–233 (2001)
6. Kashima, H., Sakamoto, H., Koyanagi, T.: Tree kernels. J. JSAI **21**, 1–9 (2006). (in Japanese)
7. Kimura, D., Kuboyama, T., Shibuya, T., Kashima, H.: A subpath kernel for rooted unordered trees. J. JSAI **26**, 473–482 (2011). (in Japanese)
8. Kuboyama, T., Hirata, K., Aoki-Kinoshita, K.F.: An efficient unordered tree kernel and its application to glycan classification. In: Washio, T., Suzuki, E., Ting, K.M., Inokuchi, A. (eds.) PAKDD 2008. LNCS (LNAI), vol. 5012, pp. 184–195. Springer, Heidelberg (2008)
9. Leslie, C.S., Eskin, E., Noble, W.S.: The spectrum kernel: a string kernel for SVM protein classification. In: Proceedings of PSB 2002, pp. 566–575 (2002)
10. Lodhi, H., Saunders, C., Shawe-Taylor, J., Cristianini, N., Watkins, C.: Text classification using string kernels. J. Mach. Learn. Res. **2**, 419–444 (2002)
11. Makino, S., Shimada, T., Hirata, K., Yonezawa, K., Ito, K.: A trim distance between positions in nucleotide sequences. In: Ganascia, J.-G., Lenca, P., Petit, J.-M. (eds.) DS 2012. LNCS, vol. 7569, pp. 81–94. Springer, Heidelberg (2012)
12. Makino, S., Shimada, T., Hirata, K., Yonezawa, K., Ito, K.: A trim distance between positions as packaging signals in H3N2 influenza viruses. In: Proceedings of SCIS-ISIS 2012, pp. 1702–1707 (2012)
13. Shimada, T., Hamada, I., Hirata, K., Kuboyama, T., Yonezawa, K., Ito, K.: Clustering of positions in nucleotide sequences by trim distance. In: Proceedings of IIAI AAI 2013, pp. 129–134 (2013)
14. Shin, K., Kuboyama, T.: Kernels based on distributions of agreement subtrees. In: Wobcke, W., Zhang, M. (eds.) AI 2008. LNCS (LNAI), vol. 5360, pp. 236–246. Springer, Heidelberg (2008)
15. Shin, K., Kuboyama, T.: A generalization of Haussler's convolutioin kernel - mapping kernel and its application to tree kernels. J. Comput. Sci. Tech. **25**, 1040–1054 (2010)

16. Sung, W.-K.: Algorithms in Bioinformatics: A Practical Introduction. Chapman & Hall/CRC, Boca Raton (2009)
17. Tai, K.-C.: The tree-to-tree correction problem. J. ACM **26**, 422–433 (1979)
18. Valiant, L.G.: The complexity of enumeration and reliablity problems. SIAM J. Comput. **8**, 410–421 (1979)
19. Zhang, K., Wang, J., Shasha, D.: On the editing distance between undirected acyclic graphs. Int. J. Found. Comput. Sci. **7**, 43–58 (1995)

A Comprehensive Study of Tree Kernels

Kilho Shin[1] and Tetsuji Kuboyama[2]([✉])

[1] Graduate School of Applied Informatics, University of Hyogo,
7-1-28 Minatojima-minamimachi, Chuo-ku, Kobe, Hyogo 650-0047, Japan
yshin@ai.u-hyogo.ac.jp
[2] Computer Centre, Gakushuin University, 1-5-1 Mejiro,
Toshima, Tokyo 171-8588, Japan
ori-dds2013@tk.cc.gakushuin.ac.jp

Abstract. Tree kernels are an effective method to capture the structural information of tree data of various applications and many algorithms have been proposed. Nevertheless, we do not have sufficient knowledge about how to select good kernels. To answer this question, we focus on 32 tree kernel algorithms defined within a certain framework to engineer positive definite kernels, and investigate them under two different parameter settings. The result is amazing. Three of the 64 tree kernels outperform the others, and their superiority proves statistically significant through t-tests. These kernels include the benchmark tree kernels proposed in the literature, while many of them are introduced and tested for the first time in this paper.

1 Introduction

Trees are an important type of data which is widely used in various application fields. Tasks to learn some hidden structures from texts are an important focus of natural language processing problems and parse trees are a typical representation of such target structures. Mark-up languages such as SGML, HTML and XML define tree generation syntax and resulting documents are naturally dealt with as trees. In biochemistry and structural biology, secondary structures of biopolymers such as proteins and nucleic acids have significant meaning, therefore, representing them as trees certainly yields practical advantages. In evolutionary biology, evolutionary trees are used to represent relationships among biological species.

On the other hand, tree kernels are useful tools to capture structural information of such kinds of tree data. The first characteristic of tree kernels is that a kernel is a two variable function used to evaluate the similarity between trees specified for the argument. More importantly, tree kernels implicitly project tree data into Hilbert spaces and, consequently, we can analyze the structural information captured through the kernels effectively and efficiently by taking advantage of a variety of multivariate analysis techniques.

In the literature, an efficient framework to design kernels for discretely abstract structured data is known and many tree kernels have been engineered

© Springer International Publishing Switzerland 2014
Y. Nakano et al. (Eds.): JSAI-isAI 2013, LNAI 8417, pp. 337–351, 2014.
DOI: 10.1007/978-3-319-10061-6_22

within this framework. The framework is known as the *convolution kernel* [6] or the *mapping kernel* [18].

In this paper, we investigate 32 tree kernels designed within this framework, some which have been presented in the literature and some which are introduced for the first time in this paper. Not only did we investigate their theoretical properties, that is, the positive definiteness and the computational complexity, but we also try to answer the question of which kernels we should use to obtain good results.

To answer this question, we ran intensive experiments using ten datasets selected from various application fields. Because we generate ten pairs of training and test data subsets from each dataset, the total number of datasets that we use is 100. On the other hand, each kernel is performed under two different settings for the parameters included in the kernel definition. Both of the settings are derived from conventional methodologies for designing tree kernels presented in the literature. One is to derive tree kernels from edit distance metrics while the other is based on the simple idea of counting substructure pairs shared between trees.

The conclusion is amazing. Three of the $32 \times 2 = 64$ kernels significantly outperformed the others and we prove the superiority statistically through pairwise t-tests. These kernels are a type of counting shared substructures. In particular, the kernel to count shared contiguous paths is included, which can be computed very efficiently and applicable to *unordered* trees. This is a clear contrast with the other kernels that are only applicable to *ordered* trees.

2   Preface

In this paper, by a tree, we mean a rooted, labeled and ordered tree. Hence, for the vertices of a tree, not only the generation (ancestor-descendant) order $<$ but also the order derived from the post-order traversal (called *post-order* for simplicity) are given. In particular, when we write $x' < y'$ for two vertices x' and y', we mean y' is an ancestor of x'.

Γ_x denotes the set of vertices of a tree x. A substructure of x is an arbitrary subset of Γ_x and it inherits both of the orders from x. In particular, if a substructure has a root, which means the maximum vertex with respect to the inherited generation order, we call the substructure a *subtree*. We also use the term *forest* to indicate a substructure because any substructure can be decomposed into one or more disjoint subtrees.

Furthermore, a substructure $\bar{x} \subseteq \Gamma_x$ is said to be *contiguous*, if $x' < z' < y' \wedge x' \in \bar{x} \wedge y' \in \bar{x}$ always implies $z' \in \bar{x}$. If $\bar{x}$ is not necessarily contiguous, we say that $\bar{x}$ is *sparse*.

For trees x and y, two substructures of $\bar{x} \subseteq \Gamma_x$ and $\bar{y} \subseteq \Gamma_y$ are said to be *isomorphic* if there exists a bijective mapping from $\bar{x}$ to $\bar{y}$ that preserves the generation order and the post-order. Moreover, if this isomorphism also preserves labels of vertices, we say that $\bar{x}$ and $\bar{y}$ are *congruent*.

3   Tree Kernels

There exist two important methodologies in the literature to design kernels for trees. One is to derive kernels from edit distance metrics and the other is to count shared substructures of focused types between trees. These methodologies can be unified into a common framework by means of the mapping kernel [18].

3.1   Deriving Kernels from Edit Distance Metrics

The most important edit distance metric for trees should be the Taï distance introduced in [19]. To define the Taï distance, we need to introduce three types of edit operations:

1. Replace the label of a vertex with a new label. When x' and y' denote the vertices with the old and new labels, we denote this operation by $\langle x' \rightarrow y' \rangle$.
2. Delete the vertex x'. The child vertices of x' if present are redefined to be children of the parent vertex of x'. We denote this operation by $\langle x' \rightarrow \bullet \rangle$.
3. Insert the vertex y' below the vertex z'. Furthermore, we can specify a subset of the child vertices of z' as the children of y'. We denote this operation by $\langle \bullet \rightarrow y' \rangle$.

Given two trees x and y, an edit script σ from x to y is a finite sequence of edit operations such that the included operations sequentially apply to x to transform x into y.

In addition, a non-negative cost is assigned to each operation (denoted by $\gamma(x' \rightarrow y')$, $\gamma(x' \rightarrow \bullet)$ and $\gamma(\bullet \rightarrow y')$), and the cost $\gamma(\sigma)$ of an edit script σ is defined as the sum of the costs of the operations that σ comprises. Finally, the Taï distance $d_T(x, y)$ between x and y is the minimum of the cost of the edit scripts that transform x into y.

The notion of *mappings*, which are also known as *traces*, is important to extend the Taï distance to computationally more contractible metrics: The mapping of a script σ is a pair of vectors of vertices $\langle \boldsymbol{x}', \boldsymbol{y}' \rangle = \langle (x'_1, \ldots, x'_d), (y'_1, \ldots, y'_d) \rangle$ such that x_i's are vertices of x, y_i's are vertices of y, and $\langle x'_i \rightarrow y'_i \rangle$'s are equivalent to the substituting operations included in σ. The importance of the mapping is partly contained in the following formula:

$$\gamma(\sigma) = \sum_{i=1}^{d} \gamma(x'_i \rightarrow y'_i) + \sum_{x' \in \Gamma_x \setminus \boldsymbol{x}'} \gamma(x' \rightarrow \bullet) + \sum_{y' \in \Gamma_y \setminus \boldsymbol{y}'} \gamma(\bullet \rightarrow y').$$

This formula indicates that the mapping $\langle \boldsymbol{x}', \boldsymbol{y}' \rangle$ uniquely determines $\gamma(\sigma)$. Therefore, we also denote the right-hand side of the formula by $\gamma(\langle \boldsymbol{x}', \boldsymbol{y}' \rangle)$. Taï has proven that $\langle \boldsymbol{x}', \boldsymbol{y}' \rangle$ is the mapping of some edit script, if, and only if, $x'_i \rightarrow y'_i$ determines an isomorphism between the substructures $\boldsymbol{x}'$ and $\boldsymbol{y}'$ of x and y. Finally, we let $M_{x,y}^T$ denote the set that comprises all such $\langle \boldsymbol{x}', \boldsymbol{y}' \rangle$.

In contrast with the Taï distance, the way to define the less constrained distance [13], the constrained edit distance [21] and the Lu distance [14] is to determine the set of mappings $M_{x,y}^{\mathsf{LC}}$, $M_{x,y}^{\mathsf{C}}$ and $M_{x,y}^{\mathsf{L}}$ first and then define $d_\$(x,y) = \min\limits_{\langle \boldsymbol{x}',\boldsymbol{y}'\rangle \in M_{x,y}^\$} \gamma(\langle \boldsymbol{x}',\boldsymbol{y}'\rangle)$ for $\$ \in \{\mathsf{LC},\mathsf{C},\mathsf{L}\}$. To define $M_{x,y}^\$$, we need to define the *least common ancestor* of two vertices in a tree.

Definition 1 (Least common ancestors). *For $v,w \in \Gamma_x$, the least common ancestor of v and w in x is the least element of $\{u \in \Gamma_x \mid v \leq u \wedge w \leq u\}$, where $v < u$ means u is an ancestor of v, and is denoted by $v \smile w$.*

Then $M_{x,y}^\$$ are defined as follows.

$$M_{x,y}^{\mathsf{C}} = \Big\{ \langle (x_1',\ldots,x_m'),(y_1',\ldots,y_m') \rangle \in M_{x,y}^{\mathsf{T}} \mid \forall(i,j,k)$$
$$\big[x_i' \nleq x_k' \wedge x_i' \smile x_k' = x_j' \smile x_k' \Leftrightarrow y_i' \nleq y_k' \wedge y_i' \smile y_j' = y_i' \smile y_k' \big] \Big\}$$

$$M_{x,y}^{\mathsf{LC}} = \Big\{ \langle (x_1',\ldots,x_m'),(y_1',\ldots,y_m') \rangle \in M_{x,y}^{\mathsf{T}} \mid$$
$$\forall(i,j,k) \big[x_i' \smile x_j' < x_i' \smile x_k' \Rightarrow y_i' \smile y_k' = y_j' \smile y_k' \big] \Big\}$$

$$M_{x,y}^{\mathsf{L}} = \Big\{ \langle (x_1',\ldots,x_d'),(y_1',\ldots,y_d') \rangle \in M_{x,y}^{\mathsf{T}} \mid$$
$$\forall(i,j,k) \big[x_i' \smile x_j' = x_i' \smile x_k' \Leftrightarrow y_i' \smile y_j' = y_i' \smile y_k' \big] \Big\}$$

Because $M_{x,y}^{\mathsf{L}} \subseteq M_{x,y}^{\mathsf{C}} \subseteq M_{x,y}^{\mathsf{LC}} \subseteq M_{x,y}^{\mathsf{T}}$ [9], $d_{\mathsf{L}}(x,y) \geq d_{\mathsf{C}}(x,y) \geq d_{\mathsf{LC}}(x,y) \geq d_{\mathsf{T}}(x,y)$ holds true.

Based on the relationship between $d_{\mathsf{T}}(x,y)$ and $M_{x,y}^{\mathsf{T}}$, for example, when we define a new kernel by

$$K_c'(x,y) = \sum_{\langle \boldsymbol{x}',\boldsymbol{y}'\rangle \in M_{x,y}^{\mathsf{T}}} e^{-c\gamma(\langle \boldsymbol{x}',\boldsymbol{y}'\rangle)},$$

$d_{\mathsf{T}}(x,y) = \lim\limits_{c\to\infty} -\frac{1}{c}K_c'(x,y)$ follows the soft-minimum approximation $\min\{a_1,\ldots,$ $a_n\} = \lim\limits_{c\to\infty} -\frac{1}{c}\ln\sum\limits_{i=1}^{n} e^{-ca_i}$. We further modify this formula, and obtain:

$$K_c'(x,y) = \prod_{x' \in \Gamma_x} e^{-c\gamma(x'\to\bullet)} \cdot \prod_{y' \in \Gamma_y} e^{-c\gamma(\bullet\to y')} \cdot K^{\mathrm{TAI}}(x,y);$$

$$K^{\mathrm{TAI}}(x,y) = \sum_{\langle \boldsymbol{x}',\boldsymbol{y}'\rangle \in M_{x,y}^{\mathsf{T}}} \prod_{i=1}^{|\boldsymbol{x}'|} \frac{e^{-c\gamma(x_i'\to y_i')}}{e^{-c\gamma(x_i'\to\bullet)}e^{-c\gamma(\bullet\to y_i')}}.$$

When we assume that the cost function γ is symmetric, $K'(x,y)$ is positive definite, if, and only if, $K^{\mathrm{TAI}}(x,y)$ is positive definite. Furthermore, we can define $K^{\mathsf{C}}(x,y)$, $K^{\mathsf{L}}(x,y)$ and $K^{\mathsf{LC}}(x,y)$ by replacing $M_{x,y}^{\mathsf{T}}$ with $M_{x,y}^{\mathsf{C}}$, $M_{x,y}^{\mathsf{L}}$ and $M_{x,y}^{\mathsf{LC}}$.

3.2   Counting Substructures

The other methodology to define tree kernels is based on the simple idea that more similar trees must share more substructures.

The *parse tree kernel* [2] should be the first tree kernel proposed using this methodology. In fact, the parse tree kernel counts the *congruent* pairs of subtrees of a certain shape (called *corooted* subtrees) shared between two trees x and y. We denote the set of the isomorphic pairs of corooted subtrees by $M^{\mathsf{P}}_{x,y}$. For $\langle x', y' \rangle \in M^{\mathsf{P}}_{x,y}$, x' and y' are corooted subtrees of x and y, and they are isomorphic to each other. Hence, the parse tree kernel can be described as

$K^{\mathsf{P}}(x,y) = \displaystyle\sum_{\langle x', y' \rangle \in M^{\mathsf{P}}_{x,y}} \prod_{i=1}^{|x'|} \delta_{x'_i, y'_i}$, where $\delta_{x'_i, y'_i} = 1$, if the labels of x'_i and y'_i are identical, and $\delta_{x'_i, y'_i} = 0$, otherwise.

On the other hand, the *elastic kernel* [7] relaxes the condition of the parse tree kernel on the substructure pairs to count up. In fact, while a corooted subtree is always contiguous, the scope of the elastic kernel contains sparse subtrees. When we let $M^{\mathsf{E}}_{x,y}$ denote the set of isomorphic subtree pairs that the elastic kernel counts up, the elastic kernel is defined by $K^{\mathsf{E}}(x,y) = \displaystyle\sum_{\langle x', y' \rangle \in M^{\mathsf{E}}_{x,y}} \prod_{i=1}^{|x'|} \lambda \delta_{x'_i, y'_i}$ with a decay factor $0 < \lambda \le 1$.

3.3   The Unified Framework

The tree kernels designed based on the aforementioned methodologies are commonly of the form of

$$K(x,y) = \sum_{\langle x', y' \rangle \in M_{x,y}} \prod_{i=1}^{|x'|} \kappa(x'_i, y'_i),$$

where $\langle x', y' \rangle \in M_{x,y}$ is an isomorphic pair of substructures. In general, kernels of this form are called *mapping kernels*. The mapping kernel was studied in [18]. The following theorem presents an important property of the mapping kernel.

Definition 2. *Let* $\mathcal{M} = \{M_{x,y} \mid (x,y) \in \chi \times \chi\}$ *with* $M_{x,y} \subseteq \chi' \times \chi'$, *where* χ *denotes the space of data objects, and* $\chi' \subseteq \chi$ *denotes the base space of data objects.* $\mathcal{M}$ *is transitive, if, and only if, the following conditions always hold true.*

- $(x', y') \in M_{x,y} \Rightarrow (y', x') \in M_{y,x}.$
- $(x', y') \in M_{x,y} \land (y', z') \in M_{y,z} \Rightarrow (x', z') \in M_{x,z}.$

Theorem 1 ([18]). *The following conditions are equivalent.*

1. $\ker xy$ is positive definite for any positive definite κ.
2. $\mathcal{M}$ is transitive.

For $\$ \in \{\mathsf{T}, \mathsf{C}, \mathsf{L}, \mathsf{E}, \mathsf{P}\}$, $\{M^{\$}_{x,y} \mid (x,y) \in \chi \times \chi\}$ are transitive. This can be easily verified from their definition (see also [9]). On the other hand, $\{M^{\mathsf{LC}}_{x,y}\}$ is not transitive, and the trees x, y and z below shows this. In fact, when $x|a$, for example, denotes the vertex of x whose label is a, the following hold.

$$\langle (x|a, x|b, x|c, x|d), (y|a, y|b, y|c, y|d) \rangle \in M^{\mathsf{LC}}_{x,y}$$

$$\langle (y|a, y|b, y|c, y|d), (z|a, z|b, z|c, z|d) \rangle \in M^{\mathsf{LC}}_{y,z}$$

$$\langle (x|a, x|b, x|c, x|d), (z|a, z|b, z|c, z|d) \rangle \notin M^{\mathsf{LC}}_{x,z}$$

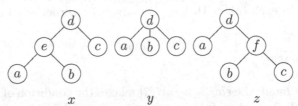

On the other hand, to have a positive definite κ, we let

$$\kappa(x', y') = \begin{cases} \alpha & \text{if } x' = y'; \\ \beta & \text{if } x' \neq y'. \end{cases}$$

with the adjustable parameters α and β such that $0 \leq \beta \leq \alpha$.

This definition covers not only κ used in the parse tree and elastic kernels but also κ derived from the cost function γ of the edit distance. In fact, assuming the common setting of $\gamma(x' \to x') = 0$, $\gamma(x' \to y') = a$, $\gamma(x' \to \bullet) = \gamma(\bullet \to y') = b$ and $0 \leq a \leq 2b$ ($a \leq 2b$ comes from the triangle inequality), we have

$$1 \leq \kappa(x', y') = \frac{e^{-c\gamma(x' \to y')}}{e^{-c\gamma(x' \to \bullet)} e^{-c\gamma(\bullet \to y')}} = e^{(2b-a)c}$$

$$\leq \kappa(x', x') = \frac{e^{-c\gamma(x' \to x')}}{e^{-c\gamma(x' \to \bullet)} e^{-c\gamma(\bullet \to x')}} = e^{2bc}.$$

Thus, in Sect. 4, we investigate two settings of the parameters: $1 \leq \beta \leq \alpha$ and $0 \leq \beta \leq \alpha \leq 1$: The former corresponds to the kernels derived from the edit distances while the latter covers the kernels that count substructures.

Finally, we determine $M_{x,y}$ that we investigate in this paper and introduce three parameters to describe these $M_{x,y}$. The parameters determine the *shape* of the substructures (see Table 1 and Fig. 1) and the corresponding $M_{x,y}$ is the set of isomorphic pairs of substructures for the determined shape. For example, because the value s-f-i indicates arbitrary substructures, $M^{\mathsf{T}}_{x,y} = M^{\text{s-f-i}}_{x,y}$ follows from Taï's theorem. It is evident that these $M_{x,y}$ meet the second condition of Theorem 1.

In addition to $M_{x,y}$ determined by Table 1, to represent the kernels presented in the literature, we let $M^{\mathsf{C}}_{x,y} = M^{\text{s-cd-i}}_{x,y}$, $M^{\mathsf{L}}_{x,y} = M^{\text{s-lu-i}}_{x,y}$, $M^{\mathsf{E}}_{x,y} = M^{\text{a-t-i}}_{x,y}$ and $M^{\mathsf{P}}_{x,y} = M^{\text{c-ct-i}}_{x,y}$.

Because s-f-r and c-f-r are identical to c-f-r and c-f-r, the total number of the kernels to investigate is $2 \times 7 \times 2 - 2 + 6 = 32$. All of them are positive definite due to Theorem 1.

Table 1. Definition of the parameters

FIRST PARAMETER	
s	Sparse substructures (See Sect. 2)
c	Contiguous substructures (See Sect. 2)
a	Agreement subtrees [7]. Used for the elastic kernels
SECOND PARAMETER	
f	Forest. A forest is an arbitrary substructures
t	Subtrees
p	Paths
w	Worms. A worm is a subtree such that only the root can be of degree two, while the others are of degree one or zero
c	Cascades. A cascade is a subtree such that only the root can be of degree greater than one, while the others are of degree one or zero
a	Amebas. An ameba is a subtree such that only one intermediate vertex is of degree two, while the others are of degree one or zero
b	Blooms. A bloom is a subtree such that only one intermediate vertex is of degree greater one, while the others are of degree one or zero
ct	Contiguous trees [2]. Used for the parse tree kernel
cd	Mappings for the constrained distance
lu	Mappings for Lu edit distance
THIRD PARAMETER	
i	Intermediate. The root can be an intermediate vertex of the parent tree
r	Root. The root must be the root of the parent tree

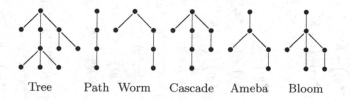

| Tree | Path | Worm | Cascade | Ameba | Bloom |

Fig. 1. Shapes of substructures

3.4   Computational Efficiency

All of these 32 kernels have efficient dynamic-programming-based algorithms to compute themselves. When trees x and y are input to the kernels, the time and space complexities of the algorithms are $O(|\Gamma_x||\Gamma_y|)$, except for the kernels whose type is one of s-f-i, s-t-i, s-t-r and c-f-i. For the excluded kernels, the complexities are known $O(|\Gamma_x|^3)$ when computed based on the decomposition strategy introduced in [4].

Table 2. Recursive formulas

$$K^{\text{c-p-i}}(x,y) = \kappa({}^{\bullet}x, {}^{\bullet}y)\left(1 + K^{\text{c-p-rr}}({}^{\circ}\blacktriangle x, {}^{\circ}y)\right) + K^{\text{c-p-i}}({}^{\blacktriangle}x, {}^{\circ}y) - K^{\text{c-p-i}}({}^{\circ}\blacktriangle x, {}^{\circ}y) + K^{\text{c-p-i}}(x, {}^{\triangle}y) + K^{\text{c-p-i}}({}^{\triangle}x, y) - K^{\text{c-p-i}}({}^{\triangle}x, {}^{\triangle}y)$$

$$K^{\text{c-p-r}}(x,y) = \kappa({}^{\bullet}x, y)\left(1 + K^{\text{c-p-rr}}({}^{\circ}\blacktriangle x, {}^{\circ}y)\right)$$

$$K^{\text{c-w-i}}(x,y) = \kappa({}^{\bullet}x, {}^{\bullet}y)\left(1 + K^{\text{c-2p-r}}({}^{\circ}\blacktriangle x, {}^{\circ}y)\right) + K^{\text{c-w-i}}({}^{\blacktriangle}x, {}^{\circ}y) - K^{\text{c-w-i}}({}^{\circ}\blacktriangle x, {}^{\circ}y) + K^{\text{c-w-i}}({}^{\triangle}x, y) + K^{\text{c-w-i}}(x, {}^{\triangle}y) - K^{\text{c-2p-r}}({}^{\triangle}x, {}^{\triangle}y)$$

$$K^{\text{c-2p-r}}(x,y) = \kappa({}^{\bullet}x, {}^{\bullet}y)\cdot\left(1 + K^{\text{c-p-rr}}({}^{\triangle}x, {}^{\triangle}y)\right) + K^{\text{c-p-rr}}(x, {}^{\triangle}y) + K^{\text{c-p-rr}}({}^{\triangle}x, y) - K^{\text{c-p-rr}}({}^{\triangle}x, {}^{\triangle}y)$$

$$K^{\text{c-p-rr}}(x,y) = \kappa({}^{\bullet}x, {}^{\bullet}y)\left(1 + K^{\text{c-p-rr}}({}^{\circ}\blacktriangle x, {}^{\circ}y)\right) + K^{\text{c-p-rr}}({}^{\triangle}x, y) - K^{\text{c-2p-r}}({}^{\triangle}x, {}^{\triangle}y)$$

$$K^{\text{c-w-rr}}(x,y) = \kappa({}^{\bullet}x, {}^{\bullet}y)\left(1 + K^{\text{c-2p-r}}({}^{\circ}\blacktriangle x, {}^{\circ}y)\right) + K^{\text{c-p-rr}}({}^{\triangle}x, y) - K^{\text{c-2p-r}}({}^{\triangle}x, {}^{\triangle}y)$$

$$K^{\text{c-c-i}}(x,y) = \kappa({}^{\bullet}x, {}^{\bullet}y)\left(1 + K^{\text{c-pf-r}}({}^{\circ}\blacktriangle x, {}^{\circ}y)\right) + K^{\text{c-c-i}}({}^{\blacktriangle}x, {}^{\circ}y) - K^{\text{c-c-i}}({}^{\circ}\blacktriangle x, {}^{\circ}y) + K^{\text{c-c-i}}(x, {}^{\triangle}y) + K^{\text{c-c-i}}({}^{\triangle}x, y) - K^{\text{c-c-i}}({}^{\triangle}x, {}^{\triangle}y)$$

$$K^{\text{c-pf-r}}(x,y) = \kappa({}^{\bullet}x, {}^{\bullet}y)\cdot\left(1 + K^{\text{c-pf-r}}({}^{\triangle}x, {}^{\triangle}y)\right) + K^{\text{c-pf-r}}(x, {}^{\triangle}y) + K^{\text{c-pf-r}}({}^{\triangle}x, y) - K^{\text{c-pf-r}}({}^{\triangle}x, {}^{\triangle}y)$$

$$K^{\text{c-c-r}}(x,y) = \kappa({}^{\bullet}x, {}^{\bullet}y)\left(1 + K^{\text{c-pf-r}}({}^{\circ}\blacktriangle x, {}^{\circ}y)\right)$$

$$K^{\text{c-a-i}}(x,y) = \kappa({}^{\bullet}x, {}^{\bullet}y)\left(K^{\text{c-w-rr}}({}^{\circ}\blacktriangle x, {}^{\circ}y) + K^{\text{c-a-rr}}({}^{\circ}\blacktriangle x, {}^{\circ}y)\right) + K^{\text{c-a-i}}({}^{\blacktriangle}x, {}^{\circ}y) - K^{\text{c-a-i}}({}^{\circ}\blacktriangle x, {}^{\circ}y) + K^{\text{c-a-i}}(x, {}^{\triangle}y) + K^{\text{c-a-i}}({}^{\triangle}x, {}^{\triangle}y)$$

$$K^{\text{c-w-rr}}(x,y) = \kappa({}^{\bullet}x, {}^{\bullet}y)\left(1 + K^{\text{c-2p-r}}({}^{\circ}\blacktriangle x, {}^{\circ}y)\right) + K^{\text{c-w-rr}}({}^{\triangle}x, y) - K^{\text{c-w-rr}}({}^{\triangle}x, {}^{\triangle}y)$$

$$K^{\text{c-a-r}}(x,y) = K^{\text{c-a-rr}}(x,y)$$

$$K^{\text{c-b-i}}(x,y) = \kappa({}^{\bullet}x, {}^{\bullet}y)\left(K^{\text{c-c-rr}}({}^{\circ}\blacktriangle x, {}^{\circ}y) + K^{\text{c-b-rr}}({}^{\circ}\blacktriangle x, {}^{\circ}y)\right) + K^{\text{c-b-i}}({}^{\blacktriangle}x, {}^{\circ}y) - K^{\text{c-b-i}}({}^{\circ}\blacktriangle x, {}^{\circ}y) + K^{\text{c-b-i}}(x, {}^{\triangle}y) + K^{\text{c-b-i}}({}^{\triangle}x, {}^{\triangle}y)$$

$$K^{\text{c-c-rr}}(x,y) = \kappa({}^{\bullet}x, {}^{\bullet}y)\left(1 + K^{\text{c-pf-r}}({}^{\circ}\blacktriangle x, {}^{\circ}y)\right) + K^{\text{c-c-rr}}({}^{\triangle}x, y) - K^{\text{c-c-rr}}({}^{\triangle}x, {}^{\triangle}y)$$

$$K^{\text{c-b-rr}}(x,y) = \kappa({}^{\bullet}x, {}^{\bullet}y)\left(K^{\text{c-c-rr}}({}^{\circ}\blacktriangle x, {}^{\circ}y) + K^{\text{c-b-rr}}({}^{\circ}\blacktriangle x, {}^{\circ}y)\right) + K^{\text{c-b-rr}}({}^{\triangle}x, y) - K^{\text{c-b-rr}}({}^{\triangle}x, {}^{\triangle}y)$$

$$K^{\text{c-b-r}}(x,y) = K^{\text{c-b-rr}}(x,y)$$

Some of the algorithms are known in the literature (c-ct-i [2]; a-t-i [7]; s-lu-i [9]; [c,s]-[f,t,p]-i [17]), while we have developed the rest. For the kernel that only counts identical contiguous path pairs, Kimura et al. showed an efficient kernel whose time complexity is $O(|\Gamma_x|+|\Gamma_y|)$ [8]. Kuboyama et al. also proposed two kernels that counts identical contiguous worm pairs. One is very fast but takes only worms of fixed length into account [12]. Although the other deals with worms of variable length like our kernels, it enumerates all the identical contiguous worm pairs and its computational complexity relies on the degrees of trees and the size of the alphabet for labels [11]. Also, Kuboyama et al. [10] and Augsten [1] proposed kernels that count identical contiguous amebas of fixed size. These kernels are faster than our kernels, but support only limited functionality. In particular, they only count identical pairs of substructures, and the substructures to be counted is required to be contiguous.

Table 2 shows a part of these recursive formulas that we have engineered to compute our kernels. It includes auxiliary kernels (for example, c-2p-r) that are necessary to compute our target kernels. The notation used in Table 2 are as follows. When x is a forest, $^{\blacktriangle}x$ is the leftmost subtree component of x, and $^{\bullet}x$ is the root of $^{\blacktriangle}x$. Also, $^{\circ\blacktriangle}x$ and $^{\vartriangle}x$ denote $^{\blacktriangle}x \setminus \{^{\bullet}x\}$ and $x \setminus {}^{\blacktriangle}x$, respectively. These recursive formulas reduce computing of the kernels for substructures to computing of the kernels for smaller substructures. By tracking this relation from the bottom to the top, we obtain dynamic-programming-based algorithms.

4   Investigation of the Kernels

4.1   Method

To investigate and compare the kernels, we use an experimental method. In fact, we choose ten datasets from various application fields and evaluate the prediction performance of the kernels by applying the kernels to the datasets.

The ten datasets that we use covers three different areas of applications, that is, bioinformatics (three), natural language processing (six) and web access analysis (one). Three (COLON, CYSTIC and LEUKEMIA) are retrieved from the KEGG/GLYCAN database [5] and contain glycan structures annotated relating to colon cancer, cystic fibrosis and leukemia cells. One (SYNTACTIC) is the dataset PropBank provided in [15]. This dataset includes parse trees labeled with two syntactic role classes for modeling the syntactic/semantic relation between a predicate and the semantic roles of its arguments in a sentence. Five (AIMED, BIOINFER, HPRD50 IEPA and LLL) are the corpora that include parse trees obtained by analyzing documents regarding protein-protein interaction (PPI) extraction [16]. PPI is an intensively studied problem of the BioNLP field. The remaining one (WEB), used in [20], consists of trees representing web-page accesses by users, and the annotation is based on whether the user is from a .edu site or not. Table 3 describes the basic features of these datasets.

For each dataset, we generate ten pairs of training and test data subsets by distributing the examples in the dataset at random. Therefore, we finally obtain $10 \times 10 = 100$ pairs of training and test data subsets. The training data subset

Table 3. Features of the datasets

Dataset Notation	#Tree	Tree (average) Size	Hight	Deg.(max)		Dataset Notation	#Tree	Tree (average) Size	Hight	Deg.(max)	
COLON	134	8.4	5.6	1.18	(3)	CYSTIC	160	8.3	5.0	1.22	(3)
LEUKEMIA	442	13.5	7.4	1.17	(3)	SYNTACTIC	225	19.7	6.5	1.42	(8)
AIMED	100	94.4	13.5	1.54	(19)	BIOINFER	100	116.4	14.1	1.55	(43)
HPRD50	100	84.4	12.7	1.54	(12)	IEPA	100	105.2	13.6	1.54	(18)
LLL	100	106.4	14.3	1.54	(13)	WEB	500	12.0	4.3	1.24	(26)

of each pair is four times larger in the number of the examples included than the corresponding test data subset.

Then, for each combination of a pair of training and test data subsets and a kernel, we train a C-SVM classifier [3] with the training data subset and measure the prediction performance of the kernel by making the trained classifier predict class labels for the examples included in the test data subset. For this evaluation of the prediction performance of the kernel, we use the Area Under Curve of Receiver Operating Characteristic curve (AUC-ROC).

When training the classifier, we simultaneously determine the optimal values for the α and β parameters of κ and the C parameter of C-SVM, and use the values we obtain when running the classifier on the test data subset. This optimization is performed based on the grid search strategy.

In the remainder of this section, we first describe the results of the experiments for the cases of $0 \leq \ln \beta \leq \ln \alpha \leq 10$ and $0 \leq \beta \leq \alpha \leq 1$ separately, and then compare the results of these two cases. The first case represents the case where the kernels are derived from edit distance metrics, while the second case abstracts the case where we use the primitive kernels κ as decay factor.

4.2   Results for $0 \leq \ln \beta \leq \ln \alpha \leq 10$

Table 10 shows the raw AUC-ROC scores obtained through the experiment on 26 kernels among the entire 32 kernels. The six kernels of the forest and tree types, which are the kernels determined by s-f-i, s-t-i, s-t-r, c-f-i, c-t-i and c-t-r, are tested separately because they require heavy computation and we cannot test these kernels on all of the datasets. The results of the experiment with these six kernels are described at the end of this subsection.

To calculate the values displayed in the bottom row headed by AVERAGE (NORMALIZED), we normalize the raw scores per dataset: With respect to the $26 \times 10 = 260$ raw scores across the 26 kernels, we calculate their average and variance, and then normalize all of these 260 scores using the average and variance we obtained. The values of the row are calculated used these normalized scores.

From Table 10, we see that the kernels of s-p-i, s-p-r, s-w-i and s-c-i outperform the others. Their superiority is also statistically significant. In fact, Table 9 shows

Table 4. The forest and tree types($0 \leq \ln \beta \leq \ln \alpha \leq 10$)

KERNEL	s-f-i	s-t-i	s-t-r	c-f-i	c-t-i	c-t-r
AUC-ROC SCORES						
COLON	0.910	0.907	0.930	0.898	0.943	0.890
CYSTIC	0.719	0.710	0.726	0.729	0.686	0.677
LEUKEMIA	0.905	0.906	0.907	0.882	0.886	0.886
SYNTACTIC	0.810	0.810	0.791	0.780	0.766	0.770
WEB	0.675	0.673	0.657	0.646	0.654	0.637
P-VALUES						
s-p-i	0.000	0.000	0.000	0.000	0.000	0.000
s-p-r	0.000	0.000	0.000	0.000	0.000	0.000
s-w-i	0.000	0.000	0.000	0.000	0.000	0.000
s-c-i	0.001	0.000	0.000	0.000	0.000	0.000

that the p-values of the pairwise t-tests that we performed to compare these kernels with the others are significantly small. Because of space limitations, only the results with the typical kernels are shown. In particular, the kernels of c-ct-i and a-t-i, namely, the parse tree and elastic kernels have been used as the benchmark kernels from the literature.

With respect to the comparison between the kernels of the sparse and contiguous types, Table 9 (right) also shows the p-values of the pairwise t-tests that compare the kernels of the path, worm and cascade types. Based on the displayed results which are shown, we can conclude that the superiority of the sparse type to the contiguous type is statistically significant.

Lastly, we investigate the kernels of the forest and tree types that are excluded in the experiment stated above because they can apply to only five of the ten datasets due to their heavy computation. From the raw AUC-ROC scores and the p-values shown in Table 4, we see that the performance of the kernels of these types is worse than the kernels that show the best performance in the previous experiment.

4.3 Results for $0 \leq \beta \leq \alpha \leq 1$

We ran an experiment for $0 \leq \beta \leq \alpha \leq 1$ in the same way as we ran for $0 \leq \ln \beta \leq \ln \alpha \leq 10$.

Table 11 shows the raw AUC-ROC scores obtained through the experiment. From the table, we see that the kernels of c-p-i, c-w i and c-c-i outperform the others, and this can be statistically verified based on Table 6, which shows the p-values of pairwise t-tests.

In contrast with the case of $0 \leq \ln \beta \leq \ln \alpha \leq 10$, the kernels of the contiguous type shows better performance than the kernels of the sparse type.

Table 5. The forest and tree types($0 \leq \beta \leq \alpha \leq 1$)

KERNEL	s-f-i	s-t-i	s-t-r	c-f-i	c-t-i	c-t-r
AUC-ROC SCORES						
COLON	0.970	0.968	0.960	0.981	0.968	0.946
CYSTIC	0.777	0.762	0.793	0.817	0.785	0.775
LEUKEMIA	0.956	0.954	0.954	0.939	0.947	0.951
SYNTACTIC	0.583	0.576	0.650	0.657	0.843	0.838
WEB	0.727	0.727	0.717	0.731	0.735	0.705
P-VALUES						
c-p-i	0.014	0.003	0.003	0.067	0.009	0.001
c-w-i	0.012	0.002	0.002	0.065	0.001	0.000
c-c-i	0.017	0.003	0.003	0.082	0.001	0.000

Table 6. P-values ($0 \leq \beta \leq \alpha \leq 1$)

c-p-i	0.401	0.988	**0.000**	**0.000**	**0.007**	**0.022**	**0.007**
	c-w-i	0.314	**0.000**	**0.000**	**0.001**	**0.007**	**0.002**
		c-c-i	**0.000**	**0.000**	**0.003**	**0.020**	**0.005**
			s-a-r	c-b-i	s-cd-i	s-lu-i	a-t-i

Table 7. Normalized AUC-ROC scores

KERNEL	$(0 < \ln \beta \leq \ln \alpha < 10)$				$(0 < \beta \leq \alpha < 1)$		
	s-p-i	s-p-r	s-w-i	s-c-i	c-p-i	c-w-i	c-c-i
AVERAGE	−0.160	−0.203	−0.176	−0.293	0.257	0.317	0.259

Also, as Table 5 shows, the kernels of the forest and tree types prove to be inferior to the kernels that show the best performance in the experiment with the entire ten datasets.

Table 8. P-values (comparison of the two cases)

	s-p-i	s-p-r	s-w-i	s-c-i
c-p-i	0.000	0.000	0.001	0.000
c-w-i	0.001	0.000	0.000	0.000
c-c-i	0.000	0.000	0.000	0.000

Table 9. P-values $(0 \leq \ln \beta \leq \ln \alpha < 10)$

s-p-i	0.225	0.695	0.279	0.000	0.000	0.000	0.000	0.000	0.000	0.002	0.000	0.031	0.001
	s-p-r	0.995	0.432	0.000	0.000	0.000	0.000	0.000	0.000	0.003	0.000	0.070	0.002
		s-w-i	0.371	0.000	0.000	0.000	0.000	0.000	0.000	0.094	0.058	0.000	0.001
			s-c-i	0.000	0.000	0.000	0.000	0.000	0.000	0.342	0.156	0.216	0.000
				s-a-i	c-b-i	s-cd-i	s-lu-i	a-t-i	c-ct-i	c-p-i	c-p-r	c-w-i	c-c-i

Table 10. AUC-ROC scores $(0 \leq \ln \beta \leq \ln \alpha \leq 10)$

Kernel	s-p-i	s-p-r	s-w-i	s-w-r	s-c-i	s-c-r	s-a-i	s-a-r	s-b-i	s-b-r	s-cd-i	s-lu-i	c-ct-i
	c-p-i	c-p-r	c-w-i	c-w-r	c-c-i	c-c-r	c-a-i	c-a-r	c-b-i	c-b-r	a-t-i	a-t-r	c-ct-r
COLON	0.954	**0.958**	0.949	0.940	0.947	0.938	0.500	0.500	0.500	0.500	0.942	0.935	0.861
	0.932	0.930	0.925	0.915	0.920	0.909	0.500	0.500	0.500	0.500	0.940	0.932	0.879
CYSTIC	0.760	**0.763**	**0.763**	0.750	0.745	0.750	0.500	0.500	0.500	0.500	0.716	0.727	0.625
	0.762	0.758	0.737	0.724	0.714	0.724	0.500	0.500	0.500	0.500	0.722	0.719	0.609
LEUKEMIA	0.936	0.935	0.935	0.922	**0.941**	0.920	0.500	0.500	0.500	0.500	0.905	0.906	0.880
	0.921	0.923	0.922	0.906	0.920	0.903	0.500	0.500	0.500	0.500	0.902	0.907	0.878
SYNTACTIC	0.870	0.857	**0.892**	0.874	0.835	0.874	0.500	0.500	0.500	0.500	0.794	0.799	0.718
	0.840	0.821	0.857	0.860	0.787	0.861	0.500	0.500	0.500	0.500	0.793	0.798	0.743
AIMED	0.606	0.606	0.524	0.606	**0.651**	0.606	0.527	0.520	0.603	0.604	0.558	0.558	0.558
	0.601	0.596	0.515	0.596	0.647	0.597	0.512	0.532	0.597	0.558	0.547	0.558	0.558
BIOINFER	0.650	0.597	0.569	0.597	0.606	0.597	0.666	**0.702**	0.615	0.611	0.585	0.598	0.633
	0.539	0.563	0.568	0.563	0.577	0.563	0.628	0.601	0.636	0.643	0.641	0.569	0.633
HPRD50	0.506	0.515	0.531	0.515	0.518	0.515	**0.556**	0.524	0.514	0.521	0.504	0.508	0.509
	0.523	0.524	0.538	0.524	0.517	0.524	0.539	0.518	0.518	0.530	0.519	0.529	0.515
IEPA	0.617	0.613	0.633	0.613	0.544	0.613	0.609	0.625	0.549	0.549	0.527	0.525	0.535
	0.594	0.601	**0.639**	0.601	0.585	0.601	0.634	0.629	0.562	0.571	0.534	0.520	0.538
LLL	0.557	0.560	**0.621**	0.560	0.613	0.559	0.580	0.575	0.583	0.574	0.550	0.565	0.570
	0.581	0.576	0.604	0.580	0.593	0.577	0.582	0.572	0.571	0.557	0.557	0.543	0.576
WEB	**0.724**	0.715	0.692	0.690	0.693	0.686	0.500	0.500	0.500	0.500	0.671	0.663	0.636
	0.706	0.690	0.678	0.676	0.671	0.661	0.500	0.500	0.500	0.500	0.669	0.656	0.621
AVERAGE	**0.478**	0.442	0.468	0.398	0.436	0.391	−0.573	−0.613	−0.701	−0.697	0.115	0.149	−0.011
(NORMALIZED)	0.372	0.360	0.385	0.330	0.322	0.308	−0.627	−0.696	−0.681	−0.691	0.201	0.129	0.006

4.4 Comparison of the Two Settings

It is interesting to note that the kernels of the path, worm and cascade types are ranked in the top three, and outperform the other types for both cases of $0 \leq \ln \beta \leq \ln \alpha \leq 10$ and $0 \leq \beta \leq \alpha \leq 1$. The contrast between the cases to note here is that the kernels of the sparse type significantly outperform the kernels of the contiguous type for $0 \leq \ln \beta \leq \ln \alpha \leq 10$, whereas this relationship is contrary for $0 \leq \beta \leq \alpha \leq 1$.

In addition, Table 7 shows the normalized averages of the AUC-ROC scores obtained through the experiments to compare the setting of $0 \leq \ln \beta \leq \ln \alpha \leq 10$ and $0 \leq \beta \leq \alpha \leq 1$: We performed the kernels of the s-p-i, s-p-r, s-w-i and s-c-i types under the condition of $0 \leq \ln \beta \leq \ln \alpha \leq 10$ while c-p-i, c-w-i and c-c-i types under the condition of $0 \leq \beta \leq \alpha \leq 1$.

From the table, we observe that the kernels performing the best under the condition of $0 \leq \beta \leq \alpha \leq 1$ consistently outperform the kernels performing the best under $0 \leq \ln \beta \leq \ln \alpha \leq 10$. The p-values obtained through pairwise t-tests also endorse this observation (Table 8).

Table 11. AUC-ROC scores ($0 \leq \beta \leq \alpha \leq 1$)

Target	s-p-i	s-p-r	s-w-i	s-w-r	s-c-i	s-c-r	s-a-i	s-a-r	s-b-i	s-b-r	s-cd-i	s-lu-i	c-ct-i
	c-p-i	c-p-r	c-w-i	c-w-r	c-c-i	c-c-r	c-a-i	c-a-r	c-b-i	c-b-r	a-t-i	a-t-r	c-ct-r
COLON	**0.980**	0.968	0.968	0.966	0.971	0.966	0.500	0.500	0.500	0.500	0.973	0.973	0.970
	0.971	0.945	0.967	0.951	0.967	0.951	0.500	0.500	0.500	0.500	0.973	0.967	0.914
CYSTIC	0.797	**0.815**	0.776	0.798	0.771	0.798	0.500	0.500	0.500	0.500	0.785	0.781	0.758
	0.797	0.805	0.796	0.775	0.798	0.775	0.500	0.500	0.500	0.500	0.768	0.777	0.709
LEUKEMIA	0.943	0.951	0.950	0.948	0.947	0.948	0.500	0.500	0.500	0.500	0.954	0.954	0.937
	0.946	0.953	0.948	**0.956**	0.947	0.955	0.500	0.500	0.500	0.500	0.952	0.955	0.906
SYNTACTIC	0.877	0.884	0.894	**0.896**	0.888	**0.896**	0.500	0.500	0.500	0.500	0.632	0.625	0.891
	0.874	0.849	0.884	0.861	0.881	0.861	0.500	0.500	0.500	0.500	0.766	0.733	0.859
AIMED	0.574	0.541	0.569	0.541	0.558	0.541	0.599	0.588	0.563	0.555	0.629	**0.641**	0.590
	0.552	0.535	0.617	0.535	0.583	0.535	0.561	0.607	0.596	0.575	0.567	0.500	0.593
BIOINFER	0.633	0.688	0.724	0.688	0.583	0.688	0.780	0.748	0.601	0.620	0.609	0.636	0.738
	0.816	0.571	**0.850**	0.571	0.836	0.571	0.731	0.671	0.803	0.715	0.595	0.602	0.570
HPRD50	0.559	0.550	0.560	0.550	0.554	0.550	0.538	0.555	**0.601**	0.583	0.537	0.589	0.527
	0.516	0.525	0.533	0.525	0.538	0.525	0.548	0.524	0.517	0.515	0.619	0.566	0.526
IEPA	0.568	0.562	0.623	0.562	0.562	0.562	0.607	0.591	0.542	0.562	0.604	0.571	0.663
	0.669	**0.688**	0.653	**0.688**	0.625	**0.688**	0.610	0.668	0.574	0.604	0.580	0.558	0.575
LLL	0.593	0.578	0.633	0.578	0.563	0.578	0.582	0.613	0.524	0.566	0.562	0.568	0.602
	0.645	0.570	0.592	0.570	0.597	0.570	0.587	0.561	0.585	0.552	0.546	0.582	0.630
WEB	0.741	0.731	0.740	0.715	0.741	0.711	0.500	0.500	0.500	0.500	0.732	0.732	0.742
	0.745	0.713	**0.745**	0.708	0.741	0.712	0.500	0.500	0.500	0.500	0.735	0.711	0.699
AVERAGE	0.329	0.304	0.475	0.282	0.223	0.279	−0.627	−0.615	−0.811	−0.769	0.187	0.256	0.443
(NORMALIZED)	**0.522**	0.250	**0.560**	0.234	**0.501**	0.238	−0.671	−0.671	−0.673	−0.760	0.260	0.105	0.149

References

1. Augsten, N., Böhlen, M.H., Gamper, J.: The pq-gram distance between ordered labeled trees. ACM Trans. Database Syst. **35**(1), 1–36 (2010)
2. Collins, M., Duffy, N.: Convolution kernels for natural language. In: Proceedings of Advances in Neural Information Processing Systems 14 (NIPS), pp. 625–632 (2001)
3. Cristianini, N., Shawe-Taylor, J.: An introduction to Support Vector Machines and Other Kernel-Based Learning Methods. Cambridge University Press, Cambridge (2000)
4. Demaine, E.D., Mozes, S., Rossman, B., Weimann, O.: An optimal decomposition algorithm for tree edit distance. ACM Trans. Algorithms (TALG) **6**(1), 2:1–2:19 (2009)
5. Hashimoto, K., Goto, S., Kawano, S., Aoki-Kinoshita, K.F., Ueda, N.: Kegg as a glycome informatics resource. Glycobiology **16**, 63R–70R (2006)
6. Haussler, D.: Convolution kernels on discrete structures. UCSC-CRL 99–10, Department of Computer Science, University of California at Santa Cruz (1999)
7. Kashima, H., Koyanagi, T.: Kernels for semi-structured data. In: Proceedings of the 9th International Conference on Machine Learning (ICML), pp. 291–298 (2002)
8. Kimura, D., Kashima, H.: Computation of subpath kernel for trees. In: Proceedings of the 29th International Conference on Machine Learning (ICML) (2012)
9. Kuboyama, T., Shin, K., Kashima, H.: Flexible tree kernels based on counting the number of tree mappings. In Proceedings of the Machine Learning with Graphs (MLG) (2006)

10. Kuboyama, T., Hirata, K., Aoki-Kinoshita, K.F.: An efficient unordered tree kernel and its application to glycan classification. In: Washio, T., Suzuki, E., Ting, K.M., Inokuchi, A. (eds.) PAKDD 2008. LNCS (LNAI), vol. 5012, pp. 184–195. Springer, Heidelberg (2008)
11. Kuboyama, T., Hirata, K., Aoki-Kinoshita, K.F., Kashima, H., Yasuda, H.: A gram distribution kernel applied to glycan classification and motif extraction. Genome Inform. Ser. **17**(2), 25–34 (2006)
12. Kuboyama, T., Hirata, K., Kashima, H., Aoki-Kinoshita, K.F., Yasuda, H.: A spectrum tree kernel. Inf. Media Technol. **2**(1), 292–299 (2007)
13. Lu, C.L., Su, Z.-Y., Tang, C.Y.: A new measure of edit distance between labeled trees. In: Wang, J. (ed.) COCOON 2001. LNCS, vol. 2108, pp. 338–348. Springer, Heidelberg (2001)
14. Lu, S.Y.: A tree-to-tree distance and its application to cluster analysis. EEE Trans. Pattern Anal. Mach. Intell. (PAMI) **1**, 219–224 (1979)
15. Moschitti, A.: Example data for Tree Kernels in SVM-light. http://disi.unitn.it/moschitti/Tree-Kernel.htm
16. Pyysalo, S., Airola, A., Heimonen, J., Bjorne, J., Ginter, F., Salakoski, T.: Comparative analysis of five protein-protein interaction corpora. BMC Bioinform. **9**(S-3), S6 (2008)
17. Shin, K., Cuturi, M., Kuboyama, T.: Mapping kernels for trees. In: Proceedings of the 28th International Conference on Machine Learning ICML (2011)
18. K. Shin and T. Kuboyama. A generalization of Haussler's convolution kernel - mapping kernel. In: Proceedings of the 25th International Conference on Machine Learning ICML (2008)
19. Taï, K.C.: The tree-to-tree correction problem. JACM **26**(3), 422–433 (1979)
20. Zaki, M.J., Aggarwal, C.C.: Xrules: an effective algorithm for structural classification of XML data. Mach. Learn. **62**, 137–170 (2006)
21. Zhang, K.: Algorithms for the constrained editing distance between ordered labeled trees and related problems. Pattern Recogn. **28**(3), 463–474 (1995)

Outliers on Concept Lattices

Mahito Sugiyama[✉]

Machine Learning and Computational Biology Research Group,
Max Planck Institute for Intelligent Systems and Max Planck Institute
for Developmental Biology, Tübingen, Germany
mahito.sugiyama@tuebingen.mpg.de

Abstract. Outlier detection in mixed-type data, which contain both discrete and continuous features, is still a challenging problem. Here we newly introduce *concept-based outlierness*, which is defined on a hierarchy of clusters of data points and features, called the *concept lattice*, obtained by *formal concept analysis* (FCA). Intuitively, this outlierness is the degree of isolation of clusters on the hierarchy. Moreover, we investigate *discretization* of continuous features to embed the original continuous (Euclidean) space into the concept lattice. Our experiments show that the proposed method which detects concept-based outliers is more effective than other popular distance-based outlier detection methods that ignore the discreteness of features and do not take cluster relationships into account.

Keywords: Outlier · Formal concept analysis · Concept lattice · Cluster · Discretization

1 Introduction

An *outlier* is typically considered as an observation which is significantly different from other observations in a numerical sense [10], and detecting such outliers is one of the central topics of data mining. Since outliers appear in many real-life situations, outlier detection techniques have lots of significant applications across a number of domains. Examples include intrusions in network traffic, credit card fraud, defective products in industry, and misdiagnosed patients.

To date, distance-based approaches, which define outliers as objects located far away from the remaining objects, have been successfully applied in various situations due to its flexibility. That is to say, we do not need to estimate an underlying probability distribution, which is often difficult in particular in high-dimensional settings. For example, LOF (local outlier factor) [6] has become one of the most popular outlier detection methods, which defines the outlierness of each object based on the difference of local densities between the object and its neighbors.

However, many datasets in real-life applications may contain not only continuous (numerical) features but also discrete (categorical) features. For example,

© Springer International Publishing Switzerland 2014
Y. Nakano et al. (Eds.): JSAI-isAI 2013, LNAI 8417, pp. 352–368, 2014.
DOI: 10.1007/978-3-319-10061-6_23

demographic data often contain continuous features such as the body height or weight and discrete features such as gender or race. One of the most famous benchmark datasets KDDCup1999, which is a network traffic dataset for intrusion detection, also contains both continuous and discrete features. Outlier detection for such *mixed-type data* is still a challenging topic since it is not trivial to construct a distance function for such mixed-type data points in order to define outliers.

Here we newly define outliers for mixed-type data, which we referred to as *concept-based outliers*, and present a new parameter-free outlier detection method. There are two key processes: (1) *discretization* of continuous features using the binary encoding scheme to convert mixed-type data into binary data, and (2) *formal concept analysis* (FCA) to produce a hierarchy of clusters, called the *concept lattice*. This technique enables us to treat the proximity of objects in an algebraic manner, that is, both objects (data points) and attributes (features) are clustered as "closed" sets, called *concepts*, which provides a higher level perspective of relationships between not objects but clusters. We then measure the outlierness of objects and features included in a cluster by counting the number of their upper and lower clusters on the hierarchy, which scores the degree of isolation of clusters. Finally, we have a real-valued score like the other distance-based approaches. We experimentally show that the proposed method is superior to other distance-based outlier detection methods which ignore the discreteness of features.

This paper is organized as follows: In Sect. 3, we describe how to measure the outlierness using FCA. In Sect. 4, we introduce a method for converting mixed-type data into binary data to apply FCA. We show experimental results in Sect. 5 and summarize this paper with future work in Sect. 6. The notation is summarized in Table 1.

2   Related Work

FCA is a mathematical discipline for data analysis introduced by Wille [27] and, to date, this has been successfully applied in many fields including data mining [25]. One of the most successful applications of FCA is for frequent pattern mining [18], where concepts containing frequent items, called frequent closed itemsets, were shown to be lossless compression of all frequent itemsets.

Recently, mining not frequent but *rare* concepts via FCA has been studied [1,16], and such concepts correspond to outliers of itemsets. The rareness of a concept is defined as the size (support) of it, that is, the number of objects contained in the concept, and small concepts are assumed to be rare. This definition, which is based on the number of objects, is essentially the same as that of distance-based outliers, where an object is considered as an outlier if it has a few objects in its ϵ-neighborhood [13].

As we will introduce in Sect. 3.2, our proposal in this paper is significantly different from the above approaches. The outlierness of an object depends only on the number of concepts produced by FCA, and hence the number of objects (and attributes) is ignored. This approach can be viewed as characterization of outliers based on the *semantics* through FCA, which is provided as subset inclusion relationships of objects and attributes on the concept lattice. Moreover, our approach also includes semantical aspects of numerical attributes into FCA, since the distance between objects on Euclidean space is preserved on the concept lattice via appropriate discretization (Lemma 2). This new approach is thus expected to find non-rare outliers which can be viewed as outliers semantically but cannot be detected by the existing rareness-based methods. More specifically, our method is able to find an outlier which belongs to a densely agglomerated cluster but the cluster itself locates sparsely if we overview the whole objects. Although some methods have been already proposed to find clustered outliers [14], they are mainly designed only for continuous data, and our method is the first to find them from mixed-type data, thanks to the formulation of FCA. Furthermore, we will show that our concept-based outlierness has a close relationship with distance-based outlierness (Theorems 1 and 2), which is defined based on the rareness of objects, when our method is combined with discretization of continuous attributes. Thus our proposal can be viewed as an extension of rareness- and distance-based outlier detection, and hence distance-based outliers are also expected to be detected.

Kaytoue *et al.* recently studied FCA for numerical attributes [11]. Although the objective is different since their method is for pattern mining, it is interesting future work to investigate the relationship to our method.

We have recently applied FCA to semi-supervised learning [22,23], where the concept hierarchy is used to characterize the *preference* of objects in classification and discretization is also applied. This paper has mainly two differences from these studies: (1) the degree of outlierness is measured in a novel way in an unsupervised manner; (2) the discretization method is further developed to appropriately embed distances into concept lattices.

3   Measuring Outlierness on Concept Lattices

3.1   Formal Concept Analysis

First we briefly introduce *formal concept analysis* (FCA) [8,9,27], an algebraic technique to construct a hierarchy of clusters from binary data. A *context* is a triplet $\mathbb{K} = (G, M, I)$ consisting of a set of *objects* G, that of *attributes* M, and a binary relation $I \subseteq G \times M$ between G and M. Let $A \subseteq G$ and $B \subseteq M$. Define

$$A' = \{\, m \in M \mid (g, m) \in I \text{ for all } g \in A \,\},$$
$$B' = \{\, g \in G \mid (g, m) \in I \text{ for all } m \in B \,\}.$$

If $A' = B$ and $B' = A$, a pair (A, B) is called a *concept*, where A is *extent* and B is *intent*. An operator $'$ is a *Galois connection* between the power set lattices on G and M. Since the mapping $''$ is a *closure operator* on the context $\mathbb{K}$, both A and B are (algebraic) *closed* sets for a concept (A, B). Thus a subset $A \subseteq G$ (resp. $B \subseteq M$) is closed if and only if $A'' = A$ (resp. $B'' = B$). This operation can be

Table 1. Notation.

x, y	Object (data point)
G	The set of objects
M	The set of attributes
I	Binary relation between G and M
$\mathbb{K}$	Context; $\mathbb{K} = (G, M, I)$
g	Object
m	Attribute
A, X	Subset of G
B, Y	Subset of M
$'$	Galois connection between power set oattices on G and M
(A, B)	Concept; $A = A''$ and $B = B''$
$\mathfrak{B}(\mathbb{K})$	The set of concepts
$(\mathfrak{B}(\mathbb{K}), \leq)$	Concept lattice
q	Concept-based outlierness
$\vee(g)$	Least upper bound of the set $\{(A, B) \in \mathfrak{B}(\mathbb{K}) \mid g \in A\}$
V	The set of features
v	Feature
$x(v)$	Value of x for feature v
Σ	Alphabet
Σ^*	The set of finite sequences
Σ^ω	The set of infinite sequences
p	Infinite sequence
ρ	Mapping from Σ^ω to $\mathbb{R}$
$\uparrow w$	Set of infinite sequences that have w as prefix
k	Discretization level
ρ_0	Modified binary representation
d	Distance function
α, δ	Parameters for distance-based outliers

viewed as *biclustering* mainly studied in machine learning, where each closed set corresponds to a cluster and objects and attributes are clustered simultaneously.

The set of concepts over a context $\mathbb{K}$ is written by $\mathfrak{B}(\mathbb{K})$, called the *concept lattice*, which is a key product of FCA for algebraic data analysis. The order $\leq$ on the lattice $\mathfrak{B}(\mathbb{K})$ can be introduced in the following manner. For a pair of concepts $(A, B) \in \mathfrak{B}(\mathbb{K})$ and $(C, D) \in \mathfrak{B}(\mathbb{K})$, $(A, B) \leq (C, D)$ if $A \subseteq C$. Then $(\mathfrak{B}(\mathbb{K}), \leq)$ becomes a complete lattice. In the following, we simply write the lattice $(\mathfrak{B}(\mathbb{K}), \leq)$ as $\mathfrak{B}(\mathbb{K})$ if it is understood from the context.

3.2 Concept-Based Outlierness Score

Here we present a new method to measure the outlierness of objects and attributes on the concept lattice $\mathfrak{B}(\mathbb{K})$. For a subset of objects $X \subseteq G$, define the *concept-based outlierness score* $q(X)$ as

$$q(X) := \left| \{ (A, B) \in \mathfrak{B}(\mathbb{K}) \mid X \subseteq A \text{ or } X' \subseteq B \} \right|^{-1}.$$

Note that q is often applied to a singleton $\{x\}$ to obtain the score of each object. This score coincides with the inverse of the number of concepts which are contained in either upper or lower sets of the least upper bound of the subset of concepts $\{ (A, B) \in \mathfrak{B}(\mathbb{K}) \mid X \subseteq A \}$. We always have $q(X) = q(X'')$ since if $X \subseteq A$ for some concept (A, B), $B \subseteq X'$ holds, hence it follows that $X'' \subseteq B' = A$. Interestingly, we can similarly define the outlierness score $q(Y)$ for a set of attributes $Y \subseteq M$,

$$q(Y) := \left| \{ (A, B) \in \mathfrak{B}(\mathbb{K}) \mid Y' \subseteq A \text{ or } Y \subseteq B \} \right|^{-1}.$$

This is achieved by the duality of FCA and is usually not possible in outlier detection methods. These definitions come from our intuition that outliers should be *isolated on the concept lattice*, that is, they have a few upper and lower sets, while inliers tend to have a relatively large number of upper and lower sets.

Every set of objects $X \subseteq G$ should be contained in some concept. We therefore have $0 < q(X) \leq 1$ for every $X \subseteq G$ and 1 is the maximum score. This property also holds for a set of attributes $Y \subseteq M$.

Example 1. Given a context $\mathbb{K} = (G, M, I)$, where $G = \{g_1, g_2, \ldots, g_5\}$, $M = \{m_1, m_2, m_3, m_4\}$, and I is given as the following table:

	m_1	m_2	m_3	m_4
g_1	×	×	×	
g_2		×	×	
g_3		×	×	
g_4				×
g_5	×	×		

Then we have seven concepts shown in Fig. 1. We have the following scores for each object and attributes:

$$q(\{g_1\}) = 1/6 = 0.17, \ q(\{g_2\}) = 1/5 = 0.2, \ q(\{g_3\}) = 1/5 = 0.2,$$
$$q(\{g_4\}) = 1/3 = 0.33, \ q(\{g_5\}) = 1/5 = 0.2,$$
$$q(\{m_1\}) = 1/5 = 0.2, \ q(\{m_2\}) = 1/6 = 0.17,$$
$$q(\{m_3\}) = 1/5 = 0.2, \ q(\{m_4\}) = 1/3 = 0.33.$$

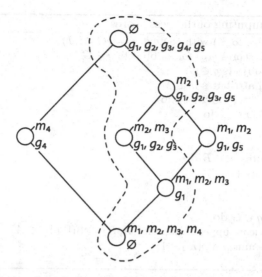

Fig. 1. Concept lattice in example 1. Dotted line shows concepts for the score $q(\{g_2\})$.

3.3   Algorithm

We focus on in this paper computing the outlierness score for every single object as it is the most typical task in outlier detection. First we use Makino and Uno's algorithm [15] (provided as LCM[1]) for constructing the concept lattice $\mathfrak{B}(\mathbb{K})$ from a given context $\mathbb{K} = (G, M, I)$, which is known to be one of the fastest algorithms for this task. Their algorithm enumerates all concepts with $O(\Delta^3)$ delay with $\Delta = \max_{x \in G \cup M} |\{x\}'|$.

Next we present an efficient algorithm, shown in Algorithm 1, for computing scores on the lattice. For each object $g \in G$, we write by $\vee(g)$ the *least upper bound* of the set $\{(A, B) \in \mathfrak{B}(\mathbb{K}) \mid g \in A\}$. This algorithm is divided into two steps: counting the number of upper concepts and that of lower concepts of $\vee(g)$ for each object $g \in G$.

Lines from 3 to 10 in Algorithm 1 are for upper concepts. Since any concept (A, B) is in the upper set of $\vee(g)$ if and only if $g \in A$, it is trivial that this part counts all upper concepts for all objects. Moreover, from the duality of concepts and attributes, each $\varphi(m)$ coincides with the number of lower concepts of the greatest lower bound of the set $\{(A, B) \in \mathfrak{B}(\mathbb{K}) \mid m \in B\}$. Lines from 11 to 14 are for lower concepts. Let $\vee(g) = (A, B)$. If there exists an attribute $m \in B$ such that $m \notin D$ for any concept (C, D) with $(A, B) \leq (C, D)$, we have $\{(C, D) \in \mathfrak{B}(\mathbb{K}) \mid m \in B\} = \{(C, D) \in \mathfrak{B}(\mathbb{K}) \mid (C, D) \leq (A, B)\}$. This means that $\min_{m \in B} \varphi(m)$ coincides with the number of lower concepts of $\vee(g)$. Later we will show that this assumption always holds in our context and hence the proposed algorithm correctly counts the number of upper and lower concepts of $\vee(g)$ for all objects $g \in G$.

---

[1] http://research.nii.ac.jp/~uno/code/lcm.html

Algorithm 1. Computing outlierness scores
***

Input: concept lattice $\mathfrak{B}(\mathbb{K})$ with context $\mathbb{K} = (G, M, I)$
Output: outlierness scores $q(g)$ for all objects $g \in G$
 1: $q(g) \leftarrow 0$ for all objects $g \in G$
 2: $\varphi(m) \leftarrow 0$ for all attributes $m \in M$
 3: **for all** concepts $(A, B) \in \mathfrak{B}(\mathbb{K})$ **do**
 4:     **for all** objects $g \in A$ **do**
 5:         $q(g) \leftarrow q(g) + 1$
 6:     **end for**
 7:     **for all** attributes $m \in B$ **do**
 8:         $\varphi(m) \leftarrow \varphi(m) + 1$
 9:     **end for**
10: **end for**
11: **for all** objects $g \in G$ **do**
12:     $(A, B) \leftarrow$ the least upper bound of $\{(A, B) \in \mathfrak{B}(\mathbb{K}) \mid g \in A\}$
13:     $q(g) \leftarrow q(g) + \min_{m \in B} \varphi(m) - 1$
14: **end for**
***

The time complexity of lines from 3 to 10 is $O(\sum_{(A,B) \in \mathfrak{B}(\mathbb{K})}(|A| + |B|))$, which is larger than $O(|\mathfrak{B}(\mathbb{K})|)$ and much smaller than $O((|G| + |M|) \cdot |\mathfrak{B}(\mathbb{K})|)$. Since the time complexity of lines from 11 to 14 is $O(|G|)$, the overall time complexity is $O(\Delta^3 \cdot |\mathfrak{B}(\mathbb{K})| + \sum_{(A,B) \in \mathfrak{B}(\mathbb{K})}(|A| + |B|) + |G|)$.

4   Context Construction

We describe how to construct a context from a given dataset to apply FCA and derive outlierness scores. Let X be a dataset, which is a set of feature vectors, and V be the set of features. For each data point $x \in X$ and a feature $v \in V$, we write as $x(v)$ the value of x for the feature v. If a feature $v \in V$ is *discrete*, the domain $\mathrm{dom}(v)$ is assumed to be a countable set. Otherwise if v is *continuous*, $\mathrm{dom}(v) = \mathbb{R}$. We always identify the set of objects G with the set X of data points (more exactly, their identifiers).

4.1   Discrete Features

Each discrete feature can be directly converted to a context using the method proposed in [22,23]. For each discrete feature $v \in V$, the corresponding set of attributes M_v is defined as

$$M_v := \{\, v.m \mid m \in \mathrm{dom}(v) \,\},$$

where each m is *qualified* as $v.m$ so that attributes are disjoint, and for each value $x(v)$,

$$(x, v.m) \in I \text{ if and only if } x(v) = m.$$

We apply this procedure for each discrete feature and concatenate all of them vertically.

Example 2. Let $X = \{x_1, x_2, x_3\}$ and $V = \{v_1, v_2, v_3\}$ with

$$(x_1(v_1), x_1(v_2), x_1(v_3)) = (2, 0, A),$$
$$(x_2(v_1), x_2(v_2), x_2(v_3)) = (3, 1, C),$$
$$(x_3(v_1), x_3(v_2), x_3(v_3)) = (1, 0, B),$$

where domains are given as $\text{dom}(v_1) = \{1, 2, 3\}$, $\text{dom}(v_2) = \{0, 1\}$, and $\text{dom}(v_3) = \{A, B, C\}$. Then

$$M_{v_1} = \{v_1.1, v_1.2, v_1.3\}, \ M_{v_2} = \{v_2.0, v_2.1\}, \ M_{v_3} = \{v_3.A, v_3.B, v_3.C\}.$$

We therefore have the following context:

	$v_1.1$	$v_1.2$	$v_1.3$	$v_2.0$	$v_2.1$	$v_3.A$	$v_3.B$	$v_3.C$
x_1		×		×		×		
x_2			×		×			×
x_3	×			×			×	

4.2   Discretization for Continuous Features

Next we *discretize* continuous features and construct a context from them. We write an alphabet by Σ and the set of finite and infinite sequences by Σ^* and Σ^ω, respectively. Let $\rho : \Sigma^\omega \to \mathbb{R}$ be a mapping from an infinite sequence $p \in \Sigma^\omega$ to a real number. We call such a mapping a *representation*, or *encoding*, of real numbers. Define

$$\uparrow w := \{ p \in \Sigma^\omega \mid w \text{ is a prefix of } p \}.$$

Then $\rho(\uparrow w) = \{\rho(p) \mid p \in \uparrow w\}$ becomes a closed interval from $\min \rho(\uparrow w)$ to $\max \rho(\uparrow w)$. In the following we always put a natural assumption that $\rho(\uparrow wa) \subseteq \rho(\uparrow w)$ for any $w, a \in \Sigma^*$. Many typically used representations satisfy this assumption, for example, binary and decimal representations.

Using such an encoding ρ, we construct a context for a continuous feature $v \in V$ as follows: For a natural number k, define

$$M_v^k := \{ v.w \mid w \in \Sigma^*, \ |w| = k \} \tag{1}$$

and for each value $x(v)$,

$$(x, v.w) \in I \text{ if and only if } x(v) \in \rho(\uparrow w).$$

We call the number k *discretization level*, which specifies the granularity of discretization.

Moreover, for each continuous feature $v \in V$, we increase k from 1 to $k_{\max}$, where $k_{\max}$ is the minimum level in which all objects are isolated. That is, for every pair of objects $x, y \in G$,

$$\{v.w \mid (x, v.w) \in I\} \cap \{v.w \mid (y, v.w) \in I\} = \emptyset.$$

This means that for every concept $(A, B) \in \mathfrak{B}(\mathbb{K})$ there always exists an attribute $m \in B$ such that $m \notin D$ for any concept (C, D) with $(A, B) \leq (C, D)$, which is the necessary condition to apply Algorithm 1. Then the resulting context is the vertical concatenation of $M_v^1, \ldots, M_v^{k_{\max}}$.

From the above construction process of the context for continuous features, we can easily check that the set inclusion relation is preserved in the order on the concept lattice. We define $\rho(B) := \bigcap_{w \in B} \rho(\uparrow w)$ for a set of attributes $B \subset M$.

Lemma 1. *Let $\mathbb{K}$ be the context constructed from a dataset with continuous features. For any pair of concepts $(A, B), (C, D) \in \mathfrak{B}(\mathbb{K})$, we have $(A, B) \leq (C, D)$ if and only if $\rho(B) \subseteq \rho(D)$.*

Thus, intuitively, sparsely located objects tend to receive higher outlierness scores than densely located objects as the number of upper and lower concepts becomes smaller.

As an example of representation, we can employ the *binary representation* of real numbers, which is defined as a mapping $\rho : \Sigma^\omega \to [0, 1]$ from an infinite sequence p with $\Sigma = \{0, 1\}$ to a real number:

$$\rho(p_1 p_2 \ldots) := \sum_{i=1}^{\infty} p_i \cdot 2^{-i}. \tag{2}$$

Thus for a finite sequence w of length k,

$$\rho(\uparrow w) = \left[\rho(w_1 w_2 \ldots w_k 000 \ldots), \rho(w_1 w_2 \ldots w_k 111 \ldots) \right]$$
$$= \left[\sum_{i=1}^{k} w_i \cdot 2^{-i}, \sum_{i=1}^{k} w_i \cdot 2^{-i} + 2^{-k} \right].$$

For example, $\rho(\uparrow 1) = [0.5, 1]$, $\rho(\uparrow 01) = [0.25, 0.5]$, and $\rho(\uparrow 10) = [0.5, 0.75]$.

Although it is a popular encoding scheme, it is lack of effectivity. For instance, different attributes are assigned to 0.4999 and 0.5001 although they are close in the geometric sense. To solve this problem, we add a simple modification as follows:

$$\rho_o(\uparrow w) := \left[\sum_{i=1}^{k} w_i \cdot 2^{-i} - 2^{-(k+1)}, \sum_{i=1}^{k} w_i \cdot 2^{-i} + 2^{-k} \right].$$

This means that two adjacent intervals overlap each other with the width $2^{-(k+1)}$. For example, $\rho_o(\uparrow 1) = [0.25, 1]$, $\rho_o(\uparrow 01) = [0.125, 0.5]$, and $\rho_o(\uparrow 10) = [0.375, 0.75]$. This modified representation ρ_o gives the following property.

Lemma 2. *Let v be a continuous feature and M_v^k be the set of attributes constructed from v at level k. For a pair of objects x, y, $\{v.w \mid (x, v.w) \in I\} \cap \{v.w \mid (y, v.w) \in I\} = \emptyset$ only if $|x(v) - y(v)| \geq 2^{-(k+1)}$.*

We call the original representation the *simple binary representation* and this modified version the *modified binary representation*. In the following we assume that all continuous values are in the interval $[0, 1]$. For real-world datasets this can be achieved by applying some normalization in advance, for example, min-max normalization.

Example 3. Let $X = \{x_1, x_2, x_3\}$ and $V = \{v_1, v_2\}$ with

$$(x_1(v_1), x_1(v_2)) = (0.3, 0.4), \quad (x_2(v_1), x_2(v_2)) = (0.18, 0.8),$$
$$(x_3(v_1), x_3(v_2)) = (0.6, 0.1).$$

Since objects are isolated at level 3 for the feature v_1 and 2 for v_2, we have the following context:

	$v_1.0$	$v_1.1$	$v_1.00$	$v_1.01$	$v_1.10$	$v_1.001$	$v_1.010$	$v_1.100$	$v_1.101$
x_1	×	×		×			×		
x_2	×		×	×		×			
x_3		×			×			×	×

	$v_2.0$	$v_2.1$	$v_2.00$	$v_2.01$	$v_2.10$	$v_2.11$
x_1	×	×		×	×	
x_2		×				×
x_3	×		×			

In this table attributes with no relation are abbreviated.

We can choose the *signed digit representation* [26] instead of the modified binary representation, which is also defined using the Eq. (2) but -1 (abbreviated as $\bar{1}$) is added to the alphabet Σ. Hence for finite sequences of length k, we have

$$\rho(\uparrow w) = \left[\rho(w_1 w_2 \ldots w_k \bar{1}\bar{1}\bar{1} \ldots), \rho(w_1 w_2 \ldots w_k 111 \ldots) \right]$$
$$= \left[\sum_{i=1}^{k} w_i \cdot 2^{-i} - 2^{-k}, \sum_{i=1}^{k} w_i \cdot 2^{-i} + 2^{-k} \right],$$

and Lemma 2 also holds in this representation. However, the signed digit representation has a significant problem in terms of efficiency: k strings could represent the same interval at level k. For example, $\rho(\uparrow \bar{1}01) = \rho(\uparrow \bar{1}1\bar{1}) = \rho(\uparrow 0\bar{1}\bar{1}) = [-0.5, -0.25]$. This means that too many unnecessary attributes could be generated if k becomes large. Thus we employ the modified binary representation and experimentally compare it to the simple binary representation.

Another possibility is to employ *Gray code* as a representation, which is theoretically known to have the same effectiveness as the signed digit representation on real number computation [24]. Formally, it is defined as an *embedding* from $[0, 1]$ to the set $\Sigma^\omega_\perp$, which is the set of infinite sequences on Σ^ω for which, in each sequence, at most one position is $\perp$. Each $x \in [0, 1]$ is mapped to an infinite sequence p such that $p_i = 1$ if

$$2^{-i} m - 2^{-(i+1)} < x < 2^{-i} m + 2^{-(i+1)}$$

for an odd number m, $p_i = 0$ if the same holds for an even number m, and $p_i = \perp$ if $x = 2^{-i} m - 2^{-(i+1)}$ for some integer m. Lemma 2 also holds in Gray code.

However, we can see that there exist $2^k + 2^k - 1 = 2^{k+1} - 1$ strings that represent intervals with the width 2^{-k}, which is larger than 2^k for the modified binary representation, hence we employ the modified binary representation as our first choice.

Example 4. Let $X = \{x_1, x_2, x_3, x_4\}$ and $V = \{v_1, v_2, v_3\}$ with

$$(x_1(v_1), x_1(v_2), x_1(v_3)) = (A, 0, 0.9), \qquad (x_2(v_1), x_2(v_2), x_2(v_3)) = (B, 1, 0.4),$$
$$(x_3(v_1), x_3(v_2), x_3(v_3)) = (B, 1, 0.17), \qquad (x_4(v_1), x_4(v_2), x_4(v_3)) = (C, 1, 0.05),$$

and $\mathrm{dom}(v_1) = \{A, B, C\}$, $\mathrm{dom}(v_2) = \{0, 1\}$, and $\mathrm{dom}(v_3) = \mathbb{R}$. We have the following context since objects are isolated at level 3 for the feature v_3.

	$v_1.A$	$v_1.B$	$v_1.C$	$v_2.0$	$v_2.1$	$v_3.0$	$v_3.1$	$v_3.00$	$v_3.01$	$v_3.10$	$v_3.11$
x_1	×			×			×				×
x_2		×			×	×			×		
x_3		×			×	×		×			
x_4			×		×	×		×			

	$v_3.000$	$v_3.001$	$v_3.011$	$v_3.111$
x_1				×
x_2			×	
x_3		×		
x_4	×			

We again abbreviate attributes with no relations. The resulting concept lattice is shown in Fig. 2. We therefore have outlierness scores

$$q(\{x_1\}) = 1/4 = 0.25, \quad q(\{x_2\}) = 1/6 = 0.17,$$
$$q(\{x_3\}) = 1/6 = 0.17, \quad q(\{x_4\}) = 1/5 = 0.2.$$

4.3 Theoretical Analysis

Here we theoretically analyze the relationship between concept-based outliers and distance-based outliers. In the following we consider only continuous (numerical) data, that is, each value is a real number.

First we introduce the formal definition of distance-based outliers, which are firstly introduced by Knorr and Ng [12,13]. Given a set of objects X, each object $x \in X$ is said to be a $DB(\alpha, \delta)$-*outlier* if

$$\left| \{y \in X \mid d(x, y) > \delta\} \right| \geq \alpha n,$$

where α and δ with $\alpha, \delta \in \mathbb{R}$ and $0 \leq \alpha \leq 1$ are parameters specified by the user. This means that at least a fraction α of all objects have a distance from x that is larger than δ. We denote the set of $DB(\alpha, \delta)$-outliers by $X(\alpha; \delta)$.

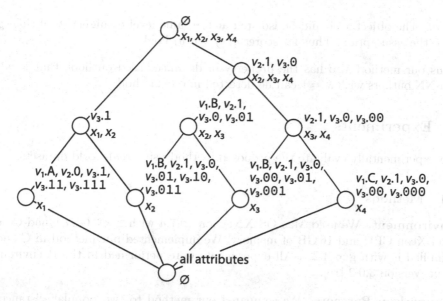

Fig. 2. Concept lattice in example 4.

Theorem 1. *Given a dataset $X \subset \mathbb{R}$. For every pair of data points $x, y \in X$, if $q(x) > q(y)$, there exist α and δ such that $x \in X(\alpha; \delta)$ and $y \notin X(\alpha; \delta)$.*

Proof. There should exist a discretization level k such that x is isolated at the level while y is not. Thus, from Lemma 2, if we set $\alpha = (n-1)/n$ and $\delta = 2^{-(k+1)}$, we have $x \in X(\alpha; \delta)$ and $y \notin X(\alpha; \delta)$. □

Thus, despite the fact of ignoring the rareness of objects in our concept-based outliers, they are actually distance-based outliers at the same time. This is why, intuitively, if objects are closely located, we need high discretization level to isolate them, resulting in larger number of concepts with high outlierness. The rareness of objects are therefore implicitly projected onto the concept lattice.

Next, let us consider κth-nearest neighbor distance, which was introduced by Ramaswamy et al. [20] and developed further [4,5,17] to overcome two drawbacks of the original distance-based outliers: the difficulty of determining the distance threshold δ and the lack of a ranking of outliers. Formally, the κth-NN score $q_{\kappa\text{thNN}}(x)$ of an object $x \in X$ is defined as

$$q_{\kappa\text{thNN}}(x) := d^\kappa(x; X),$$

where $d^\kappa(x; X)$ is the distance between x and its κth-NN in X. Notice that if we set $\alpha = (n - \kappa)/n$, the set of Knorr and Ng's DB(α, δ)-outliers coincides with the set $\{x \in X \mid q_{\kappa\text{thNN}}(x) \geq \delta\}$.

Theorem 2. *Given a dataset $X \subset \mathbb{R}$. Let $\kappa = 1$. For every pair of data points $x, y \in X$, $q_{\kappa\text{thNN}}(x) > q_{\kappa\text{thNN}}(y)$ implies $q(x) \geq q(y)$.*

Proof. The object x should be isolated at the same level or higher level than y from the assumption. Thus for scores $q(x) \geq q(y)$ holds.   □

Thus our method also has the property of distance-based outliers, that is, all κth-NN outliers with $\kappa = 1$ can be detected in our method.

5   Experiments

We experimentally evaluate the proposed method using real-world datasets.

5.1   Methods

Environment. We used Mac OS X version 10.7.4 with 2×3 GHz Quad-Core Intel Xeon CPU and 16 GB of memory. We implemented our method in C and compiled it with `gcc` 4.2.1. All experiments were performed in the R environment, version 3.0.2 [19].

Comparison Partners. We compared our method to two popular distance-based methods: *local outlier factor* (LOF) [6] and *one-class SVM* (oneSVM) [21]. Since they are designed only for numerical (continuous) attributes, we directly treat discrete values as real-valued numerical values. We used the common setting $k = 10$ for k-nearest neighbor search in LOF. In oneSVM we used a Gaussian RBF kernel and set its parameter by the popular heuristics [7]. LOF was performed by the R `DMwR` package and oneSVM by the R `kernlab` package.

Datasets. We collected six mixed-type datasets containing both discrete and continuous features from the UCI machine learning repository [3]. Their properties are summarized in Table 2. For each dataset, we assume that objects from the smallest class are outliers selected 10 data points from the class, as these datasets are originally designed for classification. Since LOF and oneSVM cannot treat missing values, we filled them in advance using the R `Hmisc` package for these two methods. Note that our method can directly treats such missing values without imputation, but we used the same imputed datasets in our method for fair comparison.

Evaluation Criteria. We used the area under the precision-recall curve (AUPRC; equivalent to the average precision), which is a typical criterion to evaluate the effectiveness of outlier detection methods [2]. It takes values from 0 to 1 and 1 is the best score, and quantifies whether the method is able to retrieve outliers correctly. These values were calculated by R `ROCR` package.

5.2   Results and Discussions

Results of AUPRCs are shown in Table 3. In the table, in our method, all results are with the modified binary representation since two discretization schemes, the simple binary representation and the modified binary representation, showed

Table 2. Summary of datasets with the ratio of outliers (corresponds to AUPRC of random scoring).

Name	# objects	# features		Ratio
		Disc.	Cont.	
Echocardiogram	60	3	9	0.17
Hepatitis	95	13	6	0.11
Heart	160	7	6	0.063
Horse colic	211	20	7	0.047
Crx	393	9	6	0.025
German	710	13	7	0.014
Annealing	774	22	10	0.013

Table 3. Area under the precision-recall curve (AUPRC) in real datasets. The best scores are denoted in bold. In our method, disc. and cont. denote the simple and the modified binary representations, respectively.

Name	AUPRC				
	Ours (disc.)	Ours (cont.)	Ours (both)	LOF	oneSVM
Echocardiogram	0.010	0.50	**0.64**	0.20	0.35
Hepatitis	**0.43**	0.20	0.30	0.14	0.17
Heart	0.12	0.54	**0.61**	0.13	0.21
Horse colic	0.05	0.13	**0.15**	0.054	0.13
Crx	0.029	0.086	**0.097**	0.028	0.029
German	0.012	**0.052**	0.036	0.028	0.021
Annealing	0.015	**0.034**	0.027	0.022	0.014
Average	0.108	0.219	**0.265**	0.086	0.131

exactly the same scores on all datasets, even though they theoretically have different effectiveness as shown in Lemma 2.

We performed our method in three ways: using only discrete attributes, only continuous attributes, and both discrete and continuous attributes. Then, the best performance is achieved on average if we use both discrete and continuous attributes. This means that our approach may appropriately take both types of attributes into account in outlier detection without any careful parameter setting. In addition, on all datasets AUPRCs from continuous attributes are much higher than those from discrete attributes. Thus we can confirm that a continuous, numerical attribute empirically has a more important role than discrete attribute. This gives an insight to the reason why most existing studies of outlier detection focus on numerical data.

Moreover, in comparison with other two distance-based methods, our methods showed the best performance on all datasets, resulting in the best average

AUPRC score across all datasets. These results indicate the power of concept-based outlierness by treating semantics via FCA, and our method has a potential to appropriately process mixed-type features without any parameter setting. Interestingly, our method is already superior to other methods when only continuous attributes are used. This means that our parameter-free method, integration of FCA with discretization, has a potential to outperform well-studied distance-based methods for numerical datasets.

6   Conclusion

In this paper we have introduced new outliers, called concept-based outliers, and proposed a detection method for mixed-type data, which contain both discrete and continuous features. The key technique is to use FCA to construct a cluster hierarchy, called the concept lattice, by discretizing continuous features and measure the outlierness by the degree of isolation on the hierarchy. We have theoretically analyzed the relationship with distance-based outliers and have been experimentally shown that it is more effective than other popular outlier detection methods.

The most important challenge is how to reduce the computational complexity. In particular, the size of the concept lattice rapidly increases as the number of object increases, which does not scale to large datasets. For example, in the dataset Annealing (774 objects) more than 10 million concepts were generated. One solution could be random sampling of objects, which we will investigate.

In addition, an interesting property of our method, which we did not analyze in detail in this paper, is the simultaneous scoring for attributes (features). This can be applied for *feature selection* for mixed-type data, which is also an important and challenging topic in machine learning.

Acknowledgments. This work is supported by the Alexander von Humboldt Foundation.

References

1. Adda, M., Wu, L., White, S., Feng, Y.: Pattern detection with rare item-set mining. arXiv:1209.3089 (2012)
2. Aggarwal, C.C.: Outlier Analysis. Springer, New York (2013)
3. Bache, K., Lichman, M.: UCI machine learning repository (2013). http://archive.ics.uci.edu/ml
4. Bay, S.D., Schwabacher, M.: Mining distance-based outliers in near linear time with randomization and a simple pruning rule. In: Proceedings of the 9th ACM SIGKDD International Conference on Knowledge Discovery and Data Mining, pp. 29–38 (2003)
5. Bhaduri, K., Matthews, B.L., Giannella, C.R.: Algorithms for speeding up distance-based outlier detection. In: Proceedings of the 17th ACM SIGKDD International Conference on Knowledge Discovery and Data Mining, pp. 859–867 (2011)

6. Breunig, M.M., Kriegel, H.P., Ng, R.T., Sander, J.: LOF: identifying density-based local outliers. In: Proceedings of the 2000 ACM SIGMOD International Conference on Management of Data, pp. 93–104 (2000)
7. Caputo, B., Sim, K., Furesjo, F., Smola, A.: Appearance-based object recognition using SVMs: which kernel should I use? In: Proceedings of NIPS Workshop on Statistical Methods for Computational Experiments in Visual Processing and Computer Vision (2002)
8. Davey, B.A., Priestley, H.A.: Introduction to Lattices and Order, 2nd edn. Cambridge University Press, Cambridge (2002)
9. Ganter, B., Wille, R.: Formal Concept Analysis: Mathematical Foundations. Springer, New York (1998)
10. Hawkins, D.: Identification of Outliers. Chapman and Hall, London (1980)
11. Kaytoue, M., Kuznetsov, S.O., Napoli, A.: Revisiting numerical pattern mining with formal concept analysis. In: Proceedings of the 22nd International Joint Conference on Artificial Intelligence, pp. 1342–1347 (2011)
12. Knorr, E.M., Ng, R.T.: Algorithms for mining distance-based outliers in large datasets. In: Proceedings of the 24th International Conference on Very Large Data Bases, pp. 392–403 (1998)
13. Knorr, E.M., Ng, R.T., Tucakov, V.: Distance-based outliers: algorithms and applications. VLDB J. **8**(3), 237–253 (2000)
14. Liu, F.T., Ting, K.M., Zhou, Z.-H.: On detecting clustered anomalies using SCiForest. In: Balcázar, J.L., Bonchi, F., Gionis, A., Sebag, M. (eds.) ECML PKDD 2010, Part II. LNCS, vol. 6322, pp. 274–290. Springer, Heidelberg (2010)
15. Makino, K., Uno, T.: New algorithms for enumerating all maximal cliques. In: Hagerup, T., Katajainen, J. (eds.) SWAT 2004. LNCS, vol. 3111, pp. 260–272. Springer, Heidelberg (2004)
16. Okubo, Y., Haraguchi, M.: An algorithm for extracting rare concepts with concise intents. In: Kwuida, L., Sertkaya, B. (eds.) ICFCA 2010. LNCS, vol. 5986, pp. 145–160. Springer, Heidelberg (2010)
17. Orair, G.H., Teixeira, C.H.C., Wang, Y., Meira Jr., W., Parthasarathy, S.: Distance-based outlier detection: consolidation and renewed bearing. Proc. VLDB Endowment **3**(1–2), 1469–1480 (2010)
18. Pasquier, N., Bastide, Y., Taouil, R., Lakhal, L.: Efficient mining of association rules using closed itemset lattices. Inf. Syst. **24**(1), 25–46 (1999)
19. R Core Team: R: A Language and Environment for Statistical Computing. R Foundation for Statistical Computing (2013). http://www.R-project.org
20. Ramaswamy, S., Rastogi, R., Shim, K.: Efficient algorithms for mining outliers from large data sets. In: Proceedings of the 2000 ACM SIGMOD International Conference on Management of Data, pp. 427–438 (2000)
21. Schölkopf, B., Platt, J.C., Shawe-Taylor, J., Smola, A.J., Williamson, R.C.: Estimating the support of a high-dimensional distribution. Neural Comput. **13**(7), 1443–1471 (2001)
22. Sugiyama, M., Imajo, K., Otaki, K., Yamamoto, A.: Semi-supervised ligand finding using formal concept analysis. IPSJ Trans. Math. Model. Appl. (TOM) **5**(2), 39–48 (2012)
23. Sugiyama, M., Yamamoto, A.: Semi-supervised learning on closed set lattices. Intell. Data Anal. **17**(3), 399–421 (2013)
24. Tsuiki, H.: Real number computation through Gray code embedding. Theor. Comput. Sci. **284**(2), 467–485 (2002)

25. Valtchev, P., Missaoui, R., Godin, R.: Formal concept analysis for knowledge discovery and data mining: the new challenges. In: Eklund, P. (ed.) ICFCA 2004. LNCS (LNAI), vol. 2961, pp. 352–371. Springer, Heidelberg (2004)
26. Weihrauch, K.: Computable Analysis: An Introduction. Springer, New York (2000)
27. Wille, R.: Restructuring lattice theory: an approach based on hierarchies of concepts. In: Ferré, S., Rudolph, S. (eds.) ICFCA 2009. LNCS, vol. 5548, pp. 314–339. Springer, Heidelberg (2009)

Author Index

Printed in the United States
by Baker & Taylor Publisher Services